GENERATING ELECTRICITY FROM THE SUN

RENEWABLE ENERGY SERIES
Editor-in-Chief: A A M Sayigh

Pergamon Titles of Related Interest

BANHIDI
Radiant Heating Systems: Design and Applications

BEI
Modern Power Station Practice, 3rd edition

GRANQVIST
Materials Science for Solar Conversion Systems

HARRISON
Geothermal Heating

HORIGOME
Clean and Safe Energy Forever

McVEIGH
Sun Power, 2nd edition

SAITO
Heat Pumps

SAYIGH
Energy Conservation in Buildings

SAYIGH
Energy and the Environment: into the 1990s, 5-vol set

SAYIGH & McVEIGH
Solar Air Conditioning and Refrigeration

STECCO & MORAN
A Future for Energy

Pergamon Related Journals (free specimen copy gladly sent on request)

Energy
Energy Conservation and Management
Geothermics
Heat Recovery Systems and CHP
International Journal of Heat and Mass Transfer
International Journal of Hydrogen Energy
Progress in Energy and Combustion Science
Renewable Energy
Solar Energy

GENERATING ELECTRICITY FROM THE SUN

Edited by

Fred C Treble
Consulting Engineer, Farnborough, UK

PERGAMON PRESS
OXFORD · NEW YORK · SEOUL · TOKYO

U.K.	Pergamon Press plc, Headington Hill Hall, Oxford OX3 0BW, England
U.S.A.	Pergamon Press, Inc., 395 Saw Mill River Road, Elmsford, New York 10523, U.S.A.
KOREA	Pergamon Press Korea, KPO Box 315, Seoul 110-603, Korea
JAPAN	Pergamon Press, 8th Floor, Matsuoka Central Building, 1-7-1 Nishishinjuku, Shinjuku-ku, Tokyo 160, Japan

First edition 1991

Library of Congress Cataloging in Publication Data

Data applied for

British Library Cataloguing in Publication Data

Treble, F. C. (Frederick Christopher), 1916–
Generating electricity from the Sun.—(Renewable energy: v. 2).
I. Title II. Series
621.47

ISBN 0 08 040936 9

Printed in Great Britain by BPCC Wheatons Ltd, Exeter

CONTENTS

FOREWORD

During the last five years, the world has awakened to the fact that fossil fuels are causing a great deal of damage to the environment. Global warming due to the emission of carbon dioxide, coupled with the depletion of the ozone layer due to CFC emissions, must be stopped or reduced. Renewable energy and primarily photovoltaic technology will offer the solution.

The first commercial use of PV was in 1954, to monitor a satellite in space. The cost of the power was $1000/Wp; today an installed system can be bought for $6/Wp. The production of PV has increased from 20 kW in 1954 to 50 MW in 1990. This is predicted to reach 200 MW in 1995 and 2000 MW in the year 2000. Amorphous silicon technology will take the lead and its cost will be about $1.5/Wp in the year 2000. A recent PV conference - Lisbon, Portugal, 8-12 April 1991 - showed the following achievements:

1. It was clear from the PV exhibition that one can buy an array for $5/Wp and the company which sells the systems is no longer being subsidised by its Government.

2. Thin film is very much in the picture with efficiency in excess of 10%.

3. Some companies are getting into consumer products, i.e. French companies (Photec) as well as the Japanese who are well in front as far as consumer products are concerned.

4. The Martin Green cell is still a luxury as far as cost is concerned, with efficiency of 23%.

5 The PV concentrator is still in the experimental stage, with efficiency up to 27%.

6. PV production for 1990 was about 50 MW with the anticipation of 200 MW for 1995 and perhaps 200 MW in the year 2000.

7. One German company (Flachglas Solartechnik GmbH - FLAGSOL) has managed to combine passive heating and PV power generation for homes which looks very promising.

8. Solar cell production machinery can be bought easily and several developing countries have it, from Spire and others.

9. CdTe panels have an efficiency of 13% and can be improved.

This book deals in a precise manner with the technology as well as the experience acquired so far in the use of photovoltaic conversion systems through various applications round the world. This book has fourteen chapters and many references. Following the introductory remarks, the questions of solar radiation intensity, orientation of panels and solar tracking are addressed. Referring to crystalline silicon, the cells, the modules, the arrays and their system performance are explained in detail. This is followed by an analysis of polycrystalline solar cells. The long term prospects for photovoltaics are thoroughly discussed, followed by battery storage: types and limitations. Finally, innovative ideas in the use of photovoltaic conversion, such as concentrators, as well as recent developments in single and polycrystalline cells, are covered. This is followed by pilot and demonstration projects round the world.

This well-written book forms part of the Pergamon Press Renewable Energy Series, and will be of interest to practising engineers, scientists and many non-technical readers.

A A M Sayigh
University of Reading

Chapter 1

INTRODUCTION

F. C. TREBLE
Consulting Engineer, Farnborough, UK

The direct conversion of daylight into electricity by photovoltaic solar cells is one of the most promising of the renewable energy options to have emerged in recent years. In terms of its potential benefits to mankind, the invention in the early 1950s of this completely new way of generating electricity may come to rank in importance with Faraday's discovery of electromagnetic induction, which led to the development of rotary electric generators and motors.

If the present downward cost trend continues, as it is expected to do, photovoltaic generation offers a way of helping to meet the increasing worldwide demand for electricity without accelerating the depletion of our finite resources of fossil fuels, adding to the contamination of the atmosphere or building hundreds of nuclear power stations. In the short run, there are many commercial applications for which solar power is already cost effective.

Solar cells today are mostly made of silicon, one of the most common elements on earth. They do their job silently and there are no moving parts to wear out. They do not pollute the atmosphere and they leave behind no harmful waste products. Their mechanical simplicity means that they can be engineered to last reliably for many years, with little or no maintenance. In fact, many existing plants operate automatically and require no attendant operators. Photovoltaic cells work effectively even in cloudy weather and, unlike solar heaters, are more efficient at low temperatures. They also respond rapidly to the sudden changes of solar input which occur when clouds pass by. These properties are of particular importance in temperate climates, where a large proportion of solar energy comes in the form of diffuse radiation from cloudy skies.

The crystalline silicon solar cell has the considerable advantage of being based on a well-established semiconductor technology, which has been developed over many years for electronic components such as diodes, transistors and microchips.

Another important advantage of the photovoltaic generator is its modularity. Arrays of any size and voltage can be constructed from standard modules. There is no scale effect, the conversion efficiency being practically independent of output. The modules can be thoroughly type tested and mass produced under close quality control, thus ensuring a reliable product. Potential users of large generators can gain experience beforehand with a smaller version. Systems can grow as more funds become available and demand increases. Repair is usually a matter of replacing a faulty module. One or more modules can fail and the system continue to operate until replacements are installed.

Photovoltaic power plants can be put up quickly and easily. The long lead times, commonly ten years or more, associated with the planning and construction of coal, oil or nuclear power stations can be avoided. Consequently, there is no need to rely on long-term forecasts of future electricity demand, which quite often turn out to be inaccurate.

However, solar power should not be thought of solely in the context of central power stations and distribution grids. Perhaps its most important characteristic is that, because sunlight is a distributed energy source, the power can be generated as and where it is needed, thus saving the cost and avoiding the losses of transmission lines. It is therefore uniquely suited to on-site generation in the many parts of the world where there is no mains supply and electricity has to be provided expensively by batteries or small diesel or gasoline generators. Because of this, photovoltaics have an important role to play in the developing countries, where most of the population live in small scattered villages and there is no lack of sunshine.

The so-called "photovoltaic effect" was first observed by Becquerel in 1839, when he directed sunlight on to one of the electrodes of an electrolytic cell. Later, Adams and Day (1877) observed the effect in selenium. The pioneering work of Lange (1930), Schottky (1930) and Grondahl (1933), with other solid-state workers in the field of selenium and cuprous oxide photocells, led to the development of the photographic exposure meter and many other useful devices. But it was not until 1954 that photovoltaic cells with an acceptably high coversion efficiency for electrical power generation were developed. In that year, Chaplin, Fuller and Pearson of Bell Telephone Laboratory reported that they had made diffused junction crystalline silicon cells with an efficiency of 6% and Reynolds et al (1954) made a similar breakthrough with a cadmium sulphide device.

With the advent of the space age in the 1950s, there came a need for a light, reliable power source for satellites, which would last several years in orbit. A small group of engineers at the US Army Signal Corps, Fort Monmouth, New Jersey recognised that solar cells could meet this requirement. As a result, the first solar-powered satellite, the American Vanguard I, was launched early in 1958. This 6 in. diameter spherical spacecraft was fitted with six small groups of silicon cells to power its 5mW transmitter. Subsequently, the satellite continued to send signals to earth over a period of eight years. Two months after the Vanguard I launch, the USSR launched a much larger solar-powered satellite, which operated for over two years. Since then, practically every one of the hundreds of satellites launched for scientific, military, meteorological, communications and other purposes have been powered by solar cells. Indeed, it can be legitimately claimed that, had it not been for the timely invention of this new power source, the exploration and exploitation of space could not have developed as they have.

Progressive improvements have raised the conversion efficiency of space solar cells to nearly 20 %. Spacecraft power has increased from the few milliwatts of Vanguard I to 10 kW for the latest communication satellite. Photovoltaic power generation in space has indeed become a well-established technology with a proven track record of reliability and durability.

Until 1973, interest in the terrestrial applications of photovoltaics was muted and sporadic. But the sudden quadrupling of oil prices in that year brought home the fact that supplies of fossil fuels are limited and are subject to political pressures. Accordingly, Governments, oil companies and utilities began to look at renewable sources of energy as a means of supplementing conventional supplies. In photovoltaics, the resulting funding of research, development and demonstration programmes, together with tax incentives and investment in production facilities, has established a new industry, which is continuing to grow.

In the 15 years up to 1975, production of crystalline silicon solar cells, mainly for space programmes, averaged about 100 kW per year. By 1987, annual shipments of terrestrial modules throughout the world had reached 28.6 MW, compared with 27.3 MW in 1986 and 23.7 MW in 1985. Among the many current applications are telecommunications, cathodic protection of pipelines and other metallic structures, water pumping, rural electrification, navigational aids, alarm systems, traffic warning lights, remote instrumentation and consumer products, such as pocket calculators. The new technology is gradually gaining the attention of rural electrification developers, the greatest progress to date having been made in the Dominican Republic and the islands of French Polynesia. Worldwide, it is estimated that more than 15000 homes now receive their electricity from solar cells.

The growth in the market has been accompanied by a gradual reduction in the price of modules. Crystalline silicon modules now cost only about one sixth of what they did in 1975 and the downward trend is expected to continue, resulting in further stimulation of the market. The recent development of thin-film amorphous silicon modules and other thin-film devices may eventually lead to solar cells being used for large-scale power plants in sunny parts of the world. Amorphous silicon cells are already being widely used in consumer products, which accounted for 37 % of world sales in 1987.

This book is intended primarily to give students, engineers and scientists entering the photovoltaics field an overview of all aspects of the subject, with pointers to further reading. However, by using simple language and avoiding jargon, an effort has been made to make the work useful and interesting to others, particularly decision makers, potential domestic, commercial and industrial users, aid agencies, investors and environmentalists.

REFERENCES

Adams, W.G. and Day, R.E. "The action of light on selenium". Proc. R. Soc. London, ser. A, 1877, 25, p. 113.

Becquerel, E. "On electric effects under the influence of solar radiation". Compt. Rend., 1839, 9, p. 561.

Chapin, D.M., Fuller, C.S. and Pearson, G.L. "A new silicon p-n junction photocell for converting solar radiation into electrical power". J. Appl. Phys., 1954, 25, pp. 676-677.

Grondahl, L.O. "The copper-cuprous oxide rectifier and electric cell". Rev. Mod. Phys., 1933, 5, p. 141.

Lange, B. "New photoelectric cell". Zeit. Phys., 1930, 31, p. 139,

Reynolds, D.C., Leies, G., Antes, L.L. and Marburger, R.E. "Photovoltaic effect in cadmium sulfide". Phys. Rev., 1954, 96, pp. 533-534.

Schottky, W. "Cuprous oxide photoelectric cells". Zeit. Phys., 1930, 31, p. 913.

Chapter 2

PHOTOVOLTAIC AND SOLAR RADIATION

A. A. M. SAYIGH
Reading University

Department of Engineering
Reading University
Whiteknights
Reading RG6 2AY UK

In the year 1839, Becquerel observed that when direct light falls on one of the electrodes of an electrolyte process, voltage was created and was called "the photovoltaic effect". Later, many scientists observed the same effect using other materials. The invention and development of photovoltaic cells was a direct result of the race to explore space. In 1954, the Russians, then in 1959, the Americans, launched their satellites in space orbiting the earth to supply electricity to their communications equipment. These satellites relied upon the photovoltaic concept. A solar cell, which will be described later, is a device to convert sunlight into direct-current electricity without any moving parts. The energy produced in this way is clean and does not have any harmful waste products. The cells are mostly made of silicon, which is one of the most common elements on earth and they have a life span of over 20 years. In 1973, the need for alternative energy also promoted great interest in the photovoltaic industry so that in 1978, the shipment of PV was 1 MW while by 1988 this had reached 40 MW and it is still progressing steadily. In this time, costs have been drastically reduced. In 1978, the cost of PV cells was $50/Wp, but by 1988 had dropped to $5/Wp.

The earth receives energy from the sun at a rate of 10^{16}KJ per minute and since the sun shines for an average of twelve hours per day, the earth receives about 7.2 x 10^{18}KJ per day. Assuming that we can utilise 1% of this energy, which is equivalent to 1.65 x 10^{2} times the projected amount of the whole world energy requirement in the year 2094 AD (1). Although solar radiation out of the earth's atmosphere has a value of 1367 W/m^2 which is called <u>Solar Constant</u>, as the solar rays enter the atmosphere a lot of it gets absorbed by Ozone, carbon monoxide and dioxide and water particles. This absorption depends upon the thickness of the atmosphere through which the ray us travelling. The thickness of the layer of the atmosphere is called <u>Air Mass</u>. Therefore, an air mass is the path traversed by the direct solar beam, expressed as a multiple of the path traversed to a point at sea level with the sun directly overhead. Thus the path length at sea level with the sun at zenith is Air Mass 1, or (AM1), while above the earth atmosphere it is (AMO). Air mass depends upon the time of day, the time of year, altitude and latitude of the place. Figure 1 shows the solar radiation at various air mass (2). About half of the scattered energy from a clear sky reaches the earth in the form of diffuse radiation. The sum of the diffuse reflected and direct radiation on a horizontal surface is called: total radiation or global radiation. When the word irradiance is used it means solar radiation intensity. The diffuse component can vary from 20% of the global on a clear day to 100% in totally overcast weather conditions.

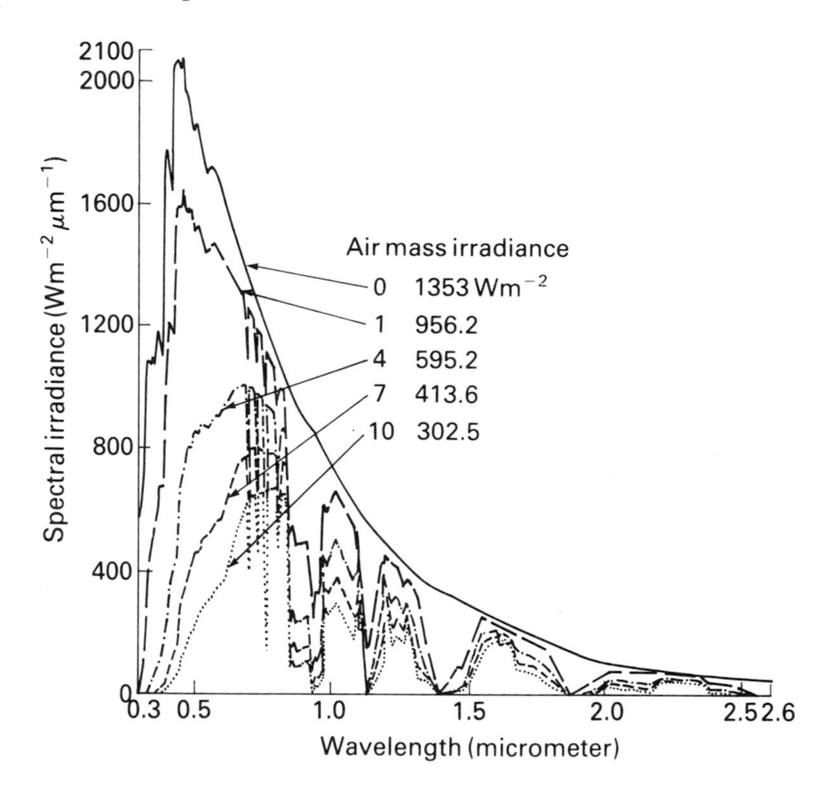

Fig. 2.1. Solar irradiance for different air mass values. U.S. Std. Atmosphere; H_2O = 20 mm; O_3 = 3.4 mm. (α = 1.3; β = 0.02).

Another very important factor dealing with PV and radiation is the response of various PV types to the solar energy wave length or what is known as the energy gapy response of photovoltaic. This concept will be discussed at a later stage, but for the purpose of this chapter Figure 2 is included which shows the maximum calculated efficiency of various PV cells versus the energy gap (2).

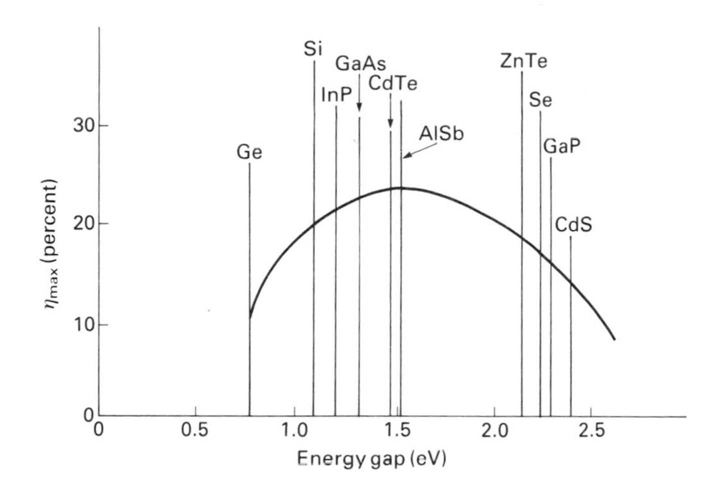

Fig. 2.2. A calculated curve of the maximum efficiency that can be obtained as a function of the energy gap of the semiconductor made into a *p-n* junction and illuminated with the solar spectrum outside the atmosphere. AMO (Loferski, 1963).

Another important factor which influences photovoltaic performance is the amount of solar energy available on a given site, since the higher this amount the greater is the amount of electricity generated by a given cell area. Figures 3 add 4 show direct and diffuse radiation at Kew, England averaged 1955 - 1970, and at Aden, averaged 1958 - 1967 respectively. Kew receives total radiation (in this case direct plus diffuse components) of 5500 MJ/m² yearly, while Aden received 10800 MJ/m² year;y. The ratio of Kew to Aden is nearly 50%. Therefore, a PV panel will generate half of the electricity of that generated in Aden.

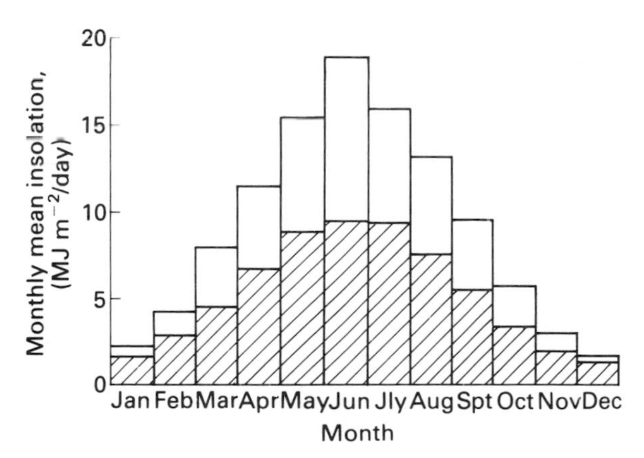

Fig. 2.3. Variation of monthly mean isolation at Kew, England (mean 1955 - 1970).

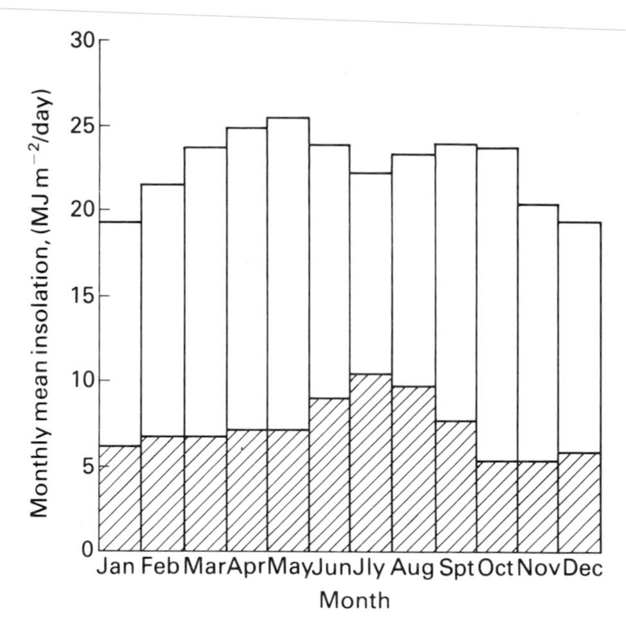

Fig. 2.4. Variation of monthly mean isolation at Aden (mean 1958 - 1967).

1. TRACKING VERSUS NON-TRACKING COLLECTORS:-

Prior to discussing the importance of tracking and non-tracking, the importance of tracking is shown in Fig. 5, which represents a daily solar radiation intensity for Kuwait on the fifth of October 1983. Figure 6 shows the different amounts of energy collected that day on an area of one square metre. They are 9.5, 6.6, 6.2, and 5.8 kWh/m^2 per day for tracking, horizontal tilt, 30° with the horizontal facing south and 45° with the horizontal facing south surfaces respectively. If one assumes that the energy collected by tracking is 100%, then in the case of the fifth of October 1983 in Kuwait the percentage of energy collected by a horizontal, 30° inclined and 45° inclined surfaces are about 69.5%, 65.3% and 61.0% respectively.

The following definitions are relevant to the solar geometry and must be realized:

i - the angle of declination "δ": It is the angle that the sun's rays make with the equatorial plane at solar noon. This angle varies from +23.45° on June 21, to 0.0° on September 21, to - 23.45° on December 21, to 0.0° on March 21. Figure 7 gives the variation of the angle declination with time during the year. Day of solar year, n, is 1 at January the 1st. The mathematical formula which expresses the declination angle at any day during the year is [1]:

$$\delta = 23.45 \text{ Sin } \frac{360}{365.24} (n + 284) \tag{1}$$

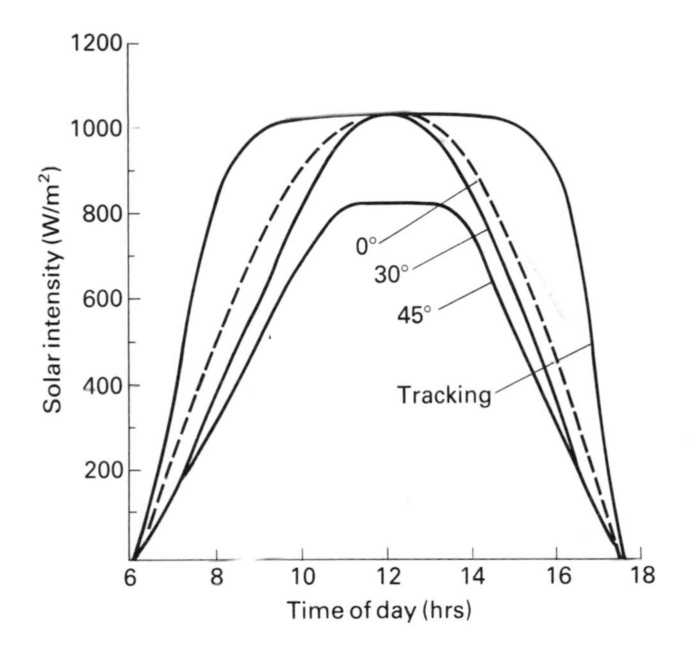

Fig. 2.5. Solar radiation intensity on various inclined surfaces in Kuwait on 5th October, 1983.

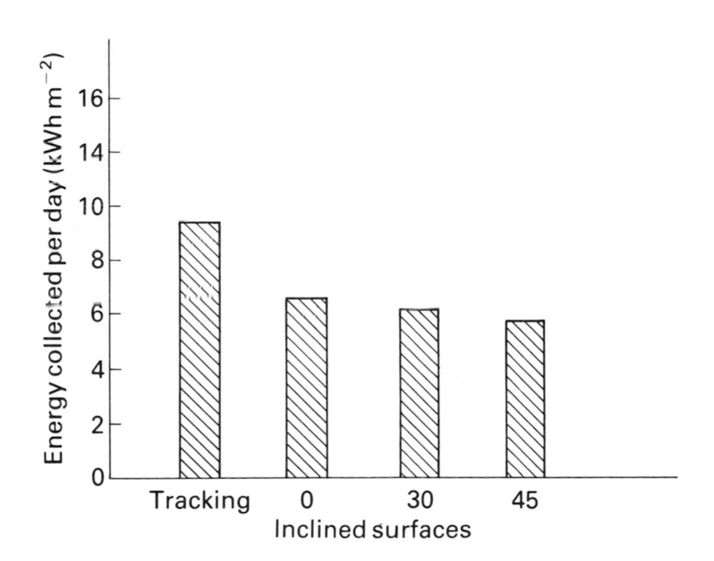

Fig. 2.6. Energy collected per day at various inclination for Kuwait on 5th October, 1983.

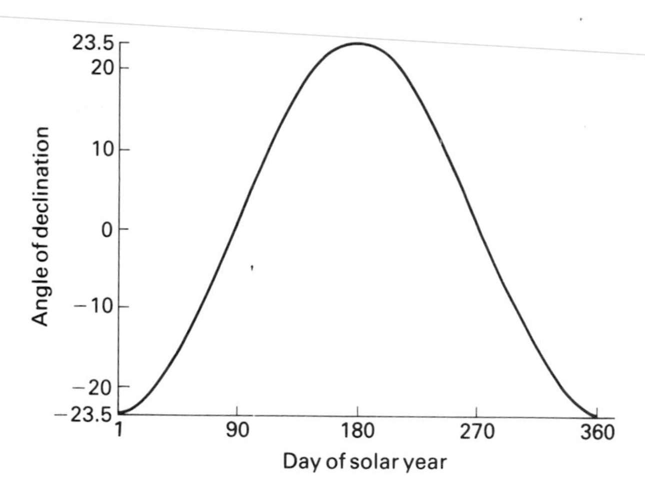

Fig. 2.7. Variation of the declination angle throughout the year according to Equation 1. The day of solar year equals one at January 1st.

ii - Solar noon: At any locality it is defined as the time when a line from the centre of the earth to the sun passes through the meridian of longitude that contains the locality.

iii - The hour angle "ω": As shown in Fig. 8(a), the hour angle, ω, is defined as the angle of arc along the equator between the meridian containing the locality, point P, and the medidian containing line OQ from the centre of the earth to the sun. The hour angle can be converted to time units at the rate of 1.0 hour per 15.0°. By convention the hour angle is positive before noon and given by the relation

$$\omega = (12.0 - t) \times 15° \quad (2)$$

where "t" is the local solar time in hours.

iv - The zenith angle "θ_z": Some times is referred to as inclination angle [1], is the angle between the sun and a vertical line at point P, Fig. 8(b). The zenith angle depends on the time of day, the day in the year and the position of the locality.

$$\text{Cos}\ \theta_z = \text{Cos}\ \delta\ \text{Cos}\ \phi\ \text{Cos}\ \omega + \text{Sin}\ \phi\ \text{Sin}\ \delta \quad (3)$$

where ϕ is the latitude angle; the angle between the equatorial plane and a line from the earth's centre passing through the point P.

v - Altitude angle: It is the angle between the direction of the sun and the horizontal. This is the compliment of the zenith angle.

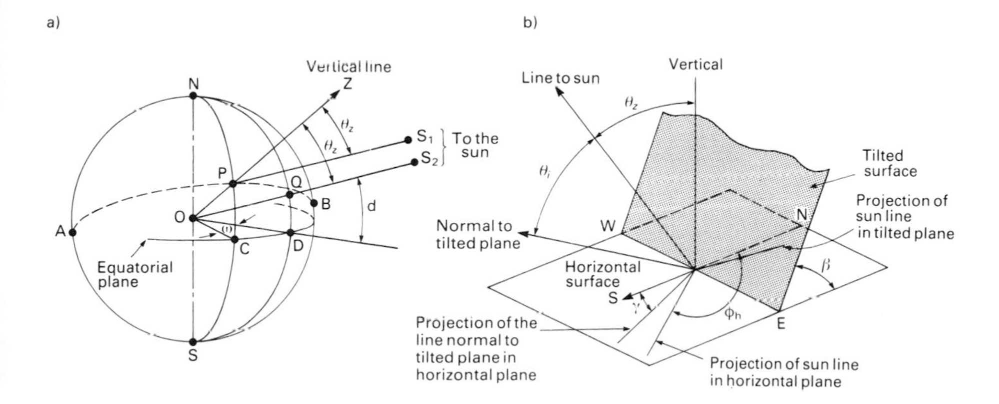

Fig. 2.8. Solar geometry with respect to a tilted plane at locality P.

vi - The sunset hour angle "ω_s": this angle indicates the length of the day and can be expresses according to the relation:

$$\cos \omega_s = - \tan \phi \tan \delta \tag{4}$$

The day length T is given by:

$$T = (2/15)\,\omega_s \tag{5}$$

2. EXTRATERRESTIAL RADIATION ON A HORIZONTAL SURFACE:

This is defined as the solar radiation outside the atmosphere incident on a horizontal plane. At any time, the extraterrestial radiation on horizontal surface, G_o, is given by [2]:

$$G_o = 1367 \left[1 + 0.033 \cos \left(\frac{360\, n}{365}\right) \right] \cos \theta_z \tag{6}$$

where n is the day of the year, 1st of January is taken as 1, and θ_z is the zenith angle.

The daily total extraterrestial solar radiation on a horizontal surface in mega Joules per square meter, H_o, is given by:

$$H_o = 2 \times 10^{-6} \int_0^{\omega_s} G_o \, dt$$

and (7)

$$t = 240\, \omega$$

where ω_s and ω are the sunset hour angle and the hour angle in degrees respectively and t is the solar time in seconds. Substituting equation (6) into (7) and replacing dt by 240 dω leads to the expression:

$$H_0 = 37.595 \left[1 + 0.033 \cos \left(\frac{360\ n}{365}\right)\right] C$$

where (8)

$$C = \cos \phi \cos \delta \sin \omega_s + \frac{2 \pi \omega_s}{360} \sin \phi \sin \delta$$

3. DAILY BEAM AND DIFFUSE COMPONENTS OF SOLAR IRRADIANCE FROM THE MONTHLY DAILY AVERAGE OF TOTAL IRRADIANCE ON A HORIZONTAL SURFACE:

In order to estimate the solar energy that can be collected by a flat plate solar collector it is necessary to know the total of the incident radiation. Knowing the total daily insolation on a horizontal surface, it is possible to estimate the beam and the diffuse solar radiation incident on the collector during the day. More over, as will be discussed in the next section, it is possible to estimate the hourly radiation in the day.

It has been found [3] that the long term average ratio of the diffuse radiation to the total, hemispherical, radiation on a horizontal surface, H_d/H_h, is strongly dependant on the sunset hour angle, ω_s, and the clearness index K_h. Periera and Rabl found that the ratio H_d/H_h can be expressed as:

$$\frac{H_d}{H_h} = a - bc$$

where (9)

$$a = 0.775 + 0.00606 (\omega_s - 90)$$

$$b = 0.505 + 0.00456 (\omega_s - 90)$$

$$c = \cos (115 K_h - 103)$$

and the clearness index K_h is defined as the ratio between the total irradiance on a horizontal surface, H_h, and the extraterrestrial insolation, H_0.

$$K_h = \frac{H_h}{H_0} \quad (10)$$

For a certain period of time K_h is:

$$K_h = \frac{1}{N} \sum_{1}^{N} \frac{H_h}{H_0} \quad (11)$$

Once H_d has been found the long term beam component can be found according to the relation:

$$H_b = H_h - H_d \quad (12)$$

4. HOURLY TOTAL IRRADIANCE ON A HORIZONTAL PLANE:

The hourly solar radiation on a horizontal surface is found to be dependant on the sunset hour angle. The ratio of hourly total to daily total is given by [3]:

$$\frac{I_{tot}}{H_h} = \frac{\pi}{24}(A + B \cos\omega)\left[\frac{\cos\omega - \cos\omega_s}{\sin\omega_s - (2\pi\omega_s/360)\cos\omega_s}\right] \quad (13)$$

where I_{tot} is the hourly total radiation on horizontal surface in mega Joules per meter square. The coefficients A and B are given by:

$$A = 0.409 + 0.5016 \sin(\omega_s - 60)$$

$$B = 0.6609 - 0.4767 \sin(\omega_s - 60)$$

5. HOURLY BEAM AND DIFFUSE RADIATION COMPONENTS ON A HORIZONTAL PLANE:

The ratio of hourly diffuse to daily diffuse solar radiation, I_d/H_d, had been developed by Liu and Jordan in 1960 [4]. This ratio is given by:

$$\frac{I_d}{H_d} = \frac{\pi}{24}\left[\frac{\cos\omega - \cos\omega_s}{\sin\omega_s - (2\pi\omega_s/360)\cos\omega_s}\right] \quad (14)$$

where I_d and H_d are the hourly and the daily diffuse solar radiation on horizontal surface respectively in MJ/m². The hourly beam component can be easily estimated according to the expression:

$$I_b = I_{tot} - I_d \quad (15)$$

6. HOURLY BEAM AND DIFFUSE COMPONENT ON AN INCLINED PLANE:

In general, any plane oriented towards the sun is making an angle with the incident solar beam. This angle, between the solar beam and the normal to the surface is called the angle of incidence, θ_i, and can be expressed as [2]:

$$\begin{aligned}\cos\theta_i = {} & \sin\delta\sin\phi\cos\beta - \sin\delta\cos\phi\sin\beta\cos\gamma \\ & + \cos\delta\cos\phi\cos\beta\cos\omega \\ & + \cos\delta\sin\phi\sin\beta\cos\omega\cos\gamma \\ & + \cos\delta\sin\beta\sin\omega\sin\gamma\end{aligned} \quad (16)$$

where:

δ = declination angle

β = tilt angle

ϕ = latitude angle

γ = the angle by which the collector is diverted from the south

Figure 8(b) shows these angles with respect to the plane under consideration. If the plane is a horizontal one then the angle of incidence is equivalent to the sum zenith angle θ_z. The expression for θ_z was introduced in section 1. In most of the applications where the solar collectors are maintained fixed, the collectors are sloped towards the equator, in this case $\gamma = 0°$ and the expression for θ_i is simplified to be:

$$\text{Cos } \theta_i = \text{Cos}(\phi - \beta) \text{ Cos } \delta \text{ Cos } \omega + \text{Sin}(\phi - \beta) \text{ Sin } \delta \quad (17)$$

It can be noticed that this incident angle is equivalent to the azimuth angle of locality of (ϕ') latitude such that, $\phi' = \phi - \beta$.

Defining R_b as the ratio of beam radiation on tilted plane to a horizontal plane then:

$$R_b = \frac{I_{bT}}{I_b} = \frac{\text{Cos } \theta_i}{\text{Cos } \theta_z} \quad (18)$$

The diffuse radiation on tilted surface is composed of two components.

i - Diffuse solar radiation from hemisphere, I_{dh}. This is given by the relation [4]:

$$I_{dh} = I_d \frac{1 + \text{Cos } \beta}{2} \quad (19)$$

where $(1 + \text{Cos}\beta)/2$ is known as the view factor of the tilted surface to the sky.

ii - Solar radiation diffusely reflected from the ground, I_{dg}. This is given by the relation [4]:

$$I_{dg} = I_{tot} \rho_g \frac{1 - \text{Cos } \beta}{2} \quad (20)$$

where ρ_g is the reflectivity of the surroundings. Reflectivities of 0.2 and 0.7 are recommended for grass and snow covered ground respectively [4]. The factor, $(1 - \text{Cos } \beta)/2$, is known as the surface view factor to the ground. The total diffuse radiation on tilted surface I_{dT} can be simply estimated by adding the two diffuse components,

$$I_{dT} = I_{dh} + I_{dg} \quad (21)$$

In conclusion, Solar Radiation and the position of the panel in regard to the incoming radiation play an important role in improving the efficiency as well as increasing the energy which can be harnessed from the Sun.

If a country has a lot of sunshine but very few labourers or expensive labourers, then an optimum tilt for panels should be used, which leads to additional energy saving of 20% compared with a position 5 degrees out of the optimum tilt. However, if we have a seasonal tracking, which can only be applied if the labour cost is not high, an additional 10% more can be achieved. If the cost is not vital, then Photovoltaic with tracking will result in almost 50% additional energy compared with an arbitrarily fixed position.

Chapter 3

THE CRYSTALLINE SILICON SOLAR CELL

F. C. TREBLE
Consulting Engineer, Farnborough, UK

3.1 Introduction

The crystalline silicon solar cell was one of the first types to be developed and it is still the most common type in use today. In this chapter, we describe its construction, how it works and its main performance characteristics. We next look at the various sources of energy loss in the device and consider what can be done to improve performance. The latest advances in cell design, including the bifacial cell, are described. Finally, the processes of manufacture are traced from the starting material, sand, to the finished product and reference is made to alternative crystal and cell fabrication processes which have been and are being developed to bring costs down.

3.2 Construction

Figure 3.1 shows the essential features of a crystalline silicon solar cell. It is made from a thin (250 - 400 µm) wafer cut from a pure silicon crystal which has been doped with boron. It can be square, as shown, square with cropped corners, circular, semi-circular or quadrant-shaped. Nowadays, it is usually 100 mm square or 100 mm diameter.

Phosphorus is diffused into the slice to form a p-n junction a fraction of a micron below the front or active surface. This is the so-called "n-on-p" cell. The alternative p-on-n type is made by diffusing boron into a phosphorus-doped wafer. The front metal contact is in the form of a narrow-fingered grid, while the back contact usually covers the entire back surface. The front surface has an antireflective coating.

3.3 How it works

To understand how a solar cell works, it is necessary to go back to some basic atomic concepts. In the simplest model of the atom,

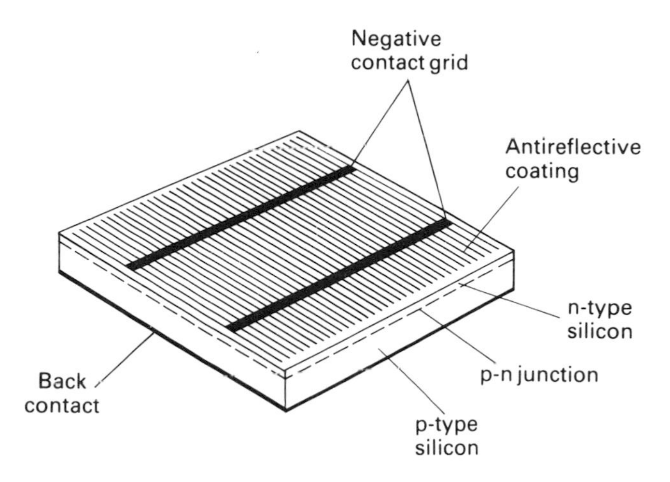

Figure 3.1. Crystalline silicon solar cell.

electrons orbit a central nucleus, composed of protons and neutrons. Each electron carries one negative charge and each proton one positive charge. Neutrons carry no charge. Every atom has the same number of electrons as there are protons, so, on the whole, it is electrically neutral. The electrons have discrete kinetic energy levels, which increase with the orbital radius. When atoms bond together to form a solid, the electron energy levels merge into bands. In electrical conductors, these bands are continuous but in insulators and semiconductors there is an "energy gap", sometimes called the "forbidden gap", in which no electron orbits can exist, between the inner or "valence" band and the outer or "conduction" band. Valence electrons help to bind together the atoms in a solid by orbiting two adjacent nucleii, while conduction electrons, being less closely bound to the nucleii, are free to move in response to an applied voltage or electric field. The fewer conduction electrons there are, the higher the electrical resistivity of the material.

In semiconductors, the materials from which solar cells are made, the energy gap E_g is fairly small. Because of this, electrons in the valence band can easily be made to jump to the conduction band by the injection of energy, either in the form of heat (thermal excitation) or light (photon excitation). This explains why the resistivity of semiconductors decreases as the temperature is raised or the material illuminated. The excitation of valence electrons to the conduction band is best accomplished when the semiconductor is in the crystalline state, i.e. when the atoms are arranged in a precise geometrical formation or "lattice".

At room temperature and low illumination, pure or so-called "intrinsic" semiconductors have a high resistivity. But the resistivity can be greatly reduced by "doping", i.e. introducing a very small amount of impurity, of the order of one in a million atoms. There are two kinds of dopant. Those which have more valence electrons that the semiconductor itself are called "donors" and those which have fewer are termed "acceptors".

In a silicon crystal, each atom has four valence electrons, which are shared with a neighbouring atom to form a stable tetrahedral structure. Phosphorus, which has five valence atoms, is a donor and causes extra electrons to appear in the conduction band. Silicon so doped is called "n-type". On the other hand, boron, with a valency of three, is an acceptor, leaving so-called "holes" in the lattice, which act like positive charges and render the silicon "p-type". The drawings in Fig. 3.2 are 2-dimensional representations of n- and p-type silicon crystals, in which the atomic nucleii in the lattice are indicated by circles and the bonding valence electrons are shown as lines between the atoms.

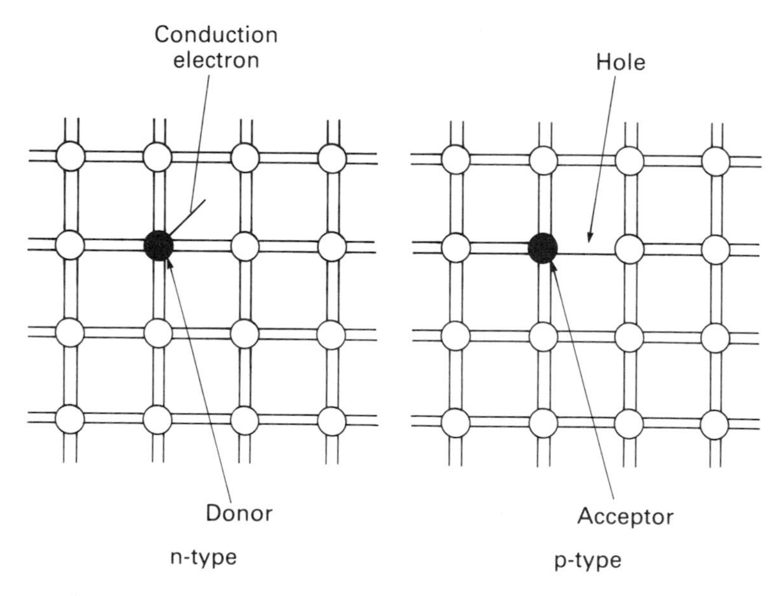

Figure 3.2. Donor and acceptor atoms in a silicon crystal.

Holes, like electrons, will move under the influence of an applied voltage but, as the mechanism of their movement is valence electron substitution from atom to atom, they are less mobile than the free conduction electrons.

As already explained, in an n-on-p crystalline silicon solar cell, a shallow junction is formed by diffusing phosphorus into a boron-doped base. At the junction, conduction electrons from donor atoms in the n-region diffuse into the p-region and combine with holes in acceptor atoms, producing a layer of negatively-charged impurity atoms. The opposite action also takes place, holes from acceptor atoms in the p-region crossing into the n-region, combining with electrons and producing positively-charged impurity atoms. The net result of these movements is the disappearance of conduction electrons and holes from the vicinity of the junction and the establishment there of a reverse electric field, which is positive on the n-side and negative on the p-side. This reverse field, as we shall see, plays a vital part in the functioning of the device. The area in which it is set up is called the "depletion area" or the "barrier layer".

When light falls on the front surface, photons with energy in excess of the energy gap (1.1 eV in crystalline silicon) interact with valence electrons and lift them to the conduction band. This movement leaves behind holes, so each photon is said to generate an "electron-hole pair". In crystalline silicon, electron-hole generation takes place throughout the thickness of the cell, in concentrations depending on the irradiance and the spectral composition of the light. Photon energy is inversely proportional to wavelength. The highly energetic photons in the ultra-violet and blue part of the spectrum are absorbed very near the surface, while the less energetic longer wave photons in the red and infrared are absorbed deeper in the crystal and further from the junction. Most are absorbed within a thickness of 100 µm.

The electrons and holes diffuse through the crystal in an effort to produce an even distribution. Some recombine after a lifetime of the order of one millisecond, neutralising their charges and giving up energy in the form of heat. Others reach the junction before their lifetime has expired. There they are separated by the reverse field, the electrons being accelerated towards the negative contact and the holes towards the positive. If the cell is connected to a load, electrons will be pushed from the negative contact through the load to the positive contact, where they will recombine with holes. This constitutes an electric current. In crystalline silicon cells, the current generated by radiation of a particular spectral composition is directly proportional to the irradiance. Some other types of solar cell, however, do not exhibit this linear relationship.

Another way of looking at the operation of a solar cell is to construct an energy diagram. Figure 3.3 is a typical one for an n-on-p crystalline silicon cell. It is a plot of the energy levels in the crystal (ordinates) as functions of distance from the front surface (abscissa).

In the n-region just below the front surface, the average energy of the charge carriers, called the "Fermi level", is near the top of the forbidden gap, with many electrons in the conduction band and few holes in the valence band. The opposite is true in the p-region, where the Fermi level is near the bottom of the forbidden gap.

The laws of thermodynamics require that the Fermi level must be the same in both regions. So, to satisfy this requirement, a potential barrier or reverse electric field is set up around the junction. Photon-generated carriers (electrons and holes) which reach the barrier are separated by it. The electrons can be visualised as rolling down towards the negative contact on the front surface and the holes floating upwards towards the positive contact.

3.4 Spectral response

The generated current is made up of contributions produced by photons with energy in excess of the semiconductor energy gap. If the incremental current density generated by unit irradiance

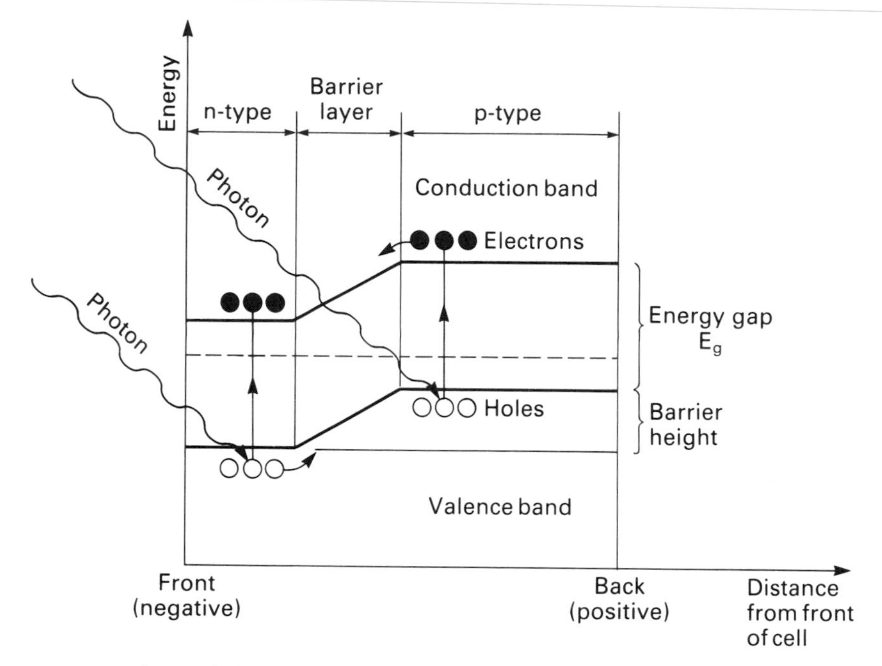

Figure 3.3. Energy diagram.

at a particular wavelength is plotted as a function of wavelength, the resulting curve is called the "absolute spectral responsivity" or, more commonly, the "absolute spectral response". Figure 3.4 shows a typical example for a crystalline silicon cell. Note that it covers the whole of the visible spectrum and part of the infrared, peaking sharply at about 900 nm.

The relative spectral response, a more easily measured parameter, is the absolute spectral response normalised to a maximum value of unity.

The generated current density of a solar cell in radiation of known intensity and spectral composition can be computed by multiplying the ordinates of the absolute spectral response by the corresponding coordinates of the absolute spectral irradiance distribution of the incident radiation and integrating the resulting products, thus:-

$$I_G = s(\lambda).E(\lambda).d\lambda \qquad (1)$$

where; I_G = the generated current density ($A.m^{-2}$)

$s(\lambda)$ = the absolute spectral response at wavelength λ ($A.W^{-1}$)

$E(\lambda)$ = the absolute spectral irradiance at wavelength λ ($W.m^{-2}.\mu m^{-1}$).

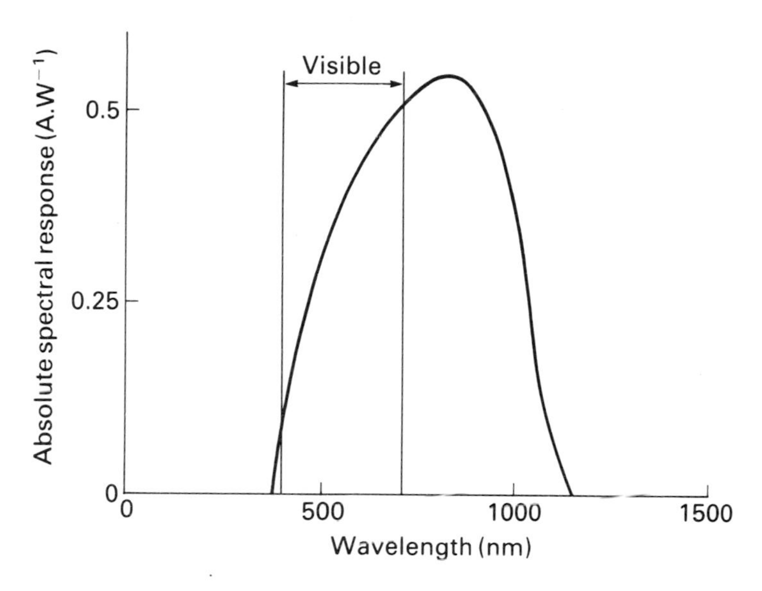

Figure 3.4. Spectral response.

As both the spectral response and the solar spectral irradiance curves are sharply peaked, it will be appreciated that a slight change in either can significantly affect the generated current. The performance of a solar cell is therefore very sensitive to the spectral content of the incident radiation.

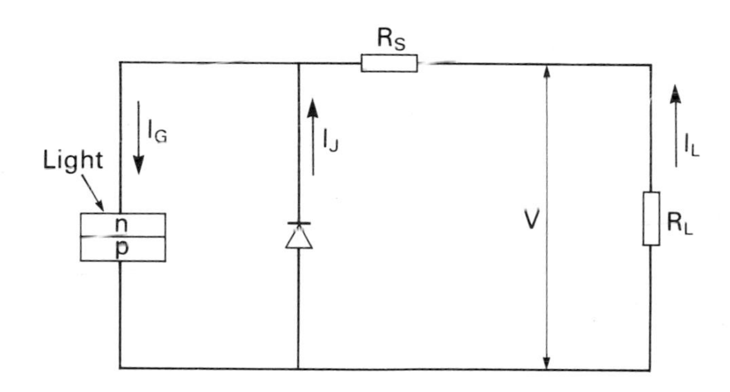

Figure 3.5. Equivalent circuit.

Figure 3.5 is the simplified equivalent circuit of a solar cell. It consists of a constant current generator shunted by the junction, which acts like a positively-biased diode. R_S is the lumped internal series resistance and R_L the load resistance. Shunt resistance and capacitance can normally be neglected.

The load current I_L is the difference between the generated current I_G and the junction current I_J:

$$I_L = I_G - I_J \quad (2)$$

$$= I_G - I_0 \left\{ \exp\left[\frac{q(V + I_L.R_S)}{A.k.T}\right] - 1 \right\} \quad (3)$$

where I_0 = the dark reverse saturation current of the diode, proportional to $\exp(-E_g/kT)$.

q = the charge on an electron (1.6×10^{-19} C).

k = the Boltzmann constant (1.38×10^{-23} J.K^{-1}).

T = the absolute temperature of the cell (K).

A = a constant between 1 and 2 (varies with type of cell)

V = the terminal voltage (V).

In the short-circuit condition, when there is no voltage across the terminals of the cell, I_J is very small and practically all of the generated current passes through the external link. The short-circuit current is therefore a useful measure of the generated current and the two terms are normally interchangeable.

From Equation (3), we can deduce the following expression for the open-circuit voltage at $I_L = 0$:-

$$V_{OC} = \frac{A.k.T}{q}. \ln\left[\frac{I_G}{I_0} + 1\right] \quad (4)$$

3.6 Current-voltage characteristic

Figure 3.6 shows how the output current and voltage of a modern crystalline silicon solar cell varies with load at a cell temperature of 25 °C and an irradiance of 1000 W.m^{-2}. The relationship with I_G and I_J (Equation 2) is indicated by broken lines. The short-circuit current I_{SC} may be read off at point A, where the curve crosses the current axis at V = 0 and the open-circuit voltage V_{OC} at point B, where the curve crosses the voltage axis at I = 0.

Maximum power P_{MAX} is represented by the area of the largest rectangle that can be fitted under the curve. It is produced at a voltage between 0.4 and 0.5V, depending on the base resistivity and the quality of the cell. In the case illustrated, P_{MAX} is 127 W.m^{-2} at 0.48V.

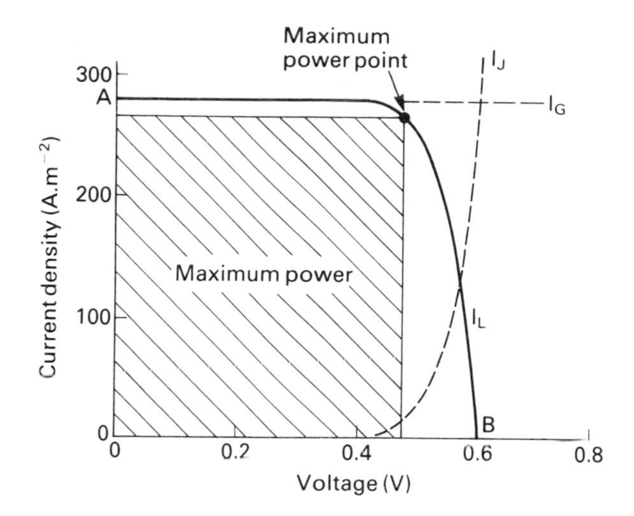

Figure 3.6. Current-voltage characteristic.

The conversion efficiency is the maximum output power, expressed as a percentage of the input power:-

$$\text{Conversion efficiency} = \frac{P_{MAX}}{\text{Irradiance x Area}} \text{ X } 100 \text{ \%} \qquad (5)$$

In efficiency calculations, the accepted convention among engineers is to use the total surface area of the cell, including the front contact and grid. But some research workers base the calculation on the active area, excluding the contact and grid. In the case illustrated in Fig. 3.6, the conversion efficiency, based on total area, is 12.7 %.

Obviously, the nearer the I-V curve is to a rectangular shape, the higher the maximum power and conversion efficiency. A measure of the rectilinearity is the "fill factor", which is defined as:-

$$\text{Fill factor} = \frac{P_{MAX}}{I_{SC} \cdot V_{OC}} \qquad (6)$$

Most modern crystalline silicon cells have fill factors exceeding 0.72.

3.7 Effects of changes of irradiance and temperature

Figure 3.7 shows the effects of changes of irradiance and temperature on the current-voltage characteristic. Changes of irradiance, such as one would experience in the course of a day,

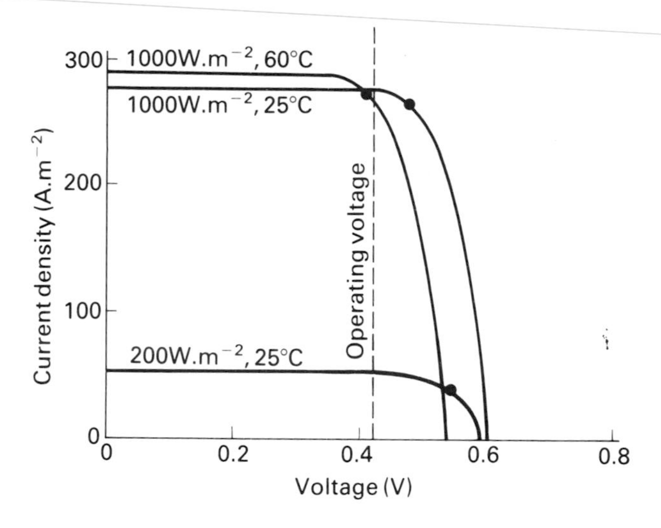

Figure 3.7. Effects of changes of irradiance and temperature.

affect the short-circuit current proportionally but have little effect on the open-circuit voltage, because of the logarithmic relationship (Equation 4). In concentrated sunlight, the short-circuit current remains proportional to the irradiance up to extremely high levels, provided the temperature is controlled at a constant value. As the open-circuit voltage also increases, albeit only slightly, one would expect the conversion efficiency to rise as the concentration ratio is increased. So it does, up to a point, but the effect of series resistance (Equation 3) progressively reduces the fill factor, offsetting the gain in efficiency and limiting the improvement that can be achieved.

Crystalline silicon cells respond very rapidly to sudden changes of irradiance, their time constant being about 20 μs.

An increase in temperature causes a slight rise in the short-circuit current but a sharp fall in open-circuit voltage and maximum power. Typical temperature coefficients for a 10 ohm.cm cell are:-

$$\frac{dI_{SC}}{dT} = 0.1 \text{ A.m}^{-2}\text{.deg C}^{-1}$$

$$\frac{dV_{OC}}{dT} = -2.2 \text{ mV.deg C}^{-1}$$

$$\frac{dP_{MAX}}{dT} = -0.5\% \text{ deg C}^{-1}$$

Because of the serious loss of power with increasing temperature, it is important in module and system design to ensure that the cells run as coolly as possible.

Note from Fig. 3.7 that the voltage for maximum power also falls with increasing temperature. However, provided the cell operating voltage does not exceed the voltage for maximum power at the highest operational temperature, the output current is little affected by changes of temperature below the maximum.

3.8 Performance improvement

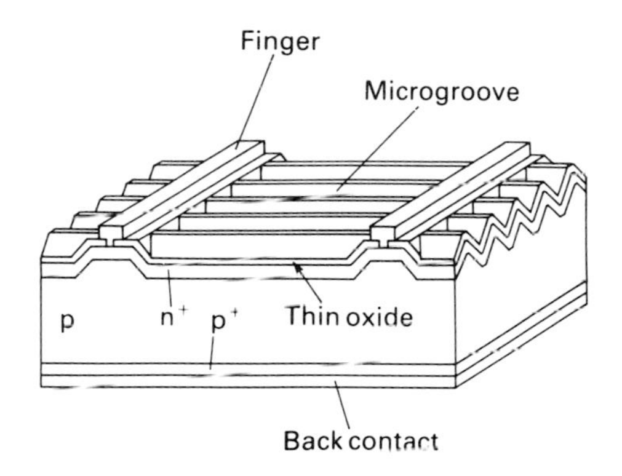

Figure 3.8. High-efficiency crystalline silicon solar cell (Green, 1985).

or rating purposes, the performance of photovoltaic devices is measured under Standard Test Conditions (STC). Insofar as flat-plate terrestrial devices are concerned, the International Electrotechnical Commission defines these conditions as:-

Reference spectral irradiance distribution (See Chapter 6).

Irradiance : 1000 $W.m^{-2}$

Cell temperature : 25 °C

The conversion efficiency of most commercial crystalline silicon solar cells is at present between 12 % and 15 %. Let us look more closely at the operation of the cell and see what can be done to improve performance beyond this level.

Energy losses in the cell can be grouped under nine headings:-

1) Front contact shading.

2) Reflection from the front surface.

3) Incomplete absorption.

4) Partial utilisation of photon energy.

5) Collection losses.

6) Voltage factor.

7) Leakage.

8) Series resistance.

9) Curve factor.

We shall discuss each of these in turn.

Front contact shading

The function of the front contact grid is to reduce the resistance encountered by the charge carriers as they move laterally through the thin highly-doped surface layer to the collection point. Unfortunately, at the same time, it prevents some of the incident radiation reaching the cell. Good design is therefore a matter of striking the best compromise between low series resistance (closely-spaced, highly conductive grid lines with good adhesion and low contact resistance) and high light transmittion (fine, widely-spaced grid lines). As will be seen, there are methods for producing extremely fine grid patterns but they are expensive and only cost-effective in special cases. Typically, the loss from front contact shading is 10 % in commercial cells.

Reflection

Some of the incident radiation is lost through reflection from the front surface of the cell. In modern cells, this loss has been cut to about 3 % by the application of a single layer antireflective coating of titanium oxide or tantalum pentoxide, one quarter of a wavelength thick, optimised to a wavelength of 600 nm. Some cells also have a textured front surface of minute pyramids, produced by a special etching process, which widens the region of low reflectivity on both sides of the optimum wavelength. A similar effect has also been achieved by micro-grooving the surface. Multilayer antireflective coatings give further marginal improvement but are generally not cost-effective.

Incomplete absorption

Photons with energy less than the energy gap are wasted, as they simply generate heat or pass right through the silicon and are absorbed in the back contact. The higher the energy gap, the greater the wastage.

Partial utilisation of photon energy

Many of the photons which generate electron-hole pairs have more energy than is needed for this operation. The excess energy is dissipated as heat. In this case, the higher the energy gap, the smaller the wastage. Combining this and the previous loss factor, it turns out that an energy gap of 0.9 eV gives maximum utilisation of the terrestrial solar spectrum in the generation of carriers. In this optimum case, the used energy amounts to about 46 % of the total incident energy. The loss is inherent and cannot be reduced in a single material cell once the semiconductor and, with it, the energy gap has been selected. However, crystalline silicon, with an energy gap of 1.1 eV, utilises about 44 % of the incident energy and so is not far from the optimum.

Collection losses

Only those electrons and holes that are separated by the barrier layer and reach the contacts contribute to the output current. The rest, which recombine in the bulk of the silicon and at the front and back surfaces, generate heat and do no useful work. The ratio of the useful charge carriers to the total generated is called the "collection efficiency". It can be increased by reducing the recombination rate and facilitating the movement of carriers towards the barrier layer. However, another important factor affecting the collection efficiency is the spatial distribution of the carriers when they are generated. This cannot be changed, as it is dependent on the absorption characteristic of the silicon and the spectral irradiance distribution of the radiation.

Collection in the base region of the cell between the junction and the back contact is optimised in modern cells by:-

(a) using good quality material with a high minority carrier lifetime (the average time elapsing between the generation and recombination of electrons in p-type material and holes in n-type),

(b) choosing processes which minimise degradation of the minority carrier lifetime in manufacture and

(c) forming an impurity concentration gradient or "back surface field", which accelerates minority carriers from the back surface towards the junction.

Another possibility, particularly in thin cells, is to use a highly-reflective material at the interface between the silicon and the back contact, so that long-wavelength photons reaching the interface before being absorbed are reflected and given another chance to generate carriers. Recombination of carriers at the back contact can be reduced by using a grid configuration, as at the front, but, in doing this, care must be taken not to cancel the gain by increased series resistance.

The surface layer of the cell, between the junction and the front contact, is very highly doped and therefore has an extremely short minority carrier lifetime. Collection of carriers generated in this region can, however, be enhanced by:-

(a) introducing the dopant in such a way as to produce a continuous impurity concentration gradient, in order to accelerate the carriers towards the junction.

(b) placing the junction very near the front surface (a fraction of a micron) and

(c) reducing the surface recombination velocity by depositing a thin passivating layer of oxide on the front surface.

In most good quality commercial cells, collection losses are now less than 5 %.

Voltage factor

The open-circuit voltage is always less than the energy gap E_g for the following reasons:-

(a) The barrier height, which is equal to the maximum applicable forward voltage on a p-n junction, is determined by the difference in Fermi levels in n- and p-regions flanking the junction. (See Fig. 3.3). The Fermi level, a function of the type of impurity, the impurity concentration and the temperature, is normally located within the forbidden gap. So the barrier height is less than E_g.

(b) A voltage equal to the barrier height can only be obtained at extremely high injection levels, which can never be reached by photon absorption from sunlight.

Since I_0 in Equation 4 is proportional to $\exp(-E_g/kT)$, an increase in E_g will raise the open-circuit voltage. But, if E_g is increased above the level for maximum generation of carriers (0.9 eV), current generation will be reduced. Hence, there is an optimum value of E_g for any particular spectral irradiance distribution, at which the product of I_{SC} and V_{OC} is a maximum. For terrestrial sunlight, the theoretical optimum is 1.4 eV. Again, silicon is near the optimum.

The Fermi level can be pushed nearer the valence band in the p-type base region by increasing the acceptor doping concentration. This increases the barrier height and with it the open-circuit voltage, while reducing the base resistivity. For this reason, terrestrial solar cell manufacturers tend to favour base resistivities in the 0.5 to 3 ohm.cm range rather than the 10 ohm.cm material which is commonly used for space applications.

Leakage

Leakage across the junction may be caused by imperfections in the junction, especially at the edges of the cell where surface conduction paths are easily formed. However, losses from this source are negligible in a well fabricated cell.

Series resistance

Series resistance in the cell causes a flattening of the current-voltage characteristic and a consequent loss of output power. It can be minimised by good contact grid geometry, good ohmic contacts and low sheet resistance in the surface layer.

Curve factor

Because the shape of the current-voltage characteristic depends on the junction characteristic (Fig. 3.6), the maximum power is always less than the product of I_{SC} and V_{OC}, even with no series resistance. The quality of the junction, like V_{OC}, depends on I_O and improves as E_g is increased.

Theoretical work by Green (1984) has suggested that the intrinsic limit of the conversion efficiency of a crystalline silicon solar cell at STC is about 30 % and a realistic experimental target is 25 %. Considerable progress towards this goal has been made as a result of research and development over many years. Indeed, efficiencies higher than 20 % have been achieved in small area devices made under laboratory conditions (Green, 1985). Figure 3.8 shows the construction of this device, which Green calls the "microgroove passivated emitter solar cell". Its main features are:-

1) Made from low-resistivity (0.2 ohm.cm), highly-refined (float zone) single crystal silicon, for high V_{OC} and high minority carrier lifetime.

2) Microgrooves in the front surface, produced by anisotropic etching through a fine photolithographic mask. These grooves improve performance by (a) reducing reflection losses to less than 1 %, (b) coupling light into the cell at an oblique angle, thus increasing the photon absorption rate and (c) reducing series resistance losses in the direction of the grooves by a factor of $\sqrt{3}$.

3) Double-layer antireflective coating.

4) Front surface covered with a thin (60 Å) passivating layer of silicon dioxide to reduce surface recombination.

5) Vacuum-deposited contact fingers and busbars of low contact area, to interfere as little as possible with the passivating layer but of adequate cross-section to keep resistive losses low,

6) A shallow junction.

7) A continuously-graded n-type doping profile in the front surface layer, to accelerate minority carriers towards the junction.

8) A back surface field to assist the movement of minority carriers in the base region towards the junction.

The performance of Green's cell in AM1.5 sunlight at an irradiance of 1000 $W.m^{-2}$ and a cell temperature of 28 °C is:-

Short-circuit current	38.3 $mA.cm^{-2}$
Open-circuit voltage	661 mV
Fill factor	82.4 %
Conversion efficiency	20.9 %

An alternative version of this cell, developed by Green (1985) to eliminate the costly photolithographic masking, vacuum deposition of contacts and 2-layer antireflective coating, has a pyramidically-textured antireflective front surface and plated contact fingers located within fine grooves formed by a laser. The latter feature reduces shading loss to 2 %. Cells fabricated in this way from commercial solar grade (0.5 - 3 ohm.cm) Czochralski silicon have achieved conversion efficiencies as high as 18.6%.

3.9 Bifacial cells

Another way of increasing the output of a solar cell is to use not only the light that falls on the front surface but also that which is reflected on the back from the ground and the surroundings. This is done in the "bifacial" cell. Many bifacial structures have been developed at the Polytechnical University of Madrid (Sala, 1983) and one version, the p^+ n n^+, is being manufactured in Spain. The starting material for this cell is a thin 10 cm diameter wafer of high resistivity n-type silicon crystal. Boron is diffused into one surface and phosphorus into the other, forming two very shallow junctions, p^+-n and n-n^+ respectively. Each surface is finished with a metallised contact grid and a single layer antireflective coating. Similar cells may be made from high-resistivity p-type material.

Cells of this type have been shown to have a conversion efficiency of about 12 % at 25 °C, when irradiated on either the front or the back. When they are installed near highly reflective surfaces, such as sand, snow or white walls, the back irradiation is commonly half of that on the front, thus yielding a 50 % gain in output over the equivalent conventional cell.

3.10 Manufacture

3.10.1 The starting material

Silicon comes from silica (silicon dioxide), which is found extensively in nature as high-grade sand and quartz. The first step in the extraction process is to melt the silica in an arc furnace and add a controlled amount of carbon. This combines with the oxygen in the silica to form carbon dioxide and leave behind "metallurgical grade" silicon, containing about 1 % of impurities.

The metallurgical grade material is refined into semiconductor grade silicon, the feedstock of the solid-state electronic industry, by the Siemens process, which reduces impurities to one atom per 10^9 atoms of silicon. This material is in the form of tiny crystals and is commonly referred to as "polysilicon". It is expensive ($ 50 - $ 70 per kg in 1984), primarily because of the large amount of thermal energy used in its production.

Fortunately, solar cells do not need starting material as pure as semiconductor grade. They can tolerate about ten times as much impurity. So, much effort has been devoted to developing a cheaper "solar grade" polysilicon specifically for solar cells (Aulich, 1986, Schwirtlich, 1986 and Callaghan, 1986). However, no production plants using these processes are yet (1988) in operation.

3.10.2 Crystal growing

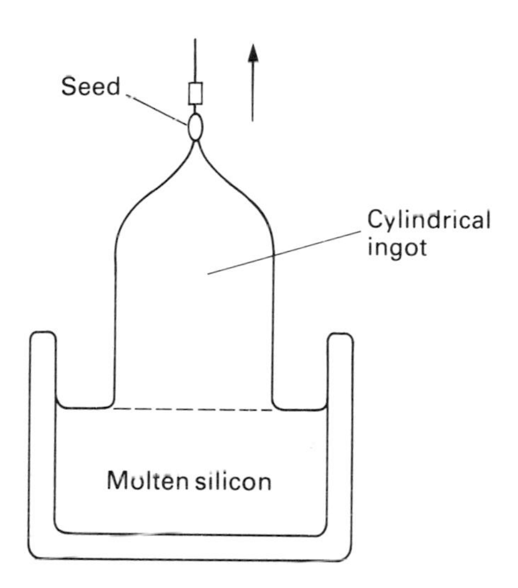

Figure 3.9. Czochralski process.

The next step is to transform the polysilicon into a single crystal. In the Czochralski process (Fig. 3.9), the polysilicon is re-melted in a special furnace with close temperature control. A trace of boron is added to make the silicon p-type with the required resistivity. Alternatively, phosphorus is added for n-type. A seed crystal on the end of a rod is dipped into the melt and then very slowly drawn upwards under close control. The molten silicon solidifies on the seed, reproducing the crystal pattern in a cylindrical ingot. In the latest production units, single crystal ingots can be formed up to 150 mm in diameter and 1 m long. Growth rates as high as 10cm/hour are possible. To reduce costs, efforts are being made to draw multiple large ingots from a single crucible at higher speeds and with continuous replenishment of the melt.

In the "float zone" process, the polysilicon feedstock is first cast into a rod of the required diameter. The rod is mounted vertically in a vacuum chamber and a molten zone, produced by radio frequency heating, is moved slowly from one end of the rod to the other, the liquid silicon in the molten zone being held in place by surface tension. After the molten zone has passed, the silicon forms into a single crystal. As the impurities are more soluble in silicon in its molten form, the molten zone sweeps most of the impurities to one end of the ingot, where they can be discarded. Higher purity can be obtained with more passes of the molten zone. This is a more expensive operation than the Czochralski process but it produces silicon with a higher minority carrier lifetime.

In recent years, a cheaper alternative to single crystal silicon, known as "multicrystalline" or "semicrystalline" material, has been developed specifically for the solar cell market. It is sold under the proprietary names "SEMIX"(Solarex), "SILSO (Wacker) and "POLYX" (Photowatt International). The pure molten silicon, suitably doped, is poured at about 1100 °C into square crucibles and there allowed to solidify under close control. The resulting ingot consists of large individual crystallites, which can be clearly discerned in the finished cell. Ingots up to 20 cm cube are produced in this way.

Another alternative, under development by Crystal Systems, is the heat exchange method (HEM), in which a 50 kg charge of silicon feedstock is melted in a 32 cm square silica crucible and then directionally solidified from the bottom under microprocessor control. This results in the formation of large vertically-orientated columnar crystals.

The multicrystalline casting process is faster and uses much less energy than the HEM process. Both use less silicon than the Czochralski method. There is evidence that less pure feedstock is needed for the multicrystalline casting process than for Czochralski crystal, because the impurities tends to get trapped in the grain boundaries of the multicrystalline material. However, in the present state-of-the-art, solar cells made from multicrystalline material are somewhat less efficient than single crystal cells.

3.10.3 Slicing

After cooling, the crystal ingots are sawn into wafers of the required size. Cast multicrystalline ingots are sawn into blocks of the required cross-section (usually 100 mm square) before wafering. Much work has gone into the development of machines to increase sawing speeds, reduce wafer thickness and reduce sawdust ("kerf") losses. The cutting medium is usually a slurry containing diamond abrasive. The saws can be annular, with the cutting edge on the inside diameter, or in the form of stretched blades or wires. Modern machines can cut up to 1000 wafers at a time in thicknesses down to 250 μm but kerf losses still amount to about 50 %.

3.10.4 Silicon ribbon

To reduce silicon consumption and avoid the two expensive steps in producing wafers (crystal growing and cutting) with the attendant kerf loss, several methods have been developed to produce crystalline silicon directly from the melt in the form of a thin ribbon. To date, the only techniques to have reached the production stage are dendritic web and edge-defined film-fed growth (EFG).

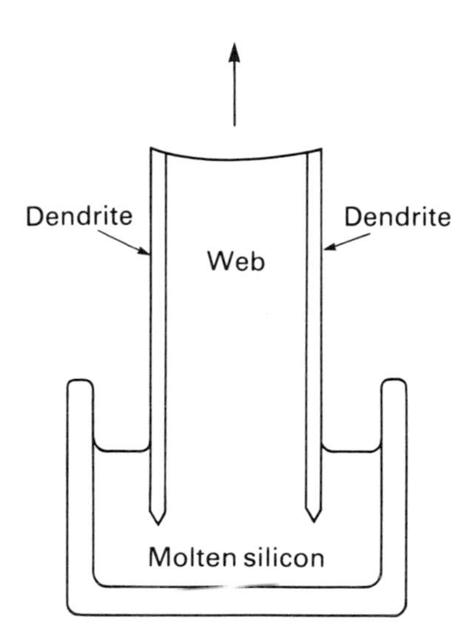

Figure 3.10. Dendritic web process.

In the dendritic web process developed by Westinghouse (Fig. 3.10), a thin sheet of single crystal silicon is produced by the solidification of a liquid film that forms between two filaments, called dendrites, that grow downwards into the molten silicon meniscus and control the width of the ribbon. Good quality ribbon 150 μm thick has been produced by this technique in widths of up to 6.9 cm and lengths of up to 17 m, at a growth rate of about 6 $cm^2.min^{-1}$. The process is automatically controlled, with continuous melt replenishment. Solar cells with an efficiency of 16 % have been made from dendritic web material.

The EFG process was developed by Mobil Solar Energy Corporation. Here, (Fig. 3.11) silicon ribbon is pulled from the top of a die through which molten silicon is fed by capillary action. Early attempts to produce single and multiple ribbons at economic rates were abandoned because it proved to be impossible to control inbuilt stresses. The stress problem was eventually solved by growing the ribbon under close automatic control in the form of a nine-sided prism and cutting the prism afterwards into nine 5 cm wide strips. This technique is capable of producing ribbon at the rate of nearly 150 $cm^2.min^{-1}$. However, EFG ribbon is not yet as good as dendritic web, the best cells so far produced from it being about 11.5 % efficient.

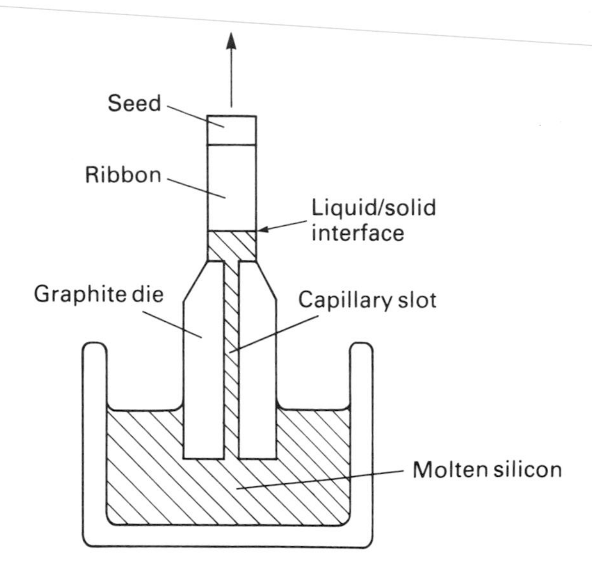

Figure 3.11. EFG ribbon process.

3.10.5 Cell processing

The fabrication of a solar cell from a crystalline silicon wafer can be divided into the following steps (van Overstraeten, 1980):-

Surface preparation

Junction formation

Back surface field formation

Contact metallisation

Antireflective coating

The aim of the surface preparation is to remove work damage caused by the wafering saw and leave clean surfaces for the subsequent operations. This is done by removing 10 - 20 µm in an acid or alkaline etch. Acid etchants (e.g. 4 % nitric or 16 % hydrofluoric acid) give good results whatever the crystal orientation but are expensive. Alkaline etchants, such as 30 % hot sodium hydroxide, are about ten times cheaper but are orientation-dependent. They can be used to produce an antireflective texturised surface on <100> material.

The most widely used method of forming the p-n junction is by diffusion of the dopant out of the gaseous phase at 850 - 950 °C. The prepared wafers are loaded into a quartz holder and placed in a temperature-controlled electric furnace, where they are exposed to the gas. For n-type diffusion, the gas is commonly phosphorus oxy-chloride, $POCl_3$. This technique can give good results but uniformity may be a problem for very shallow junctions. Other disadvantages are that it is a batch process and the diffusion affects both sides of the wafer, so that the back parasitic junction has to be removed.

An alternative method is to deposit a solid or liquid dopant layer on the front surface by chemical vapour deposition or by a spin-on, spray-on or screen-printing process and then drive in the dopant by heating to 850 - 950 °C. The CVD/drive-in process yields uniform dopant concentration profiles but uses highly pure and relatively expensive gases. Spin-on and spray-on processes are easily automated but result in a rather low fill factor, probably because of non-uniformities. They also produce a parasitic junction on the edges of the wafer, which must be removed by etching or laser scribing. Diffusion out of screen-printed doped paste is also easy to automate. The equipment is cheap and the edge junction can be avoided by suitable masking.

Yet another method of producing the junction is by ion implantation, as described by Kirkpatrick (1980). The dopant ions are implanted at an energy of the order of 10keV, followed by thermal, laser or electron annealing. This process gives excellent reproducible control of the doping profile and is suitable for automation. But the equipment is expensive and is only cost-effective on a large scale.

Other methods, such as laser-induced diffusion from a vacuum-deposited thin layer of dopant, epitaxial growth, etc. are being developed but have not yet been adopted for production.

In the processing of multicrystalline cells, junction formation is often followed by a hydrogen passivation step, as this has been found to reduce grain boundary losses.

All these methods for p-n junction formation can, in principle, be used to produce a back surface field. The most common method in n-on-p cells is diffusion from an aluminium film deposited on the back surface by evaporation or screen printing. If a similar method is used for the front junction, the drive-in of both dopants can be carried out simultaneously. BSF formation by diffusion from a gas is undesirable, as it entails another high temperature step, with consequent degradation to the minority carrier lifetime.

The metallisation must make good ohmic contact with the silicon and have low series resistance, good adherence and good solderability. The front grid must, at the same time, provide optimum transmission of light. There are several methods of applying the contacts.

Vacuum evaporation of aluminium/silver or titanium/palladium/silver through metal masks yields efficient cells but it is a batch process which uses expensive materials rather wastefully.

Electroless nickel plating through a mask etched in photoresist or in the antireflecting coating, followed by copper plating or solder dipping is another batch process, which can give good results if carefully controlled. Problems due to poor adhesion of the nickel, caused by an oxide layer on the silicon, can be overcome by first applying an electroless plated layer of palladium and heat treating it to form palladium silicide. Photomasking produces fine grid lines but it is not a particularly cheap process.

In recent years, screen-printed contacts have been developed to a high standard of reliability. The technique is well known in the elctronics industry and lends itself to automated mass production. The equipment is cheap and easy to maintain. Little of the material is wasted. The process can be combined with screen-printing techniques for junction and BSF formation. The front metallisation is usually done with a silver paste, which is dried and sintered at about 700 °C after application. If desired, the sintering can be done through the antireflective coating. Although fine grid lines are not possible, 150 µm being the usual width, good adherence and low contact resistance can be achieved. The back contact is formed with aluminium paste, followed by a second screening of solderable silver/palladium paste. Alternatively, a one-step process using a mixture of aluminium and silver pastes can be used. Copper and nickel pastes are being investigated as cheaper alternatives to silver.

The single layer antireflective coating of silicon monoxide, titanium oxide or tantalum pentoxide can be applied by vacuum evaporation, sputtering, spin-on, spray-on or screen-printing techniques. Vacuum evaporation and sputtering produce excellent, well-controlled layers but are expensive batch processes. Spin-on and spray-on are cheap continuous processes but thickness control is more difficult. Spin-on is not suitable for textured surfaces or square cells.

After manufacture, all cells are visually inspected for faults and then tested under a solar simulator. The test may consist of the measurement at STC of a complete current-voltage characteristic or it may be confined to measuring the short-circuit current, load current at a selected operating voltage near the maximum power point and the open-circuit voltage. For matching purposes, the cells are usually graded according to their load current. Testing and grading are often automated.

Chapter 3 References

Aulich, H.A. "Removal of carbon and silicon carbide from silicon prepared by advanced carbothermic reduction". 7th EC Photovoltaic Solar Energy Conference, Sevilla, Spain, October 1986, pp. 731-735.

Callaghan, W.T. "Crystalline silicon photovoltaic arrays: A final summary of Flat-Plate Solar Array (FSA) Project work, (1975-1986)". ibid, pp. 792-799.

Green, M.A. IEEE Trans. Electron Devices, ED-31, 1984, p. 671.

Green, M.A. "Improvements in silicon solar cell efficiency". 18th IEEE Photovoltaic Specialists' Conference, Las Vegas, Nevada, USA, October, 1985, pp. 39-49.

Kirkpatrick, A.R. et al. "Low-cost ion implantation and annealing technology for solar cells". 14th IEEE Photovoltaic Specialists' Conference, San Diego, CA., USA, January 1980, pp 820 - 824.

Overstraeten, R.J. van, "Advances in silicon solar cell fabrication". 3rd EC Photovoltaic Solar Energy Conference, Cannes, France, October, 1980, pp. 257 - 262.

Sala, G. et al. "Albedo-collecting photovoltaic bifacial panels". 5th EC Photovoltaic Solar Energy Conference, Kavouri, Athens, Greece, October 1983, pp. 565 - 569.

Schwirtlich, I.A. "Solar grade silicon". 7th EC Photovoltaic Solar Energy Conference, Sevilla, Spain, October 1986, pp. 736 - 739.

Chapter 4

CRYSTALLINE SILICON PHOTOVOLTAIC MODULES

F. C. TREBLE
Consulting Engineer, Farnborough, UK

4.1 Introduction

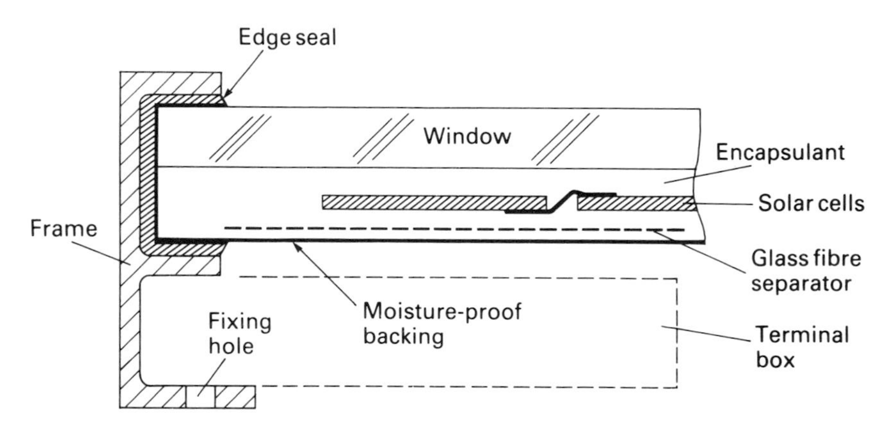

Fig. 4.1. Module construction.

A module is the basic building block of a photovoltaic solar array. Figure 4.1 illustrates its principal features. It consists of one or more solar cells, interconnected to produce the required power and voltage. The cells are encapsulated between a transparent window and a moisture-proof backing to insulate and protect them from the weather and accidental damage. Fixing holes or clamps are provided for securing the module to a supporting structure and a terminal box or pigtail leads for connecting it to other modules or the load.

4.2 Design requirements

Modules must be capable of reliable maintenance-free operation for many years in the environmental conditions for which they are intended. The current target is a lifetime of 30 years.

Window and encapsulating materials must be highly transparent to radiation over the response range of the solar cells, 350 nm to 1200 nm in the case of crystalline silicon. The transparency must not be unduly affected by prolonged weathering and exposure to sunlight. The window must have good impact resistance against hailstones and accidental knocks. The surface must be abrasion-resistant, non-staining, hard, smooth and flat to promote self-cleaning by wind, rain or spray. It should be free of projections which might provide lodgement for water, dust and other foreign matter.

To keep operational temperatures as low as possible and thereby maximise performance, the module should be designed to absorb the minimum of unused solar energy and take full advantage of radiative, convective and conductive cooling.

The encapsulation system must be highly resistant to the permeation or ingress of gases, vapours and liquids, as condensation on the cells and circuitry may cause short-circuits or galvanic corrosion. Failure of adhesive bonds in the assembly will increase the rate of moisture absorption and chemical reactions at the interfaces. It will also affect adversely the cooling of the module under operational conditions. Delamination between the window and encapsulant or between the encapsulant and the front surfaces of the cells will increase reflection losses. Therefore, the materials in the assembly must be compatible and the bonds between them must be capable of withstanding the extremes of temperature and the repeated thermal cycling that will be experienced in use.

For the best performance, series-connected cells should be matched in terms of short-circuit current within, say, ± 2 %, otherwise the cell with the lowest short-circuit current will limit the output of the module under certain load conditions. Similarly, if two or more series strings are connected in parallel, they should be voltage matched.

Electrical connections between the cells should preferably be duplicated in the interests of reliability. The cells must be well insulated against the high voltages possible when several modules are connected in series.

The module must be strong and rigid enough to support the fragile cells before, during and after installation in the array. It should be capable of accommodating slight imperfections or distortions in the support structure, withstand wind-induced vibrations and take the loads imposed by high winds, snow and ice.

The module should be easy to mount, interconnect and replace. Mountings, terminals and connectors should be non-corrosive.

As yet, there is no consensus on the optimum size of module, although most nowadays are in the 30 W to 50 W range. Small modules are cheaper to replace and have advantages in automated manufacture and testing but large ones (100 W - 150 W) yield a higher module efficiency because there is proportionally less frame area, need simpler support structures and require less on-site labour. Pre-assembled panels of small modules may prove to be the best compromise.

Finally, production costs must be kept as low as possible. Since the cost of the materials is, by and large, proportional to the area of the module, the cell packing factor, i.e. the ratio of cell area to total area, should be as high as is consistent with adequate spacing. Obviously, square or rectangular cells are better than circular or semicircular ones in this respect. As many of the production processes as possible should be capable of automation.

4.3 Encapsulation systems

The encapsulation system consists of the window, the backing and the encapsulant. Let us consider each of these in turn.

Tempered low-iron glass is now the most popular window material. It is highly transparent, stable and impervious to moisture and gases. It is strong enough to act as a mechanical support and protection for the cells. In fact, a thickness of 3.2 mm has been found sufficient to withstand the impact of 32 mm diameter hailstones at their terminal velocity. Glass has a high thermal emittance, an important property for a cool-running module. It has a hard, smooth, abrasion-resistant, easily cleaned surface. But it is rather heavy and not particularly cheap.

Plastic windows, although lighter and sometimes cheaper, are generally less stable and not so resistant to abrasion and contamination. Moreover, they are all to some extent permeable to moisture and gases. Silicone elastomers are among the most permeable but polyesters and acrylics are reasonably good in this respect. Unlike glass, transparent plastics suffer some transmission loss after prolonged exposure to ultra-violet radiation. Epoxy resins are the worst affected and acrylics, polyesters and uv-stabilised polycarbonate (Lexan) among the least affected.

Among backing materials may be listed glass, aluminium, stainless steel, glass-reinforced polyester sheet and plastic-coated metal foils. Glass, although heavy, is the most moisture-proof. It also enables most of the unused radiation to pass through the module, thereby reducing the operating temperature. This is, of course, less an advantage with square or rectangular cells than with round ones. Aluminium and stainless steel are strong and facilitate the conduction of heat from the cells to the back surface. However, bare polished metal is a poor thermal emitter and great care must be taken to ensure that the solar cells are adequately electrically insulated from it. Polyester sheet is light and cheap and is a good electrical insulator but it is not a perfect barrier against moisture. Plastic-coated metal foil combines cheapness and good thermal conductivity with good insulation and weather resistance. It is widely used in the latest modules, although glass backing is still favoured by some manufacturers. If the plastic is white, light reflected from the spaces between circular cells and re-reflected by the window has been found to give a slight gain in performance.

Transparent silicone elastomer used to be the most commonly used encapsulant because of its high transparency and resistance to fatigue failure after repeated heating and cooling. However, it has now largely been superseded by polyvinylbutyral (PVB) and ethylene vinyl acetate (EVA), which are cheaper, easier to apply and give a better bond to the window, cells and backing.

To minimise the possibility of moisture ingress at the edges, the most vulnerable parts of the module, most designs incorporate a separate edge seal under the frame.

Figure 4.2 shows two examples of modern (1988) design. Both modules have a window of toughened high-transmission glass, EVA encapsulant and a white Tedlar/aluminium foil/Tedlar laminate backing. The assembly is mounted in a frame of anodised aluminium alloy, with a silicone edge seal to provide shock resistance and an additional moisture barrier. Module A has 34 100 mm diameter circular monocrystalline silicon cells in series and produces 39W at 16.5V under Standard Test Conditions. Module B, with 35 100 mm square multicrystalline silicon cells in series, produces 44 W at 16.5 V under the same conditions. The positive and negative connections to the cell string are brought out to a sealed terminal box fixed to the frame at the back of the module. Module A is 1073 mm long, 415 mm wide and 38.5 mm deep while Module B is 982 mm long, 436 mm wide and 40 mm deep. Each weighs 6.5 kg.

4.4 Manufacture

Modern modules similar to those shown in Fig. 4.2 are made in the following way:-

1) The required number of current-matched cells is selected.

2) Interconnects are attached by soldering or ultrasonic welding to the back contacts of the cells ("cell tabbing").

3) The cells are located in a positioning jig to ensure accurate spacing and the interconnects from the back contact of one cell are attached to the front contact of the next, thereby connecting them in series.

4) Bus ribbons are soldered to the end cells for connection to the output terminals.

5) The connected cell circuit is visually inspected and its electrical performance is checked under a simulator to ensure that it is within the prescribed limits.

6) The prepared glass window is washed, rinsed in de-ionised water and dried.

7) The EVA, which comes in the form of a soft, pliable lamina, is cut to size. Two sheets are required.

Fig. 4.2.

8) The plastic-coated foil backing and a piece of open-weave fibreglass cloth are cut to size.

9) The components of the assembly are laid up in the following order from the bottom: backing laminate, fibreglass cloth, EVA, cell circuit, EVA, glass window.

10) The assembly is subjected sequentially to vacuum (1 mbar), heat (200 °C) and pressure (1 atmosphere) in a laminator. This extracts the air, softens the EVA and forces it between the cells and through the fibreglass to form a strong bond. The fibreglass cloth serves to protect the backing from damage by protrusions from the back side of the cell circuit. The whole laminating cycle takes under 10 minutes and is automatically controlled.

11) The assembly is placed in an oven to complete the curing of the EVA.

12) The edges of the laminated assembly are trimmed, sealed with adhesive tape and fitted with a rubber gasket.

13) An extruded aluminium alloy frame is fitted around the laminated assembly.

14) The positive and negative bus ribbons are connected to output leads or to a terminal box fixed to the frame at the back of the module.

15) The completed module is visually inspected and subjected to performance and insulation tests.

All the equipment necessary for a module production line, comprising cell tabber, circuit assembly station, glass preparation station, module lay-up station, laminator, curing station, final assembly station and solar simulator, is obtainable from specialist suppliers.

4.5 Power rating

The rated or peak power of a photovoltaic solar cell, module or array is defined as its power output at the rated voltage under Standard Test Conditions (STC), as defined in Chapter 3. The rated voltage is the nominal operating voltage at which the device is designed to produce near maximum power at STC.

The rated power is normally expressed in terms of "peak" watts, kilowatts or megawatts (Wp, kWp or MWp).

4.6 NOCT and energy rating

For much of its operational life, a module will be working in irradiances lower than 1000 Wm^{-2} and at temperatures higher than 25 °C. The rated or peak power is therefore a limited criterion for comparing the performance of competing designs. What is needed is a rating more closely related to the electrical energy the module will yield in an average day, month or year in the chosen location. This, after all, is the most important consideration from the user's point of view.

As a first step in tackling this problem, the concept of a "nominal operating cell temperature" (NOCT) has been established. This is defined as the equilibrium mean solar cell junction temperature within an open-rack mounted module in the following Standard Reference Environment:-

Module tilt angle:	at normal incidence to the direct solar beam at local solar noon.
Total irradiance:	800 Wm^{-2}
Ambient temperature:	20 °C
Wind speed:	1 ms^{-1}
Electrical load:	Nil (open circuit)

NOCT can be used by the system designer as a guide to the temperature at which a module will operate in the field. It is therefore a useful parameter, which should be taken into consideration, together with rated power, when comparing different modules. However, the actual operating temperature at any particular time is affected by the mounting structure, irradiance, wind speed, ambient temperature, sky temperature and reflections and emissions from the ground and nearby objects. For accurate performance predictions, all these factors must be taken into account.

As a further advance, it has been suggested (Gay, 1982) that an energy rating be instituted, based on the maximum energy output of the module in the course of a "standard solar day". The standard day would be specified in terms of profiles showing how the global irradiance, ambient temperature and air mass vary hour by hour. The maximum energy output would be computed from these profiles, a model relating the global irradiance to the irradiance on the module, the NOCT of the module and its I-V characteristics at various irradiances and temperatures. This suggestion is still under consideration by standards authorities. One problem is which "standard day" to specify. It might be necessary to specify more than one, to suit different regions of the world.

4.7 Hot-spot effect

In the early days of photovoltaics, modules sometimes failed through what is known as the "hot-spot" effect. This phenomenon can be provoked by partial shadowing or soiling, cracked or mismatched cells or interconnect failures. In extreme cases, it can lead to solder melting and damage to the encapsulant.

To understand the hot-spot effect, consider a series connected string of s matched cells, as in Fig. 4.3 (a). (For simplicity, only three cells are shown). If one of the cells, Y, is shadowed, soiled or damaged so as to reduce the current it can generate to a value below that of the others, it will be forced into reverse bias. This is because all the cells must carry the same current and the shadowed or damaged cell can only do this under negative voltage.

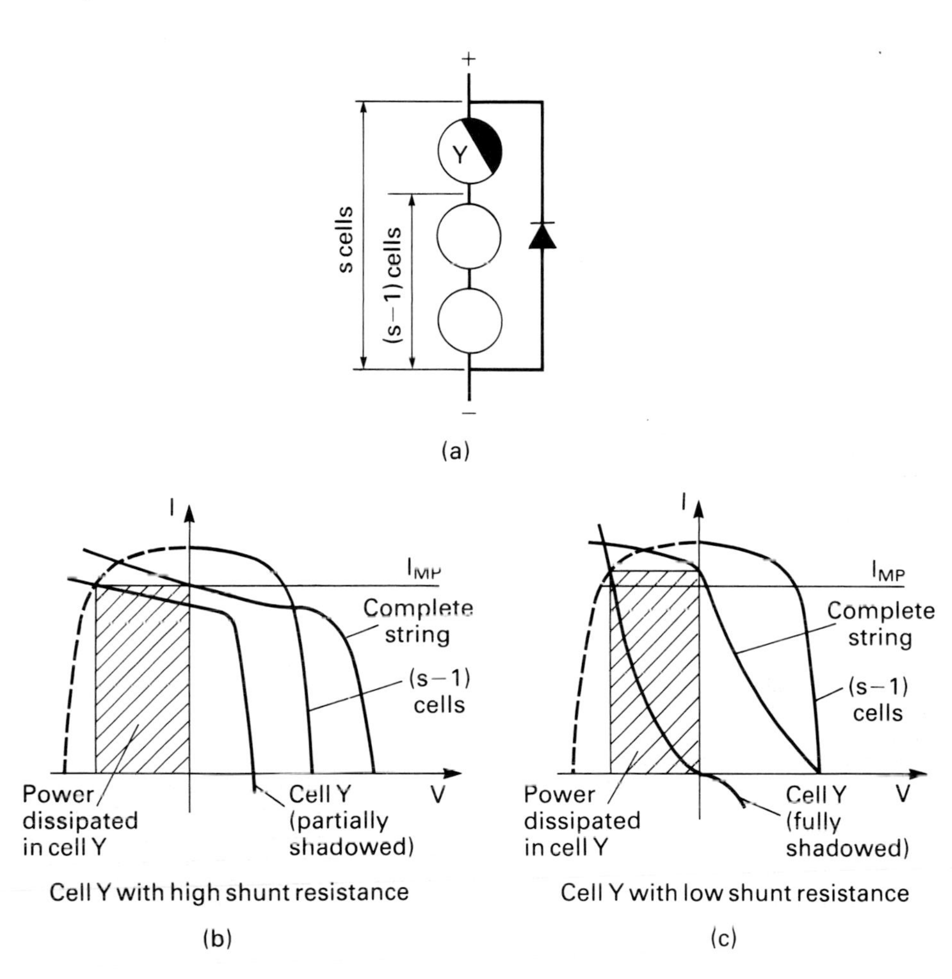

Fig. 4.3. Hot-spot effect.

In this condition, power is dissipated in cell Y and the amount is equal to the product of the string current and the reverse voltage developed across Y. For any irradiance level, maximum power will be dissipated in the short-circuit condition, when the reverse voltage across Y is equal to the voltage across the other (s-1) cells in the string. This condition is illustrated for two types of cell in Figs. 4.3 (b) and (c). The power dissipated in cell Y is shown by the hatched rectangle constructed at the intersection of the reverse I-V characteristic of Y with the mirror image of the forward I-V characteristic of the (s-1) cells. In the case of cells with a high shunt resistance (Fig. 4.3 (b)), the condition of maximum dissipation occurs when cell Y is partically shadowed, to the extent that its reverse characteristic intersects the image of the (s-1) characteristic at its maximum power point. In contrast, with cells of low shunt resistance (Fig. 4.3 (c)), the condition of maximum dissipation occurs when cell Y is fully shadowed. Note in both cases how the I-V characteristic of the complete string is distorted. Such distortions provide important clues in fault detection.

In cases where the cell string extends along several modules in series, the reverse voltage developed across Y can amount to hundreds of volts. To prevent this happening and thereby limit the power that can be dissipated in a damaged or shadowed cell, it is now common practice to connect a by-pass diode across each module or, in some cases, across sections of the module, as indicated by the broken line in Fig. 4.3 (a). These diodes are commonly housed in the terminal box but sometimes they are integrated in the module during fabrication. The modules shown in Fig. 4.2 each contain two by-pass diodes in the terminal box.

4.8 Design qualification

Design qualification or type approval tests are carried out on samples of a new product to ensure, as far as is possible within reasonable constraints of cost and time, that the design requirements have been met and that the product will continue to function satisfactorily throughout its lifetime under the environmental and operational conditions for which it is intended.

Of course, no single set of accelerated environmental tests in a laboratory will guarantee a particular lifetime, since this will depend not only on the design and workmanship of the product but also on the severity of its actual working environment. Nevertheless, type approval tests, augmented by real life experience, provide essential data on which the manufacturer can base his warrantee and the purchaser can base the amortisation period in estimating the life-cycle cost of his installation. Such tests are particularly important for products like photovoltaic modules, where the user relies on a long, trouble-free life to offset the high capital cost.

The International Electrotechnical Commission is at present (1988) engaged in the preparation of an international standard specification for the design qualification of terrestrial flat-plate pv modules. They are taking into account the existing specifications published by authorities in USA, Europe, Australia and Japan. Of these, the most widely used are the Jet Propulsion Laboratory's Block V Specification and the Commission of the European Communities' Specification No. 502.

As an example, Fig. 4.4 shows the qualification test sequence laid down in CEC Specification No. 502. They require eight samples, one of which is used as a control. The test comprises:-

<u>Visual inspection</u>

Careful inspection under a bright light for the detection of

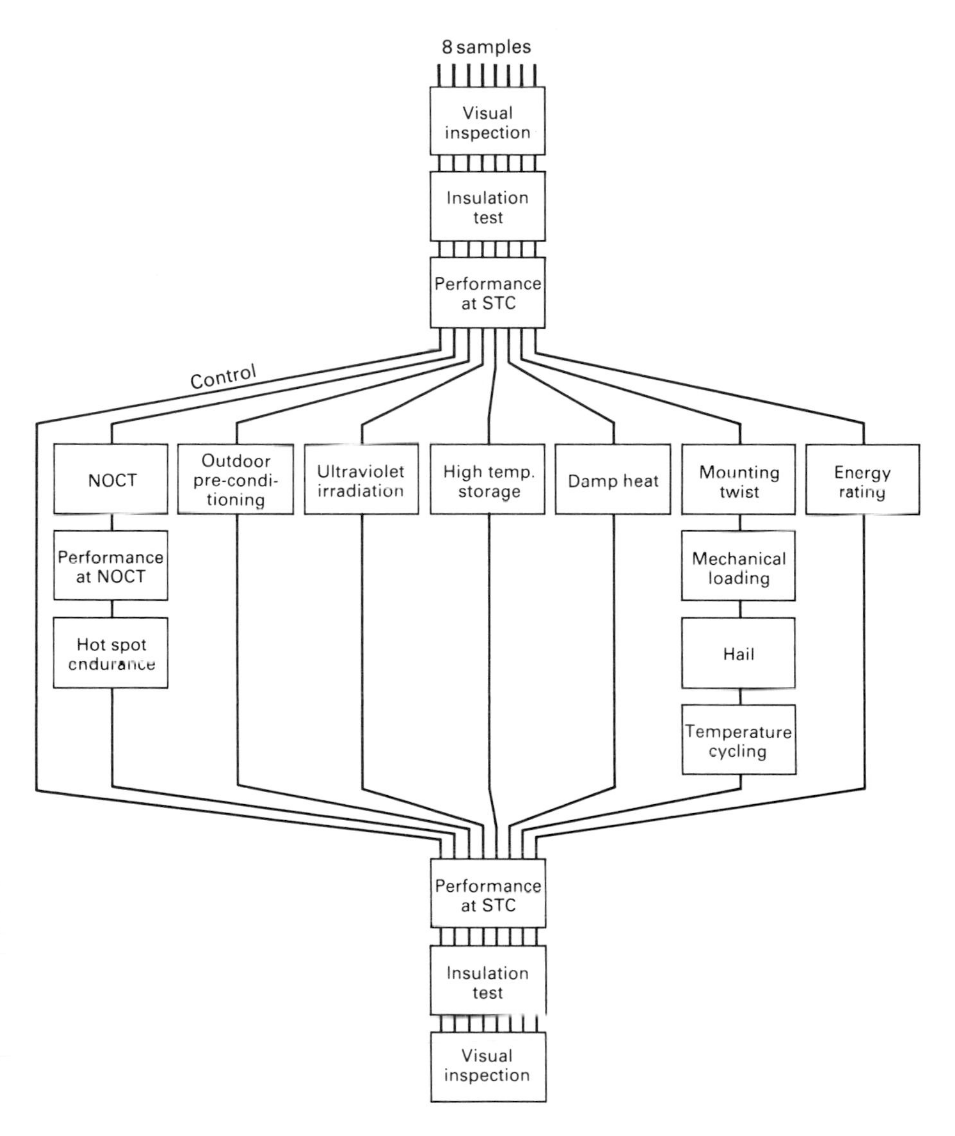

Fig. 4.4. Qualification tests (EUR. SPEC. 502).

faults such as cracks, faulty interconnections, bubbles and delamination.

Insulation test

The insulation resistance between the shorted output terminals and the frame, or simulated support structure, is measured at a voltage of 1000 Vdc. It must not be less than 100 megohms.

The module is then subjected between these points to a voltage gradually increased to 1000 Vdc plus twice the maximum system voltage.

Performance at STC

The I-V characteristic is measured under Standard Test Conditions.

NOCT

The NOCT may be determined, as in the original JPL procedure, by mounting the module outdoors in an open rack and monitoring its temperature by means of attached thermocouples under a range of environmental conditions. After excluding data recorded when the wind is too strong or the ambient temperature not sufficiently stable, the temperature rise of the module above the ambient temperature (T_J - T_{AMB}) is plotted against irradiance and a best fit line drawn through the points. The interpolated value of (T_J - T_{AMB}) at 800 $W.m^{-2}$ is then added to 20 °C to give the NOCT. If the average wind speed and ambient temperature during the measurements differ from the Standard Reference Environment values of 1 ms^{-1} and 20 °C, small corrections are applied. Because this procedure can involve waiting days for suitable weather, the CEC Joint Research Centre, Ispra, developed an indoor procedure for Specification No. 502. In this, the test module with a reference module of known NOCT are exposed in still air to radiation from a steady-state solar simulator. The irradiance is adjusted until the reference module stabilises at a temperature of (NOCT + T_{AMB} - 20 °C). The temperature of the test module is then noted and its NOCT determined by subtracting T_{AMB} and adding 20 °C.

Performance at NOCT

The I-V characteristic is measured at the determined NOCT and an irradiance of 800 $W.m^{-2}$.

Hot-spot endurance

To determine the ability of the module to withstand hot-spot heating effects, it is exposed in the short-circuited condition to natural sunlight or a good steady-state simulator at an irradiance of not less than 700 $W.m^{-2}$ and the hottest cell is found, using an approprite temperature detector, for example, an infra-red camera. After procedures for determining the condition of maximum power dissipation in this cell, the module is kept in this condition, exposed to an irradiance of 1000 $W.m^{-2}$ for 1 hour.

Outdoor pre-conditioning

To determine the stability of the module in natural sunlight, it is mounted normal to the sun at noon and exposed to a total irradiation of 60 $kWh.m^{-2}$, as measured with a pyranometer in the same plane.

Ultra-violet irradiation

To check its ability to withstand UV irradiation, the module, maintained at 60 °C ± 5 °C, is exposed to a lamp which simulates reference sunlight in the waveband 280 nm to 400 nm, until the total UV fluence is 15 $kWh.m^{-2}$. This is the equivalent of 40 clear summer days in the Mediterranean.

High temperature storage

To reveal any encapsulation deficiencies, the module is placed in an oven and subjected to a temperature of 90 °C for 20 days.

Damp heat storage

To test the ability of the module to withstand the penetration of moisture over long periods, it is subjected for 20 days to a temperature of 90 °C and a relative humidity of 95 %.

Mounting twist

To detect defects which might be caused when the module is mounted on an imperfect structure, one corner of the module is displaced from the plane of the other three by a distance corresponding to a deformation angle of 1.2°.

Mechanical loading

This test is to determine the ability of the module to withstand high winds, snow and ice. It is in three parts. First, the front of the module is subjected to a uniform pneumatic load, which is increased in steps up to 1200 Pa. At each step, the load is released and a measurement is made to detect any permanent deformation. This is repeated on the back of the module. Next, the load on the front is increased gradually to 2400 Pa, simulating wind gusts up to 130 $km.h^{-1}$, or 5400 Pa, simulating snow and ice loads. Any permanent deformation is detected after the load has been removed. This again is repeated on the back. Finally, alternating positive and negative loads of 600 Pa are applied for 10 cycles.

Hail

Using a pneumatic or spring-actuated gun, ice balls are fired at various points on the window of the module so as to impact at the terminal velocity a hailstone of the same size would attain. Unless otherwise specified by the manufacturer, 25 mm diameter iceballs are used at a terminal velocity of 23 $m.s^{-1}$.

Temperature cycling

To test the ability of the module to withstand thermally-induced stresses, it is subjected to ten temperature cycles between -20 °C and +80 °C.

Energy rating

Although included in Specification No. 502, this test has not yet been fully developed.

The JPL Block V specification has no outdoor pre-conditioning, UV or long-term storage tests. But it has a more stringent temperature cycling test, consisting of 200 cycles from -40 °C to +90 °C, and this is followed by a humidity/freeze test consisting of 10 cycles from 85 °C and 85 % RH to -40 °C. The mechanical loading test is more severe in terms of fatigue-inducing stresses. An important difference is that the mechanical tests are carried out after temperature and humidity/freeze cycling on one set of samples, instead of on separate modules.

The IEC is seeking agreement on a schedule which combines the best features of these and other current specifications.

Chapter 4 References

Block V solar cell module design and test specification for residential applications. Publication 5101-162. Published by Jet Propulsion Laboratory, California Institute of Technology, Pasadena, CA., USA, Jan. 1981.

EUR Specification No. 502, Issue 1. Qualification test procedures for photovoltaic modules. Published by the Commission of the European Communities, Joint Research Centre, Ispra Establishment, Italy, May 1984.

Gay, C.F., Rumburg, J.E. & Wilson, J.H. "AM/PM - All-day module performance measurements". 16th IEEE Photovoltaic Specialists' Conference, San Diego, CA., USA, September 1982, pp. 1041-1046.

Chapter 5

FLAT-PLATE ARRAYS

F. C. TREBLE
Consulting Engineer, Farnborough, UK

5.1 Introduction

A flat-plate photovoltaic array is an assembly of modules mounted on a support structure and electrically connected to form a dc power producing unit. Some photovoltaic generators may comprise more than one array, in which case the aggregate is termed the "array field". Within the array field, there may be groups of arrays associated by a distinguishing feature such as physical arrangement, electrical connection or common power conditioning. Such groups are referred to as "array sub-fields".

Within an array, there may be groups of modules that, by virtue of physical layout or series or parallel connection, can be considered a unit. Such groups are called "subarrays". Groups of modules that are pre-assembled at the factory and transported to the site in this condition to save on-site labour costs are termed "panels".

This nomenclature is illustrated in Fig. 5.1.

For the best performance, series-connected modules in an array should be current-matched, for the same reason as it is desirable to match series-connected cells in a module. Similarly, series-connected strings of modules connected in parallel should be matched in terms of open-circuit voltage, if the maximum output is to be obtained under all load conditions.

5.2 Orientation

In most arrays, the modules are supported at a fixed inclination facing the Equator. This has the virtue of simplicity, no moving parts and low cost. But, as the solar beam is seldom at normal incidence to the modules and may indeed be behind them in the early morning and late evening in summer, the daily energy output from a fixed array is not as high as it could be.

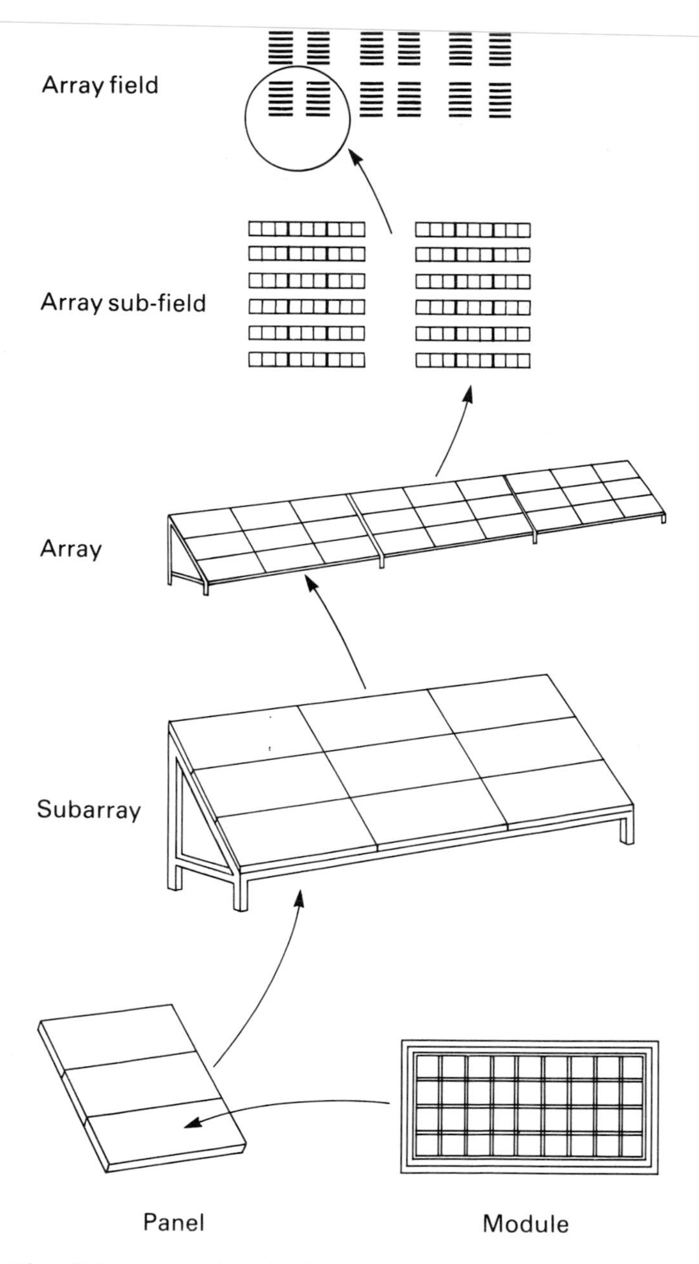

Fig. 5.1. Array terminology.

The optimum angle in a fixed tilt installation depends on the latitude and the proportion of diffuse radiation at the site, as well as on other factors such as the load profile, seasonal weather variations and storage battery capacity. In an ideal situation, with perfectly clear sunny weather all the year round and a flat load profile, maximum annual output is obtained with the modules inclined to the horizontal at the angle of latitude. But a smaller inclination will be better for places where there is a high proportion of diffuse radiation and a steeper inclination will increase output during sunny periods in winter and thus help to reduce the necessary storage capacity. In many places, the weather in the afternoons is often sunnier than in the morning. In these situations, it can be advantageous to make the modules face slightly west of the direct line to the Equator.

By mounting the array on a 2-axis sun tracker, up to 40 % more solar energy can be collected over the year than in a fixed tilt installation (Gay, 1982). Moreover, the gain is mostly in the early morning and late evening, when it can be particularly valuable in meeting peak loads. For flat-plate arrays, the tracking accuracy need not be as high as that needed for solar concentrators. An error of 5° reduces the irradiance by only 0.4 %. Nevertheless, the additional complexity of full 2-axis tracking and the introduction of moving parts is not generally worthwhile, except in large generators in the megawatt range.

Single-axis tracking, in which the array, at a fixed tilt, is swung about a vertical or polar axis to follow the sun, is less complex but yields a smaller gain.

Where labour is available, a better way of increasing output may be manual adjustment of the orientation, which can be surprisingly effective. In fixed arrays, it is a fairly simple matter to make the tilt angle adjustable, so that it can be changed at monthly or quarterly intervals. In small portable arrays, azimuth adjustments can also be made. It has been estimated that, in sunny climates, a flat plate array moved to face the sun twice a day (at mid-morning and mid-afternoon) and given a quarterly tilt adjustment can intercept nearly 95 % of the energy collected with full 2-axis tracking.

5.3 Array layout

The land area required for an array can be minimised by mounting all the modules in a single plane. For large installations, however, this can entail a costly structure and render the modules less accessible for inspection and replacement. Furthermore, the siting of large single plane arrays in some locations may be objected to on aesthetic and environmental grounds.

The more common approach is to arrange the modules in easily-accessible rows, as shown in Fig. 5.1. When this is done, care must be taken to space the rows so as to prevent too much shadowing of one row by another at the beginning and end of the day. The land area required for fixed tilt arrays of this type can be from 1.5 to 4 times the total module area, depending on the tilt angle. The smaller the tilt angle, the less the margin required to avoid shadowing.

For array fields with 2-axis tracking, even more space is required. As an example, the first 1 MWp array to be installed in Southern California (Arnett, 1983) occupies 8 hectares of land - about 7.8 times the total module area of 10285 m^2.

5.4 Support structures

The main criteria in the design of an array structure are (a) it must provide a non-distorting support for the modules at the required orientation, it must be capable of withstanding the strongest winds likely to be experienced at the chosen site and (c) it must cost as little as possible.

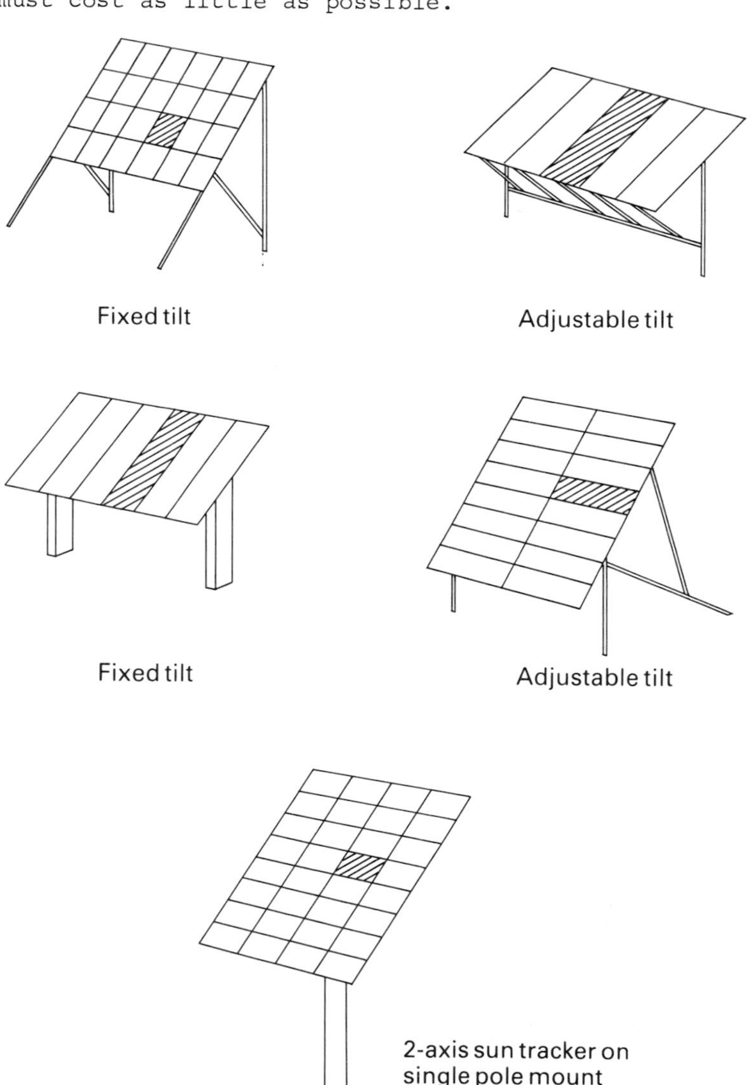

Fig. 5.2. Support structures.

Some of the great variety of designs in existing arrays are illustrated in Fig. 5.2 as thumbnail sketches. In each case, the hatched area indicates a module. No standard designs have yet been formulated. The predominant construction material is galvanised steel but aluminium, reinforced concrete and wood are also used. One problem with wood is that it shrinks as it ages, so fixings tend to loosen and have to be retightened periodically. Foundations are usually of concrete, either pre-cast or cast in situ.

The cost of support structures in European plants varies from \$ 55 to over \$ 200 per square metre, so there is considerable scope for cost reduction. Generally speaking, the larger the module, the simpler the structure and the lower the cost. One cheap solution might be a simple pole-supported structure for large modules or pre-assembled panels, using mass-produced galvanised steel pipes and clips. The supporting poles would be sunk in holes drilled in the ground and anchored in concrete. Such a structure might cost as little as \$ 30/m². However, in considering cheaper structures, one must bear in mind the appearance of the completed array and how well it will fit in with its surroundings. For large installations, an architect should always be employed to deal with this aspect.

On windy sites, it may be cost-effective to reduce structure and foundation costs by erecting wind breaks around the array field.

5.5 Site preparation

In choosing the site for an array field, the cost of preparing it for the installation should be carefully considered, as this can amount to a surprisingly high proportion of the total outlay, particularly in remote situations. Trees and shrubs may have to be removed, uneven sites flattened, sloping ones terraced and supported by retaining walls. Afterwards, the disturbed earth will have to be grassed or otherwise covered to prevent dust rising and settling on the modules. If possible, therefore, the chosen site should be fairly flat and be covered with grass or low scrub. The depth of soil should be sufficient for the array foundations. On such a site, cheap pole-mounted structures, like that described in the previous paragraph, can be installed with the minimum of ground disturbance and labour costs can be kept low.

At some sites, a second use can be found for the ground below and around the arrays, such as grazing for sheep or growing low crops. This will, of course, enhance the economic viability of the installation.

5.6 Roof-mounted arrays

For some applications, such as domestic electricity supply, it may be possible to mount the array on an existing roof and thereby avoid land and site preparation costs altogether. However, there are problems in doing this. The roof may have to be strengthened to take the weight of the modules and support structures. The pitch of the roof may be inappropriate and the available area may impose unwelcome design constraints. Arrangements must be made for the free circulation of air behind the modules, otherwise they will get too hot in full sunlight. In large single-plane arrays, some means must be provided to facilitate access for module inspection and replacement, such as laddered gangways between sections.

In new buildings, the modules can be designed to take the place of the normal roof covering and thus reduce costs by serving two purposes. One approach being investigated by some manufacturers is to design the modules in the form of traditional roofing tiles. This may be aesthetically pleasing but it aggravates the problems of cooling and electrical interconnection. A better solution may be to mount large modules as one would the panes of glass in a greenhouse roof, taking care to provide adequate sealing around the edges and sufficient ventilation underneath.

5.7 Diodes

As we have seen in Chapter 4, it is common practice to connect a diode across each module or, in some cases, across sections of the module, to limit the power that can be dissipated in a shadowe or damaged cell due to the hot spot effect. These by-pass diodes are either housed in the module terminal boxes or, less commonly, integrated in the module during fabrication.

A diode (or two in parallel for redundancy) is also inserted between each series string of modules and the connecting dc bus in order to prevent leakage of current from the battery back through the array during the hours of darkness and also to protect the battery should a short-circuit develop in one of the strings. These blocking diodes also prevent the reverse currents which would otherwise flow under some load conditions in parallel-connected strings which are imperfectly matched in terms of voltage.

Figure 5.3 shows how the by-pass and blocking diodes are connected The voltage drop across the blocking diodes (about 1 V in the case of silicon diodes) represents a power loss which must be taken into account by the array designer.

5.8 Module interconnection and cabling

All cables used in photovoltaic arrays must comply with local and national standards for flammability and fire safety. They must be sufficiently rugged and durable to withstand all operating conditions during the lifetime of the array and must not be adversely affected by exposure to moisture in the form of rainfall, melting ice or condensation. They should be adequately supported and protected against damage from the sun, the wind and ice loading. Protection is also needed against the depredations of birds, animals and insects, which have been known to cause considerable damage. Care should also be taken to preven excessive electromagnetic interference radiating from the cable system.

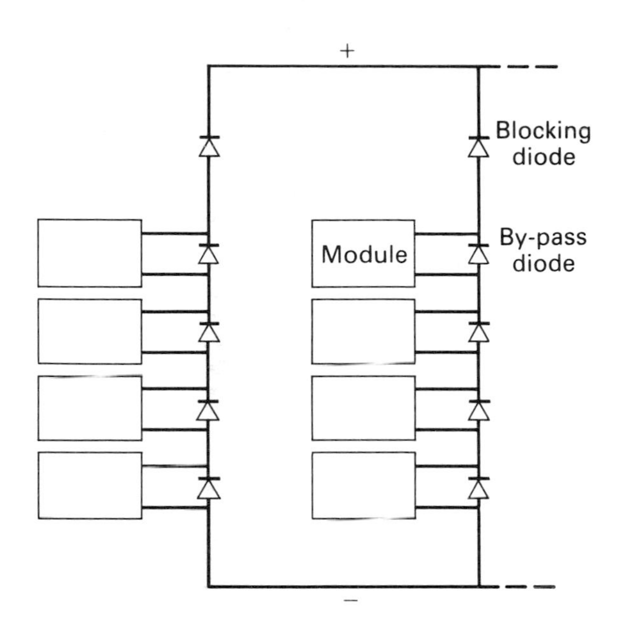

Fig. 5.3. Protective diodes.

In selecting conductor sizes, the effects of resistive loss must be weighed against cost. However, as a general rule, the total losses in an array from cable resistance, diode voltage drop and module mismatch should not exceed 5 % of the maximum power output under Standard Test Conditions.

5.9 Lightning protection

Photovoltaic arrays are often installed on mountain tops and other exposed places where there is a severe risk of lightning.

As yet, there is no general agreement amoung designers on how best to protect an array field against this hazard. Reliance is usually placed on a good earthing system with surge protectors such as varistors to conduct any lightning-induced or fault-induced current surges to ground. Varistors are semiconductor devices whose resistance drops suddenly when the applied voltage exceeds a certain critical limit. In selecting them for this purpose, care must be taken to ensure that they are capable of withstanding, with an adequate safety margin, the maximum system voltage that can be experienced in normal operation. The consequences of failure to take this precaution can be catastrophic.

Experience to date indicates that lightning is not a serious problem in well-earthed arrays. If lightning does strike, it usually damages only one module, which can be readily replaced. Advice on earthing systems is given in Chapter 7.

Chapter 5 References

Arnett, J.C. et al. "Design, installation and performance of the ARCO-Solar One Megawatt Power Plant". 5th EC Photovoltaic Solar Energy Conference, Kavouri (Athens), Greece, October 1983, pp. 314-320.

Gay, C.F. et al. "Performance advantages of two-axis tracking for large flat-plate photovoltaic energy systems". 16th IEEE Photovoltaic Specialists Conference, San Diego, CA, USA, September 1982, pp. 1368-1371.

Chapter 6

PERFORMANCE MEASUREMENT

F. C. TREBLE
Consulting Engineer, Farnborough, UK

6.1 Introduction

Reference was made in Chapter 4 to the performance rating of modules. Let us now look at the whole subject of performance measurement in more detail.

In current practice, the performance of a photovoltaic device, be it a cell, a module, a subarray or an array, is determined by exposing it at a known temperature to stable sunlight, natural or simulated, and tracing its current-voltage characteristic, while at the same time measuring the incident irradiance. If desired, the measured characteristic may then be transposed to Standard Test Conditions or some other irradiance and/or temperature. The power output at the rated voltage and STC is commonly referred to as the rated or "peak" power.

As we have seen in Chapter 3, the performance of a crystalline silicon solar cell is very sensitive to the spectral irradiance distribution of the incident radiation. The same applies to all other types. With natural sunlight, the spectral content varies with location, weather, time of year and time of day. With a solar simulator, it depends on the light source and the optical system and can be affected by ageing.

If the irradiance is measured with an instrument that is not spectrally selective, such as a pyranometer or other thermopile-type radiometer, spectral changes can affect outdoor measurements by as much as 15 %. Moreover, the measurements cannot be related to a reference spectral irradiance distribution for rating purposes. This problem is overcome by measuring the irradiance with a reference device that has essentially the same relative spectral response as the test specimen and has been calibrated in terms of the short-circuit current per unit of irradiance ($A.W^{-1}.m^2$) which it would generate in radiation of the reference spectral distribution. Such a device automatically measures the irradiance in terms of the reference distribution, insofar as it affects the test specimen. Because of this, location, time and weather conditions are not so critical in outdoor measurements, neither is the quality of the simulator so critical indoors. Furthermore, since the reference device, unlike a radiometer, has the same time constant as the test specimen, some fluctuation in the irradiance can be accepted.

If the measured performance is related in this way to a reference spectral irradiance distribution and the spectral response of the test specimen is known, it is possible to compute with reasonable accuracy the performance in radiation of a different spectral content.

6.2 Reference spectral irradiance distribution

Until recently, the most widely used reference solar spectral irradiance distribution was that published in an American official publication ERDA/NASA 1022/77/16, "Revised Terrestrial Photovoltaic Measurement Procedures" (June 1977). It was based on direct sunlight on a horizontal plane at AM 1.5. However, the International Electrotechnical Commission is proposing to replace this, insofar as flat-plate terrestrial pv devices are concerned, by a new international standard based on total (direct + diffuse) sunlight at AM 1.5. The new reference distribution, shown in Fig. 6.1, was computed from the latest AMO data by Bird, Hulstrom and Lewis (1983) for an irradiance of 1000 $W.m^{-2}$ on a plane surface tilted at 37° to the horizontal, under the following meteorological conditions:-

Atmospheric water content - 1.42 cm

Atmospheric ozone content - 0.34 cm

Atmospheric turbidity - 0.27 at 0.5 μm

Ground reflectance - 0.2

The new reference results in more realistic ratings for flat-plate modules and arrays, because they are exposed to diffuse as well as direct sunlight in operation. However, it should be noted that ratings of crystalline silicon modules based on the new reference are 2 to 3% lower than those based on the old reference. This is because the diffuse light from the sky makes total sunlight bluer than the direct beam and therefore less well matched to the spectral response of the solar cells.

6.3 Measurement of relative spectral response

The relative spectral response of a solar cell is measured by placing it on a temperature-controlled mount, irradiating it uniformally from a monochromatic source and measuring the short-circuit current and irradiance at fixed wavelength intervals over the response range. The current density (i.e. current divided by the total area of the cell) at each wavelength is then divided by the irradiance or a proportional parameter and plotted as a function of wavelength. Alternatively, the irradiance may be kept constant, in which case the relative spectral

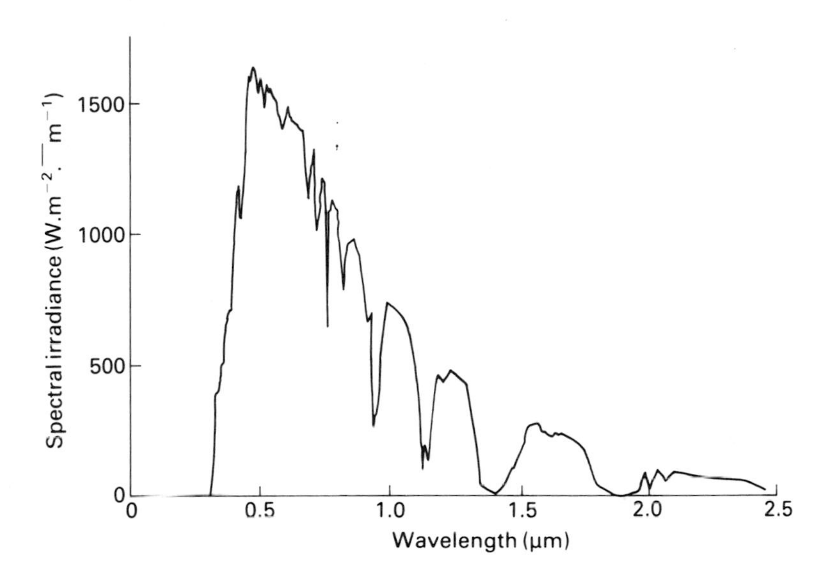

Fig. 6.1. Reference solar spectral irradiance distribution (AM 1.5 total irradiance).

response is obtained directly from the current density readings.

The irradiance monitor may be a vacuum thermopile, a pyroelectric radiometer or other suitable detector. Another alternative is a reference cell whose relative spectral response has been previously measured and found to cover the required range. In this case, the ratio of the current densities at each wavelength is equal to the ratio of the spectral responses. Knowing one spectral response, the other can be calculated.

Figures 6.2 and 6.3 show two possible test arrangements. The first uses a quartz prism monochromator and the second a filter wheel as the monochromatic source. In both cases, the light source is a 1000 W tungsten halogen lamp operated from a stable supply at a colour temperature of 3200 K. The test cell and irradiance monitor are mounted on opposite sides of a rotatable temperature-controlled block, so that either may be presented to the beam in precisely the same place. Alternatively, they may be mounted on a slide with suitable positioning stops.

The filter wheel holds a number of narrow-band filters to cover the response range of the cell in wavelength steps not exceeding 50 nm. The filters are arranged so that each can be indexed in turn to intercept the beam. It is important that the filters should have negligible (under 0.2%) side bands. The monochromator is normally used with fixed slits and is manually set to wavelengths at 50 nm intervals.

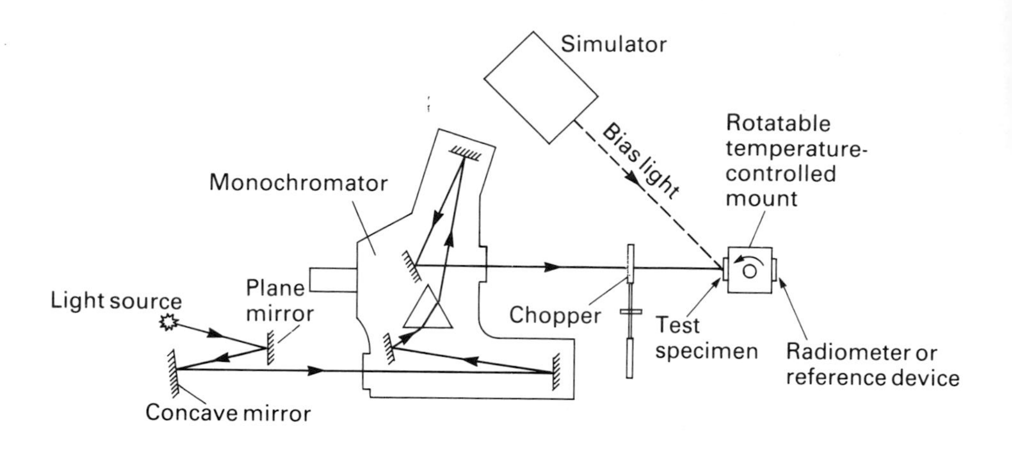

Fig. 6.2. Spectral response measurement using a monochromator.

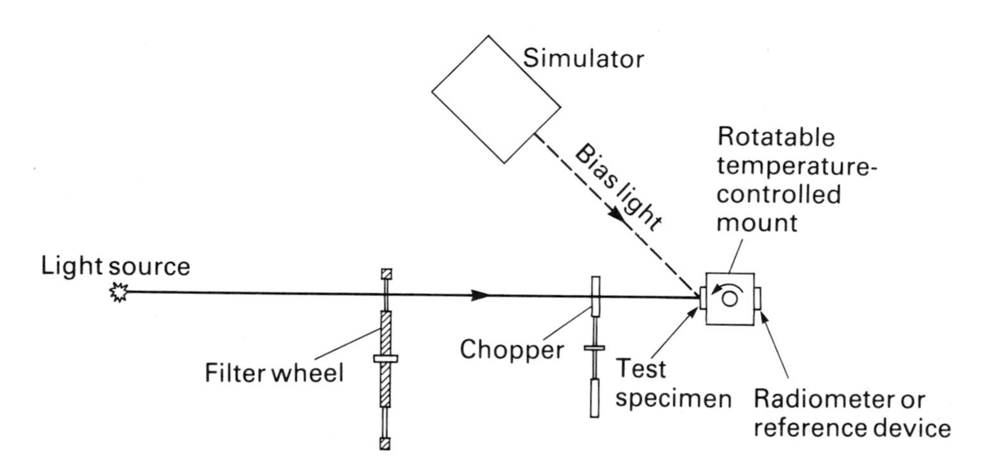

Fig. 6.3. Spectral response measurement using a filter wheel.

With crystalline silicon and other cells, where the response has been shown to change linearly with irradiance, the short-circuit current of the cells (voltage drop across a standard 4-terminal fixed resistor) and the open-circuit voltage of the irradiance monitor may be measured directly with a dc digital voltmeter or potentiometer. If this is done, the exit beam, test cell and irradiance monitor must be completely enclosed in an antireflective light-tight box and meticulous precautions must be taken to avoid thermoelectric and other randomly-generated voltages which would cause errors. Alternatively, the exit beam may be chopped at a low frequency (e.g. 13 Hz) and the output voltages amplified and rectified. In this case, it is important to ensure that the amplifiers are linear and drift-free.

With non-linear devices, such as most thin-film solar cells, it is necessary to use a chopped monochromatic beam and increase the total irradiance to the operational level of interest (e.g. 1000 W.m^{-2}) with unmodulated bias light from a suitable steady-state solar simulator, as indicated in Figs. 6.2 and 6.3.

6.4 Measurement of current-voltage characteristics.

The current-voltage characteristic of a solar cell, a sub-assembly of solar cells or a module may be measured either in natural sunlight or with a solar simulator. International standard procedures for this are laid down in IEC Publication 904-1. Subarrays and full arrays are usually measured outdoors.

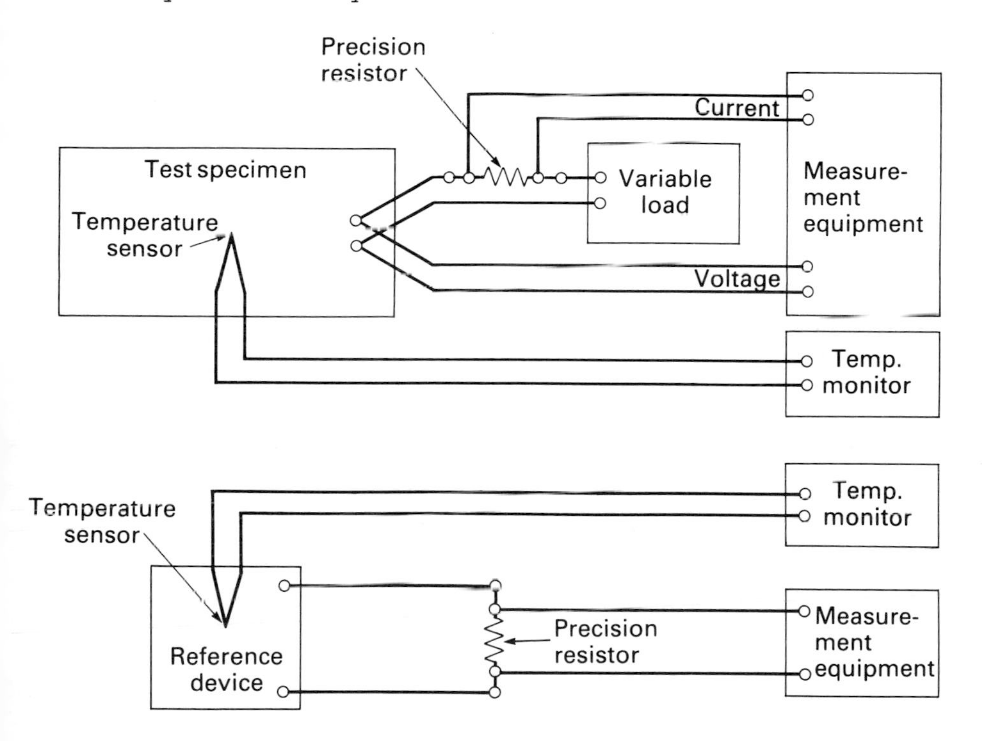

Fig. 6.4. Test connections for I-V characteristic measurement.

The test instrumentation is shown in Fig. 6.4. The variable load should preferably be an electronic circuit, which sweeps the load from the short-circuit to the open-circuit condition slowly enough for the characteristic to be measured either digitally or with an x-y plotter. Such equipment is commercially available. For small powers (under 1 kW), a manually-adjusted rheostat may be used instead. If digital measurements are made, they should consist of a minimum of 30 evenly-distributed points. The temperature-measuring equipment should be accurate to ± 1°C and voltages and currents should be measured by an accuracy of ± 0.5%. Short-circuit currents should preferably be measured at zero voltage, using the electronic load to offset the voltage drop across the external series resistance. Alternatively, the short-circuit current may be measured at a voltage not exceeding 3% of the open-circuit voltage of the test specimen. If this is done, the characteristic should be extrapolated at zero voltage.

In natural sunlight

Measurements may be made outdoors on sunny days, when the total irradiance is not fluctuating by more than ± 1% during a measurement. If the measured characteristics are intended to be transposed to STC, the irradiance should be at least 800 $W.m^{-2}$, although the limit may be reduced to 700 $W.m^{-2}$ for fixed arrays.

The test procedure is as follows:-

1) Mount the reference device as near as possible to and in the same plane as the test specimen (cell, sub-assembly, module, subarray or array) and connect to the test instrumentation, as in Fig. 6.4.

2) If possible, orientate the test specimen and reference device so that they are normal to the direct solar beam, within ± 10°.

3) Record the current-voltage characteristic and temperature of the test specimen and, at the same time, the short-circuit current of the reference device. When testing a module, a more uniform temperature may be obtained by shading it from the sun, waiting for the temperature to stabilise at the ambient air temperature and making the measurement immediately after removing the shade.

4) Transpose the measured characteristic to STC or other desired irradiance and temperature, in accordance with the procedure in Section 6.5 below.

In sunlight, the temperature can vary considerably over an array and even within a module. It is therefore difficult by direct measurement to determine a reliable average value. Experience of on-site array tests has shown that the best solution to this problem is to derive the average temperature from open-circuit voltage readings, thus:-

$$T_{AV} = \frac{V_M - V_{STC}}{\beta} + 25 \tag{1}$$

where: T_{AV} = average cell temperature (°C)

V_M = the measured open-circuit voltage of the test specimen

V_{STC} = its open-circuit voltage at STC (from a typical module value, as measured in the laboratory)

β = temperature coefficient of voltage (negative).

With a steady-state solar simulator

If a steady-state solar simulator is used, it should meet the requirements listed in Section 6.9 below. The test procedure is as follows:-

1) Connect the reference device and test speciment to the test instrumentation, as in Fig. 6.4.

2) Mount the reference device with its active surface in the test plane, normal to the centre-line of the beam, within ± 5°.

3) Note the temperature of the reference device and set the irradiance at the test plane so that the reference device produces its calibrated short-circuit current (corrected to that temperature) at the desired level.

4) Remove the reference device and mount the test specimen in its place. If the beam is sufficiently wide and uniform, the test specimen may be mounted beside the reference device and the two irradiated together.

5) Without changing the irradiance setting, record the current-voltage characteristic and temperature of the test specimen. As in the outdoors procedure, a more uniform temperature may be obtained by using the shading technique.

6) If necessary, transpose the measured characteristic to the desired cell temperature, using the procedure in Section 6.5.

With a pulsed solar simulator

The procedure with a pulsed simulator (see Section 6.9 for requirements) is the same as that with a steady-state source, except that the reference device is always irradiated with the test specimen, after a preliminary check with the reference device to ensure that the irradiance is at the desired level. Also, since the test specimen is not heated by the simulator, its temperature may be assumed to be equal to the ambient air temperature. The simulator equipment includes a computer which triggers the load sweep when the irradiance reaches the present level, records the current and voltage data and automatically transposes them to any other irradiance and temperature when required.

6.5 Transposition of I-V characteristics

A measured current-voltage characteristic may be transposed to

Standard Test Conditions or other desired irradiance and temperature by applying two simple equations, providing that the test specimen and reference device meet the following conditions of linearity over the irradiance and temperature ranges of interest:-

(a) The ratio of short-circuit current to irradiance remains constant within ± 1.5%,

(b) the current and voltage temperature coefficients remain constant within ± 10% and

(c) the relative spectral response is not affected by voltage, temperature or irradiance.

More crystalline silicon solar cells fulfil these conditions.

The transposition equations, as laid down in IEC Publication 891, are:-

$$I_2 = I_1 + I_{SC}\ \frac{I_{SR}}{I_{MR}} - 1 + \alpha(T_2 - T_1) \qquad (2)$$

$$V_2 = V_1 - R_s(I_2 - I_1) - KI_2(T_2 - T_1) + \beta(T_2 - T_1) \qquad (3)$$

where:

I_1 , V_1 — are coordinates of points on the measured characteristic

I_2 , V_2 — are coordinates of the corresponding points on the transposed characteristic

I_{SC} — is the measured short-circuit current of the test specimen

I_{MR} — is the measured short-circuit current of the reference device

I_{SR} — is the short-circuit current of the reference device at the standard or other desired irradiance

T_1 — is the measured temperature of the test specimen

T_2 — is the standard or other desired temperature

α & β — are the current and voltage temperature coefficients of the test specimen in the standard or other desired irradiance and within the temperature range of interest (β is negative)

R_S — is the internal series resistance of the test specimen

K — is a curve correction factor.

To minimise errors, these transpositions should be carried out only within an irradiance range of ± 30% of the level at which the measurements were made.

6.6 Determination of temperature coefficients

The temperature coefficients of current (α) and voltage (β) are best measured in simulated sunlight, using a minimum of two representative solar cells. A pulsed simulator is preferred, as it does not heat the cells during the measurement. The procedure is as follows:-

1) Attach a suitable temperature sensor to the test cell so that the temperature can be measured to an accuracy of ± 0.5°C.

2) Mount the test cell with good thermal contact to a temperature-controlled block and use the attached sensor to provide the control signal.

3) Mount the test cell as near as possible to a suitable reference device, with their active surfaces in the test plane and normal to the centre-line of the beam, within ± 5°.

4) Set the irradiance at the test plane so that the reference device produces its calibrated short-circuit current (temperature-corrected as necessary) at the desired level.

5) With the test cell stabilised at or near the minimum temperature of interest, measure its short-circuit current (I_{SC}) and open-circuit voltage (V_{OC}). At sub-ambient temperatures, precautions may be necessary to prevent condensation on the active surfaces of the test cell and reference device, e.g. by passing dry nitrogen over the surfaces.

6) Stabilise the test cell at a temperature about 10°C above the previous level and repeat the I_{SC} and V_{OC} measurements. Repeat this procedure at approximately 10°C increments up to the maximum temperature of interest.

7) Repeat Steps 1 to 6 on each of the other test cells.

8) From each set of data, plot I_{SC} and V_{OC} as functions of temperature and construct least squares fit curves through the points.

9) From the slopes of the current and voltage curves at a point midway between the minimum and maximum temperatures of interest, calculate α_C and β_C, the temperature coefficients for single cells.

10) For a module or other assembly of cells, calculate the temperature coefficients thus:

$$\alpha = n_p \cdot \alpha_c$$

$$\beta = n_s \cdot \beta_c$$

where n_p is the number of cells in parallel and n_s the number in series.

6.7 Determination of internal series resistance

R_S may be determined in simulated sunlight by the following procedure:-

1) Trace the current-voltage characteristics of the test specimen at room temperature and at two different irradiances, as in Fig. 6.5. During the two measurements, the temperature must not change by more than 2°C.

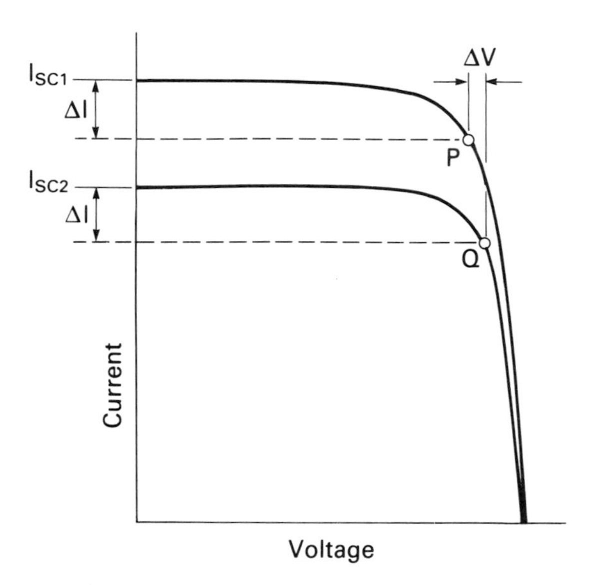

Fig. 6.5. Determination of R_S.

2) Choose a point P on the higher curve, at a voltage slightly higher than that for maximum power. Measure ΔI, the difference between the current at this point and I_{SC1}.

3) Determine point Q on the lower curve at which the current is equal to $I_{SC2} - \Delta I$.

4) Measure the voltage displacement ΔV between points P and Q.

5) Calculate R_{S1} from: $$R_{S1} = \frac{\Delta V}{I_{SC1} - I_{SC2}}$$
where I_{SC1} and I_{SC2} are the two short-circuit currents.

6) Repeat Steps 3 to 5, using a characteristic taken at a third irradiance level and the same cell temperature, in combination with each of the first two curves, to determine two further values of series resistance R_{S2} and R_{S3}. R_S is the mean of the three calculated values R_{S1}, R_{S2} and R_{S3}.

6.8 Determination of curve correction factor

The procedure for detemining K in simulated sunlight is as follows:

1) Trace the current voltage characteristic of the test specimen at an irradiance within ± 30% of the selected level and at three different temperatures (T_3, T_4 and T_5) over an interest range of at least 30°C. If the test specimen is a solar cell, it should be mounted on a temperature-controlled block. If it is a module, it should be enclosed in a glass-fronted temperature-controlled chamber to ensure uniformity of the cell temperature within ± 2°C of the intended level.

2) Using an assumed value of K (say 1.25 x 10^{-3} ohm/°C, which is typical for a crystalline silicon cell), transpose the characteristic measured at temperature T_3 to T_4 by applying the following equations:

$$I_4 = I_3 + \quad (T_4 - T_3) \tag{4}$$

$$V_4 = V_3 - KI_4(T_4 - T_3) + \quad (T_4 - T_3) \tag{5}$$

where: I_3, V_3 are coordinates of points on the T_3 characteristic

and I_4, V_4 are coordinates of the corresponding points on the T_4 characteristic.

3) If the transposed T_4 characteristic does not coincide to the desired accuracy with that obtained by measurement, repeat Step 2 by inserting different values for K until the transposed and measured characteristics coincide.

4) When the proper value of K has been determined, transpose the T_3 and T_4 characteristics sequentially to match the characteristic at temperature T_5. If the transposed and measured curves do not coincide, repeat the transposition using a slightly different value of K until the value for a correct fit is determined in each case.

5) Use the mean of the three values of K thus determined.

6.9 Simulator requirements

Two types of solar simulator are commercially available for photovoltaic performance measurements. The steady-state type of, for example, filtered xenon, dichroic filtered tungsten (ELH) or modified mercury vapour with tungsten electrodes, is suitable for measurements on single cells and modules. It can also be used for bias lighting in spectral response measurements and for tests, such as hot spot endurance, where the heating effect is an essential element. The pulsed type, consisting of one or two long-arc xenon flash lamps, has simpler optics, needs less power and can produce uniform irradiance over large areas. There is negligible heat input to the test specimen, so it remains uniformly at room temperature, which can be easily and accurately measured. The pulse-forming network and the data acquisition and processing computer are generally supplied with the simulator.

The IEC is proposing to group photovoltaic solar simulators in three classes, A, B and C, according to their uniformity, temporal stability and spectral match, i.e. how well the spectral irradiance distribution matches the reference distribution. Class A simulators are intended for accurate performance measurements in the laboratory and reference device calibration, Class B for production acceptance testing in factories and Class C for bias lighting and other purposes where lower quality is acceptable. For Class A, the non-uniformity of irradiance in the test plane, as measured with a small scanning detector, must not exceed ± 2% of the mean value. The temporal instability, i.e. the variation of the irradiance during the time of data acquisition, must also not exceed ± 2% of the mean value. As regards the spectral match, the percentage of the total radiant energy in each of six 100 nm wavelength intervals between 0.4 μm and 1.1 μm must be within ± 25% of the reference value. The corresponding tolerances for Class B are ± 5%, ± 5% and ± 40% respectively and for Class C ± 10%, ± 10% and $^{+100\%}_{-60\%}$.

In a pulsed simulator, it is important to ensure that the pulse characteristics and the time interval between successive data points are consistent with accurate measurement.

6.10 Spectral mismatch error

The spectral mismatch error in photovoltaic performance rating measurements is caused by the interaction of two mismatches. First, the spectral irradiance distribution of the sunlight, be it natural or simulated, does not perfectly match the reference spectral irradiance distribution. Second, the relative spectral response of the reference device may not exactly match that of the test specimen. If the test specimen and reference device are linear according to the definitiion in Section 6.5, the error can be computed from their relative spectral responses and the relative spectral irradiance distribution of the simulator in the following manner:

Let:

J_1 = short-circuit current density of the reference device in sunlight having the reference spectral distribution and sn irradiance of 1000 $W.m^{-2}$. ($A.m^{-2}$)

J_2 = short-circuit current density of the reference device, as measured in the natural or simulated sunlight. ($A.m^{-2}$)

$s_1(\lambda)$ = absolute spectral response of the reference device at wavelength λ. ($A.W^{-1}$)

$k_1 s_1(\lambda)$ = relative spectral response of the reference device at wavelength λ.

J_3 = short-circuit current density of the test specimen in sunlight having the reference spectral distribution and an irradiance of 1000 $W.m^{-2}$. ($A.m^{-2}$)

J_4 = short-circuit current density of the test specimen, as measured in the natural or simulated sunlight. ($A.m^{-2}$)

$s_2(\lambda)$ = absolute spectral response of the test specimen at wavelength λ. ($A.W^{-1}$)

$k_2 s_2(\lambda)$ = relative spectral response of the test specimen at wavelength λ.

$E_S(\lambda)$ = absolute spectral irradiance at wavelength λ of the reference spectral irradiance distribution. ($W.m^{-2}.\mu m^{-1}$)

$k_3 E_S(\lambda)$ = relative spectral irradiance at wavelngth λ of the reference spectral irradiance distribution.

$E_T(\lambda)$ = absolute spectral irradiance at wavelength λ of the natural or simulated sunlight. ($W.m^{-2}.\mu m^{-1}$)

$k_4 E_T(\lambda)$ = relative spectral irradiance at wavelength λ of the sunlight.

Then:

$$J_1 = \int s_1(\lambda).E_S(\lambda).d\lambda$$

$$J_2 = \int s_1(\lambda).E_T(\lambda).d\lambda$$

$$J_3 = \int s_2(\lambda).E_S(\lambda).d\lambda$$

$$J_4 = \int s_2(\lambda).E_T(\lambda).d\lambda$$

Integration of the products of the measured relative spectral responses and the relative spectral irradiances yields the following parameters:

$$A_1 = \int k_1 s_1(\lambda).k_3 E_S(\lambda).d\lambda = k_1 k_3 J_1$$

$$A_2 = \int k_1 s_1(\lambda).k_4 E_T(\lambda).d\lambda = k_1 k_4 J_2$$

$$A_3 = \int k_2 s_2(\lambda).k_3 E_S(\lambda).d\lambda = k_2 k_3 J_3$$

$$A_4 = \int k_2 s_2(\lambda).k_4 E_T(\lambda).d\lambda = k_2 k_4 J_4$$

The error in measuring the short-circuit current density of the test specimen is then given by:

$$\text{Spectral mismatch error (\%)} = \frac{J_4 - J_3}{J_3} \cdot 100$$

$$= \left[\frac{A_1 \cdot A_4}{A_2 \cdot A_3} \cdot \frac{J_2}{J_1} - 1 \right] \cdot 100 \qquad (6)$$

The calculated spectral mismatch error may be applied as a correction to measured data.

6.11 Reference devices

6.11.1 Selection and packaging

There are two categories of reference device:-

Primary - a reference device whose calibration is based on a radiometer or standard detector conforming to the current World Radiometric Reference (WRR).

Secondary - a reference device calibrated against a primary reference device.

The selection, packaging and calibration of reference devices are of the utmost importance if unacceptable discrepancies in photovoltaic performance rating are to be avoided.

At least two cells or modules should be selected for calibration as reference devices. They must be stable and linear, according to the definition in Section 6.5. Their relative spectral responses should match that of the product to be tested to the extent necessary to reduce spectral mismatch errors to less than ± 1%. (If this is not practicable, then the appropriate correction, computed as in Section 6.10, must be applied). Means must be provided for measuring the cell junction temperature to an accuracy of ± 1°C.

The packaging of the reference device should suit the application. When it is intended to be used only in natural sunlight at normal incidence or with a Class A simulator, such as in the pulsed type, where the light source is not close to the test plane, the device can be a single solar cell, either unpackaged (Fig. 6.6A) or encapsulated behind a protective window with a field of view of at least 160 ° (Fig. 6.6 B and C). Types A and B can be mounted on a temperature-controlled block. Type C, in which the cell is encapsulated and framed in the same way as the test specimen, cannot easily be temperature-controlled.

When measuring the performance of fixed arrays in situ, it may not always be possible or convenient to work with the solar beam at normal incidence. In such cases, it has been found that internal reflections can cause errors, particularly with round cells on a white background. An unpackaged reference cell or

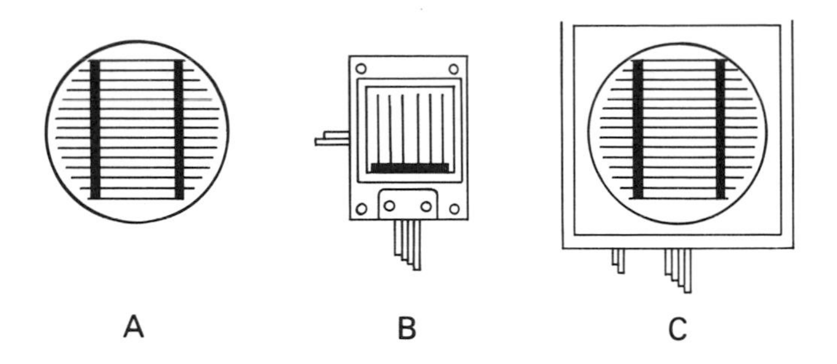

Fig. 6.6. Single reference cells.

one in a single cell package will not have the same reflective properties and therefore will not be subjected to exactly the same irradiance as the modules being tested. One solution to this problem is to enclose the reference cell in a multicell package, in which the frame, the encapsulation system and the shape, size and spacing of the surrounding cells are the same as in the modules in the array. Figure 6.7 shows the minimum acceptable size. Reference cells in multicell packages may be used as alternatives to those in single cell packages. The cells surrounding the reference cell may be real ones or dummies having the same optical properties.

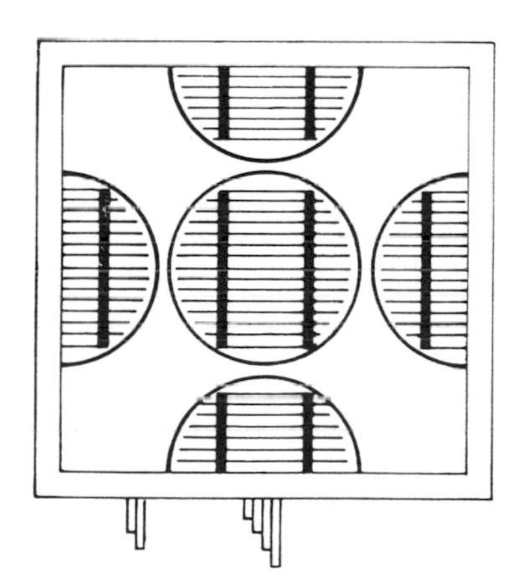

Fig. 6.7. Multicell package for reference cell.

Another problem may arise in indoor measurements, if the simulator has a light source close to the test plane. With such equipment, the irradiance may be enhanced by light reflected back and forth between the simulator and the module under test. Because of the restricted area of the test plane, the usual procedure with such simulators is first to set the irradiance with the reference device, then replace the reference device by the test module and finally carry out the measurement at the same setting. The problem is that, unless the reference device covers the same area as the test module and has the same reflective properties, the irradiance on the module may not be the same as that measured by the reference device. The answer is to use a reference module of the same size and shape as the module under test, embodying the same type of cell and encapsulation system. If the simulator is Class B, the series-connected cells in the reference module should be current-matched to ± 2%, as this can help to reduce the effects of the ± 5% non-uniformity of irradiance. Reference modules may also be used as alternatives to single cells in single- or multicell packages.

To sum up:-

1. Single reference cells, whether unpackaged or mounted in an individual package, should be used only at normal incidence in natural sunlight or with a Class A simulator with no multiple reflection problem.

2. Reference cells in multicell packages may be used instead of individually packaged cells and also for outdoor tests at sloping incidence.

3. Reference modules with matched cells (± 2%) are essential for Class B simulators. They may also be used for all other applications.

6.11.2 Calibration

At present, the calibration of primary reference devices is carried out by specialist agencies. The principal methods are as follows:-

Global Sunlight Method

The calibration is carried out on a clear day, in stable sunlight with a global irradiance of at least 800 $W.m^{-2}$. The diffuse irradiance must be less than 25% of the global irradiance and the solar elevation not less than 54° (equivalent to AM1.24 at sea level). The reference device is mounted on a horizontal platform co-planar with a pyranometer which has been calibrated to the World Radiometric Reference. Simultaneous measurements of short-circuit current, cell temperature and irradiance are taken, until the ratio of short-circuit current (corrected to 25 °C) to irradiance varies by less than 1% over five successive readings. The mean of these ratios is taken as the uncorrected calibration value CV_U in $A.W^{-1}.m^2$. Concurrent with or immediately after the calibration, the relative spectral irradiance distribution of the global sunlight is measured with a specially-adapted spectroradiometer. The calibration value CV, corrected to the reference spectral irradiance distribution, is then computed from the measured data, the relative spectral response of the reference device and the reference spectral irradiance. The procedure is carried out on at least three days and a mean is taken of the calibration values thus determined.

The spectral irradiance measurements require special expertise and equipment, together with very stable conditions. However, once the spectral correction factor k, the ratio of CV to CV_U, has been determined for a particular site and reference spectrum, there is no need to repeat the spectral irradiance measurements. One simply determines CV_U and multiplies by k. Experience over many years has shown that this method gives very consistent results.

Total Sunlight Method

This is derived from the Global Method but differs in that the reference device and pyranometer are tilted normal to the direct solar beam, thus allowing readings to be taken up to AM2 and easing the environmental conditions. The solar spectral irradiance distribution is not measured but, instead, a sun photometer is used to check that the atmospheric turbidity and water vapour content are within prescribed limits. Tilting the platform introduces some ground reflection, which must be kept to less than 3% of the total irradiance. The calibration value is calculated from a number of short-circuit current/irradiance ratios, as in the Global Method, but there is no spectral correction. The Total Sunlight Method has not acquired the same background of experience as the Global Method but the indications are that the results are equally consistent. It may prove possible to eliminate the sun photometer measurements and replace them by a limit on the diffuse irradiance, as in the Global Method.

Solar Simulator Method

This is basically the same as the Global Sunlight Method but it is carried out indoors, using a Class A solar simulator (steady state) instead of natural sunlight. Like the Global Method, it involves spectral irradiance measurements but, in this case, because of the strong lines in the xenon spectrum, the measurements are more subject to error and the mismatch between the simulator and reference spectra causes errors to have a greater effect on the resulting calibration value. The consistency of this relatively new method has yet to be established.

Differential Spectral Response Method

In this method, the calibration value CV is computed from a direct measurement of the absolute spectral response of the cell and the reference spectral irradiance distribution, thus:-

$$CV = \frac{s(\lambda).E_R(\lambda).d\lambda}{E_R(\lambda).d\lambda} \qquad (7)$$

where:

$s(\lambda)$ = the absolute spectral response of the reference cell at wavelength ($A.W^{-1}$).

$E_R(\lambda)$ = the reference spectral irradiance at wavelength ($W.m^{-2}.\mu m^{-1}$).

This is the most scientifically elegant of all calibration methods, as it does not rely on the weather nor require an expensive simulator. But there are many pitfalls in measuring spectral response in absolute terms and small errors can significantly affect the result. With the present limitations of the available equipment, this method, unlike the others, cannot be applied to reference modules.

Efforts are being made by the International Electrotechnical Commission to select one, or at most two methods for future standardisation. The most likely candidate is one derived from the Global and Total Sunlight Methods. Uncorrected calibration values would be simply determined in natural sunlight, with the reference cell or module and pyranometer either horizontal or facing the sun in appropriate environmental conditions. The spectral correction factor k would be established once-and-for-all for each location and type of cell by specialist agencies, using on-site solar spectral irradiance data. The separation of the basic simple calibration from the spectral correction in this way would encourage more laboratories and manufacturers to undertake their own calibrations, thus relieving the current bottleneck and helping the many parts of the world that lack calibration facilities.

Once a reliable primary reference device has been acquired, any number of secondary reference devices can be calibrated against it. This is done by exposing both devices at normal incidence to natural sunlight or a Class A simulator at an irradiance of at least 800 $W.m^{-2}$ and measuring the short-circuit currents and cell temperatures simultaneously. The ratio of the short-circuit current (corrected to 25 °C) of the secondary reference device to that of the primary reference device is calculated. After repeating this procedure until five successive ratios do not vary by more than ± 1%, the mean ratio is multiplied by the calibration value of the primary reference device to give the required calibration of the secondary reference device.

6.11.3 Care

The windows of reference devices should be kept clean and scratch-free. Unpackaged reference cells should be kept, when not in use, in a box or drawer to protect them against damage, contamination and degradation. The calibration of reference devices in frequent use should be cross-checked every month by comparing their short-circuit currents in the same irradiance. If the current ratios change by more than 1%, the devices should be re-calibrated. In any case, re-calibration should be carried out at least once a year.

Chapter 6 References

Anon "Revised terrestrial photovoltaic measurement procedures" Publication ERDA/NASA 1022/77/16, June 1977.

Bird, R.E., Hulstrom, R.L. & Lewis, L.J.	"Terrestrial solar spectral data sets". Solar Energy, Vol. 30, No. 6, 1983.
IEC Publication 904-1	"Measurement of photovoltaic current-voltage characteristics". Published by the International Electrotechnical Commission, 3, rue de Varembe, Geneva, Switzerland.
IEC Publication 891	"Procedures for temperature and irradiance corrections to measured I-V characteristics of crystalline silicon photovoltaic devices". Ibid.

Chapter 7

PHOTOVOLTAIC SYSTEMS

F. C. TREBLE
Consulting Engineer, Farnborough, UK

Introduction

A voltaic system is an integrated assembly of modules and other components, designed to convert solar energy into electricity to provide a particular service, either alone or in conjunction with a back-up supply. It can vary greatly in size and complexity, from a solar-powered pocket calculator to a multimegawatt power station delivering electricity to a grid. In between, there is a wide range of small and medium power applications.

There are two main categories - standalone and grid-connected. Grid connected systems are sub-divided into those in which the grid merely acts as an auxiliary supply (grid back-up) and those in which it may also receive excess power from the photovoltaic generator (grid interactive).

In this chapter, we shall discuss the main types of system, review some aspects of system design and give some information on array and battery sizing, with a worked-out example. But we begin with a description of the main components, excluding the array, which has been dealt with in Chapter 4, and storage batteries, which are reviewed in Chapter 18.

DC/DC Converters and MPPTs

As its name implies, this device converts the dc output from the array to a voltage or voltages suited to the requirements of the battery or load(s). In modern solid-state converters, the transformation is usually achieved by high frequency chopping, using transistors. Efficiencies of over 95% at full load and 90% at 10% load can be expected. Of course, converters are not required if the desired dc voltage or voltages can be provided directly from the array, by suitable arrangement of the module connections.

A maximum power point tracker (MPPT) is a special type of converter, whose function is to keep a pv generator operating at or near its maximum power point. It has built-in control logic, usually operated by a micro-processor, that senses the array voltage and current at frequent intervals (typically every 30ms), computes the power output and compares it with the previous value. If the power output has increased, the array voltage is stepped in the same direction as the last step. If the power has decreased, the array voltage is stepped in the opposite direction. Eventually, the array voltage reaches and is kept near the maximum power point, irrespective of changes in irradiance, temperature and load. The power consumption of MPPTs is typically 3 to 7% of the array output. Some systems have several MPPTs, each serving a part of the array field. The system designer has to decide whether the added complexity, cost and power loss of these devices are worth the gain in energy output.

7.3 DC/AC Inverters

An inverter converts the dc from the array or battery to single- or three-phase ac to suit load requirements. In grid interactive systems, the inverter output must meet the often stringent requirements of the utility in terms of voltage, frequency and the harmonic purity of the sine wave. The voltage requirement may necessitate an additional transformer and the harmonic requirement special filtering, both of which add to the losses in the device. But in standalone systems and for certain non-critical loads, such as motors, the requirements may be eased. For example, higher harmonic distortion - even a square or trapezoidal wave - may be acceptable, in which case filters may be omitted, with a consequent gain in efficiency.

Two techniques are mainly used in modern solid-state inverters to construct a sinusoidal ac output:-

> Waveform synthesis. The phased outputs of several inverter stages, each producing a square wave by chopping the dc input, are combined by switching at the fundamental frequency to construct a stepped output waveform approximating to a sine wave. The more inverter stages used, the better the approximation and the lower the harmonic distortion.
>
> Pulse-width modulation (PWM). An approximate sine wave, free of the main harmonics, is generated by switching square-wave inverter stages at a rate higher than the fundamental frequency. The output voltage at any instant is controlled by varying the conduction time of the power switches, i.e. the pulse width. The total harmonic distortion is inversely proportional to the switching rate.

The process by which the forward current is interrupted or transferred from one switching device to another is called "commutation". A "self-commutated" inverter, as used in standalone systems, is one in which the switching is performed wholly within the unit, using power from the dc input. On the other hand, in a "line-commutated" inverter, commonly used in grid-interactive systems, the switching is triggered by the ac system to which power is being supplied or by reactive elements connected to the output side of the unit.

Many commercially-available inverters embody thyristors or silicon controlled rectifiers (SCRs) as the switching devices. A thyristor is a semiconducter device in which one or more control electrodes initiate but do not limit the anode current. As collector current is increased to a critical value, the gain rises above unity to produce a high-speed triggering action, allowing the anode current to rise rapidly. The anode current must be arrested ("forced commutation") in order to turn off the device. Thyristor inverters are available for power ratings up to several thousand kVA. They are commonly used for uninterruptible power supply systems (UPS) for computers and other specialised equipment, where reliability is more important than efficiency. By modifying standard units for pv applications, efficiencies of over 95% at full load can be achieved. But part-load efficiency is not so good because of the relatively high internal power consumption associated with commutation. Thyristors have good overload characteristics.

In recent years, power transistors have become available with power ratings as high as 50kVA. Transistors need very little power to operate as switches and, unlike thyristors, do not need forced commutation to turn off the current. The efficiency of transistor inverters at low loads can therefore be higher. However, transistors are sensitive to overloads.

Inverters for pv applications usually incorporate the following protective features:-

- Automatic switch-off if the dc input voltage is too high or too low.
- Automatic re-start when the dc input voltage reaches a set minimum.
- Protection against short-circuits and overloading.

Some types embody a maximum power point tracker.

7.4 Battery Controllers

The simplest form of battery controller is a charge regulator, which regulates the charge current at the optimum rate and prevents over-charging. Some regulators also embody a control to prevent excessive discharge. In a pv system, such devices can regulate the charge current by interrupting the current from the array or a subarray (series type) or by short-circuiting sections of the array (shunt type). However, short-circuiting is usually considered inadvisable, because it aggravates any tendency to hot-spot failure. A number of regulators can be used in parallel. Power consumption is very small - typically under 0.2% of the power being controlled.

A more sophisticated type embodies a microprocessor, which monitors current, voltage and temperature, computes the state-of-charge and regulates the input and output currents so as to avoid overcharging and excessive discharge. Attempts to design controllers of this type for lead-acid batteries have so far proved unsuccessful, as the relationship between the state-of-charge and the measured parameters changes with age and is affected by the type of usage. Furthermore, the capacities of individual battery cells of the same construction vary and there is at present no way in which series-connected cells can be matched in this respect. So, even if a method of measuring state-of-charge could be perfected and applied to one cell, it would not necessarily indicate accurately the state-of-charge of the other cells in the string and there would remain the possibility of the cell with the smallest capacity becoming over-discharged. The accurate modelling of batteries for control purposes remains an outstanding problem.

7.5 Power Management and Control

In addition to battery controllers, most pv systems have means for automatically controlling the flow of energy from the array and, where applicable, from the grid or back-up generator, in accordance with load requirements and the prevailing conditions of irradiance, temperature and battery state-of-charge. Sometimes, the battery control is integrated with other controls into a master control unit. The energy balance is achieved in various ways, according to circumstances. In some cases, the load is varied to suit the available power, e.g. by using excess power for secondary loads such as immersion heaters. In others, when the load requirements are being fully met and the battery is fully charged, the available power is reduced by disconnecting all or part of the array. In grid-interactive systems, excess power is fed into the grid.

Simple control systems are generally reliable, as there is less to go wrong. More complex systems with microprocessors give better power management and the software can be improved in the light of experience. However, special expertise is needed for programming and fault detection and this is not always available.

7.6 Types of System

In its simplest form, a standalone pv system consists of the array supplying the load directly, as shown in schematic form in Fig. 7.1. Such a system can be used for battery charging (with a simple charge regulator) or for water pumping, where the storage medium is a water tank.

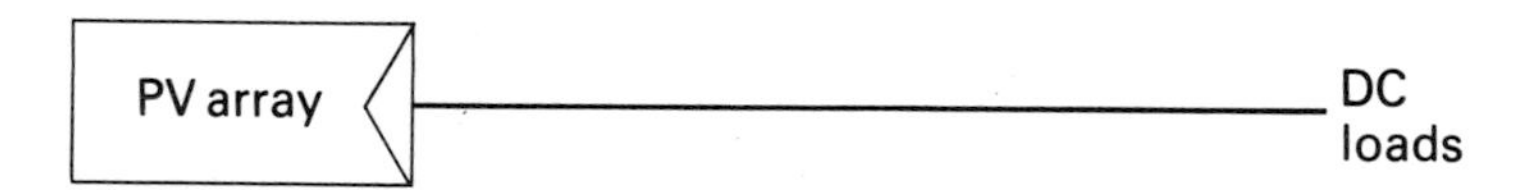

Fig. 7.1. Standalone dc system without battery.

The addition of a self-commutated inverter (Fig. 7.2) makes it suitable for an ac pump or other ac loads.

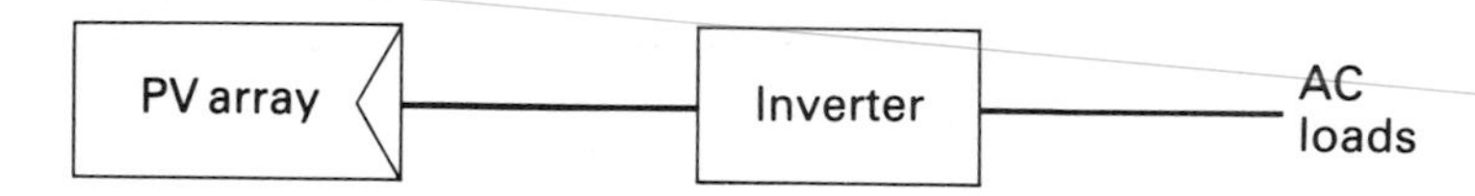

Fig. 7.2. Standalone ac system without battery.

For dc loads such as telecommunication equipment, cathodic protection, navigational aids, traffic control warning signs, dc lighting, refrigerators, TV, radio, etc., where the load has to be supplied overnight and during periods of low irradiance, a storage battery with charge regulator must be added to the basic dc system (Fig. 7.3).

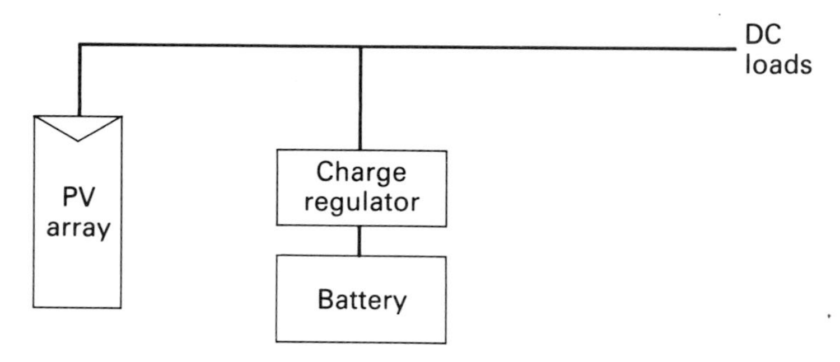

Fig. 7.3. Standalone dc system with battery.

Combining systems of the types shown in Figs. 7.2 and 7.3 gives a mixed ac/dc system with battery (Fig. 7.4). This type is appropriate for domestic supplies in remote houses and villages.

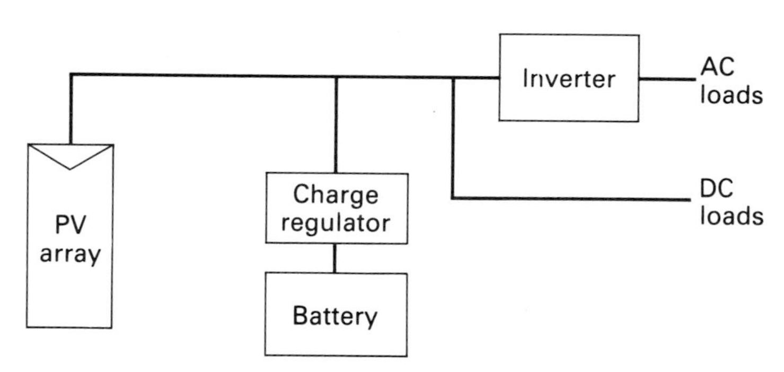

Fig. 7.4. Standalone ac/dc system with battery.

A back-up generator is commonly used to improve the security of supply (Fig. 7.5). In a hybrid system, an auxiliary source, such as a wind turbine, hydro-electric generator or diesel generator, is integrated with the pv system to reduce the required storage capacity and lower the capital cost.

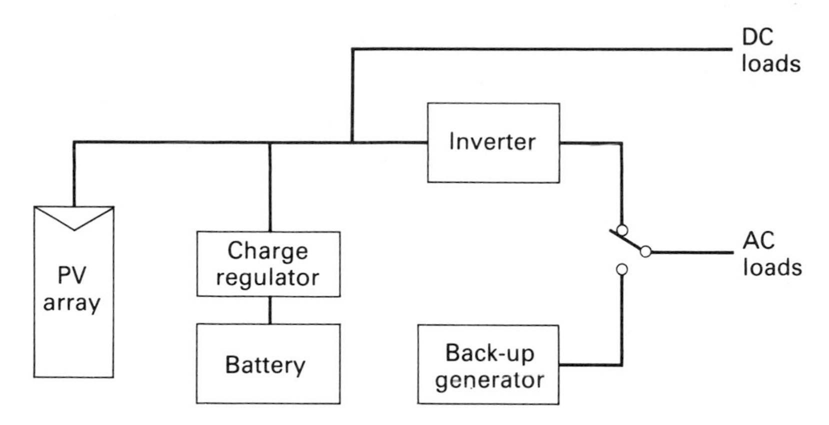

Fig. 7.5. Standalone ac system with battery and back-up.

Turning to the grid-connected category, Fig. 7.6 illustrates a simple ac system with self-commutated inverter and grid back-up. Figure 7.7 shows a grid-interactive system, where surplus power is fed to the grid through a second inverter, which is usually line-commutated. This inverter can also serve as a rectifier for charging the battery from the grid.

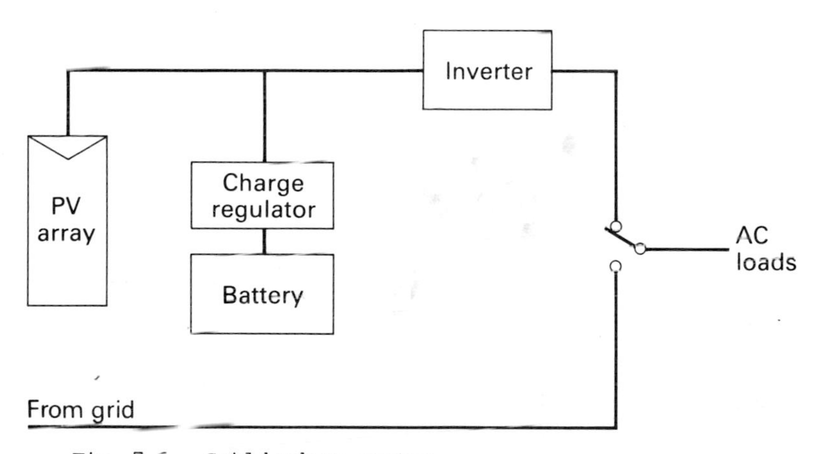

Fig. 7.6. Grid back-up system.

The basic schemes illustrated in Figs. 7.1 to 7.7 can be varied to suit particular applications. For example, a MPPT may be inserted between the pv array and the rest of the system, or it may be embodied in the inverter, in order to enhance the output. The array field may be divided into sub-fields or subarrays, each with its own MPPT, to serve different parts of the load. To improve the low-load efficiency of dc to ac conversion, multiple inverters, switched in and out to suit load requirements, may be used instead of a single unit. Back-up from the grid or an auxiliary generator may be effected through a battery charger instead of by direct connection to the load.

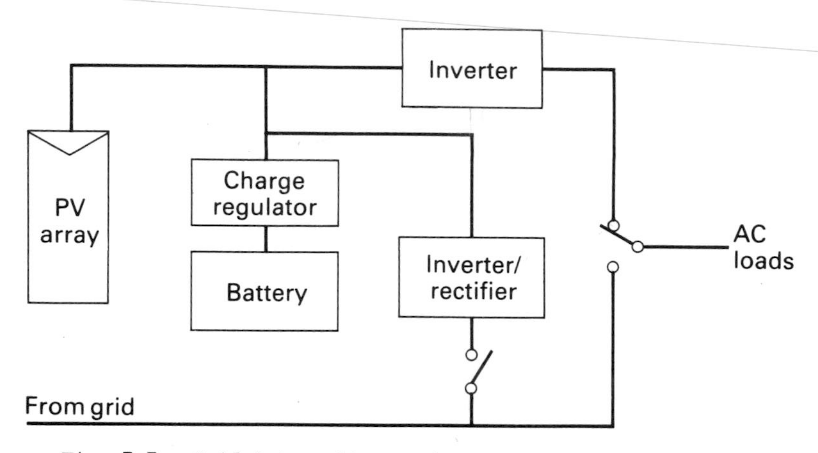

Fig. 7.7. Grid-interactive system.

7.7 General Design Considerations

The basic objective of the system designer is to deliver the required electrical energy or final product, such as pumped water, at the minimum unit cost. The unit cost is calculated by dividing the estimated life-cycle cost of the whole system, including load appliances, by the estimated useful output over the anticipated life of the system. The life-cycle cost depends, inter alia, on the capital cost (including transport, site preparation and installation), discount rate, the amortisation period and the costs of operation, maintenance and repair. The useful output is the product of the solar input, the system efficiency and the utilisation factor, (i.e. how much of the potential output is actually used). All these factors must therefore be taken into account when comparing design options.

Reliability is vital to minimise maintenance and repair costs, particularly in remote areas, where there may be a lack of spare parts and the necessary technical skills. So, before opting for complexities such as sun-trackers. MPPTs or multiple inverters, the designer should carefully consider the likely effects on reliability. A good general rule for remote installations is "keep it simple", even though this may mean some loss of efficiency.

All components should be of proven design and of the best available quality. Modules should have passed as internationally-recognised type approval test. (See Chapter 4). Power conditioning components should have demonstrated their capability to withstand the environmental conditions to which they will be subjected by passing appropriate climatic tests.

Batteries should preferably be of the sealed type, requiring minimal maintenance, and should be protected against temperature extremes, over-charging and excessive discharge. It may well prove cost-effective to pay a premium for modules of the highest efficiency, since the cost of many of the other components, such as land, fencing, support structures and cabling, are area-dependent. The smaller the array area, the lower will be the cost of these so-called "balance-of-system" (BOS) items.

Where there is a choice of load appliances, the most efficient and reliable option should be selected. Other factors that should be considered are the minimum start-up power required and the degree to which the load can be matched to the pv array under variable operating conditions.

In some standalone applications, like water pumping and domestic electricity supply, there is a choice between ac and dc. Direct current pv generators are cheaper, simpler and more efficient, because there is no inverter, but dc appliances may be more costly and less readily available than their ac counterparts. Moreover, for water pumping, submerged water-filled ac induction motors are proving to be more reliable than sealed dc motors. The balance of advantage will depend on particular circumstances.

Another choice that must be made at the outset is the dc bus voltage. In small domestic dc systems, 12V or 24V may be chosen for safety reasons but, in larger systems, it can vary between 100V and 500V. High voltage has the advantage of minimising transmission costs and losses and this factor will become of increasing importance as pv generators get bigger. Lower voltages may be considered safer but even at 100 V adequate safety precautions must be taken. Other factors influencing the selection of the dc bus voltage are (a) the use of standard commercially-available equipment such as motors and inverters, (b) the input voltage requirements of load appliances and (c) the fact that the cost of power distribution devices increases sharply at switching voltages above 120Vdc.

In rural electrification, there is often the choice between individual household systems and a centralised generator serving the whole village. Individual systems have a number of advantages. They can be roof-mounted, safely out of the reach of livestock and people. They do not take up valuable space near the houses, nor do they require a distribution network - an expensive item if the houses are widely separated. With no grid, the problems of unauthorised connection and theft of electricity are avoided. No metering system is needed. The householder has a personal incentive to keep his equipment in good condition and perhaps modify his load profile to match more closely the solar input. Lastly, the breakdown of a household system affects only one consumer, whereas the failure of a centralised system affects everyone. On the other hand, it must be said that a centralised system may have a better utilisation factor and a lower initial cost, despite the extra expense of the distribution network.

7.8 Safety Aspects

The design, construction, operation and maintenance of the system must be in accordance with applicable safety codes and regulations. It is wise to keep this in mind from the early stages of the design process. For large systems, it may be advisable to employ a safety consultant to provide an independent viewpoint. The maximum array voltage under open-circuit conditions must be within the voltage isolation capabilities of the modules and other system components. All terminals must be adequately shielded against accidental handling. All equipment must be properly earthed, for safety reasons and as a protection against lightning and the voltage surges that may arise from switching transients or the failure of components.

Protection is provided in the array field by connecting the positive or negative dc bus or a centre tap bus to a common earth terminal by a sufficient number of conductors to create an equipotential environment. All non-current-carrying metal frames, supports and enclosures for circuit conductors and equipment must be similarly earthed. The conductors should be as short as possible, avoiding sharp bends. They must be mechanically strong and their current-carrying capacity must be sufficient to limit surge voltages to a safe level and ensure fast operation of circuit protective equipment without being damaged by overheating. It may be necessary in large installations to provide an earth mat as a common earth terminal.

The system must be provided with means for interrupting the current between subsystems, for the protection of personnel during servicing and repair operations or in the event of a failure.

All materials must comply with applicable codes relating to flammability, flame spread and the release of excessive levels of smoke and toxic gases.

It is most important that system installers be given clear safety instructions, particularly as many may be unfamiliar with photovoltaics. The installation must be planned so that hazardous conditions do not exist at any time. To this end, lightning and other fault protection components should be installed as soon as possible. The designer should remember that photovoltaic modules are energised when illuminated, whether or not they are connected into the system - a fact that may be easily overlooked by the installer. It is a good idea to fit modules with integral protection, such as short-circuiting jumpers, that are removed only when installation is complete.

There are particular hazards in the installation of storage batteries, because of their high fault current capability and the possibility of chemical burns from the electrolyte. Existing codes and standards for battery installations should therefore be strictly observed.

7.9 Electromagnetic Interference

Studies have shown that a pv array field can act as a transmitting radio frequency antenna when the dc cabling is stimulated with rf energy. The inverter can add ripple to the cabling, which may be of sufficient magnitude to cause electromagnetic interference. Conversely, the solar cells can receive rf energy, which can be conducted through the dc bus to the inverter.

Suitable filtering and screening measures should be taken in cases where such interference may cause problems.

7.10 Maintenance

The whole system should be designed for simple operation and maintenance. Replacement of modules and other components, such as circuit boards, should be an easy matter and should not require special tools. The supplier should provide a maintenance manual, with clear instructions for servicing, fault detection and safety observance. He should also ensure that spare parts required for repair or replacement are readily available. Hazard warning labels should be posted where necessary.

Instrumentation should normally be kept to the minimum necessary for an unskilled operator to keep an eye on the performance of the system and be aware of faults as and when they arise.

7.11 System Sizing

Photovoltaic system sizing consists of working out the cheapest combination of array size and storage capacity that will meet the anticipated load requirements with the minimum acceptable level of security. By "security" is meant the probability that the system will always satisfy the load. The minimum acceptable level has a very important influence on cost. It can vary widely, depending on the application and the wishes of the user or procurement agency. For example, for standalone household systems in rural areas, a security level of 95% may be considered sufficient but, for telecommunications systems, a level of 99.99% or more is generally required. If the system is grid-connected or has some other back-up, a lower security level will be acceptable. The final design is always a compromise between cost and security. Unless the purchaser specifies what level of security he wants, the manufacturer, in a highly competitive market, will tend to favour a low cost option.

The following input data are required for system sizing:-

1) The daily or hourly load requirements during a typical year.
2) The required security of supply, taking into account the back-up source, if any.
3) The mean daily irradiation in the plane of the array at the chosen site for every month of a typical year.
4) The maximum number of consecutive sunless days likely to be experienced.
5) The mean daily ambient temperature for every month of a typical year.
6) The estimated cell temperature rise above ambient of the modules in the array.
7) Typical current-voltage characteristics of the selected type(s) of module at various irradiances and temperatures. Alternatively, the nominal I-V characteristics at STC, with the temperature coefficients of current and voltage and other parameters necessary to transpose it to other conditions.

8) The selected dc bus voltage.
9) The maximum allowable depth of discharge of the battery.
10) The estimated percentage energy losses in the battery, power conditioning equipment and control system.
11) The estimated losses in the array from module mismatch, cables and voltage drop across blocking diodes.
12) The estimated losses from dust and shading.

In some applications, such as navigation aids, the load profile is easily obtained. In others, for example rural electrification, it is more difficult to estimate and it may change after the system has been installed. However, it is most important that the load be estimated as accurately as possible, as it can have a profound effect on the system efficiency and the security of supply.

The designer normally uses published radiation data based on measurements over several years. He has to use figures from the meteorological station nearest the proposed site or construct data from a number of such stations. There are two major uncertainties. One is that the site may experience a micro-climate markedly different from that at the nearest meteorological station particularly if it is on a mountain or in a coastal area. The other is the variation, both annual and seasonal, from the published average figures. This can amount to several percent. As very few stations measure the irradiation on inclined surfaces, this has to be computed from records of global (horizontal) irradiation or sunshine hours. Programmes for doing this have been devised by Kondratyev (1977) and Rodgers (1979).

If local global irradiation data are not available, estimtes may be made from worldwide radiation maps, for example those produced by Lof (1966) and Major (1981). European data can be obtained from the European Solar Radiation Atlas (1984). Volume I contains maps with isopyrs (lines of equal irradiation) indicating the mean daily global irradiance in $kWh.m^{-2}$ for every month of the year in an area stretching from Scotland to Saudi Arabia. Tables list the mean data, with maxima and minima, for 340 stations in the area. These data are based on 10-year means over the period 1966-75. Volume II contains similar maps of an area covering 19 European countries, indicating the mean daily global irradiances on vertical surfaces and on South-facing surfaces tilted at 30°, 60° and the angle of latitude. Tables list estimated monthly means of daily global and diffuse irradition on inclined planes at 102 sites in the area.

Mean ambient temperatures are available from meteorological authorities. The temperature rise of the solar cells will depend on how and where the modules are mounted, the irradiance and, most importantly, the wind speed. In the absence of direct experimental data, the designer must rely on an estimate. The NOCT, if available, may be a useful guide in this respect.

The estimation of energy losses in the battery, power conditioning equipment and control system presents further problems. Battery losses depend on the charge/discharge efficiency, which varies with state-of-charge, and how much of the generated power passes through the battery. The energy loss in an inverter depends not only on the efficiency/load characteristic but on how the load varies with time. This is difficult to predict.

Because of the variability of the input data, sizing is necessarily an approximate calculation. There is at present no generally accepted procedure. Most suppliers have developed their own proprietary models and computer programmes, the details of which they are not prepared to divulge. However, for a standalone system without back-up or MPPT, a logical sequence of design steps is as follows:-

1) From the ambient temperature data and the estimated cell temperature rise, determine the maximum operating temperature the solar cells are likely to reach, on a probability based on the required level of security.

2) From the I-V characteristic of the selected module, determine the voltage for maximum power at the maximum operating temperature. Select a working voltage just below this point and from this determine the number of series-connected modules required for the selected dc bus voltage, making due allowance for the voltage drop across the blocking diodes.

3) From the I-V characteristics of the module and the monthly mean daily irradiation data at an inclination equal to the angle of latitude, calculate the mean daily energy output of the series string of modules for every month of the year. Reduce these figures by an appropriate safety margin and an allowance for losses in the array, battery, power conditioning equipment and control system. The safety margin is based on the required security level, the tolerance on the module performance, the variability of the irradition data and the expected loss from dust and shading effects on the array. For a ± 10% tolerance on module performance, it could be between 15 and 35%.

4) Divide the annual mean daily load by the annual mean daily output from the module string and round up to give the minimum number of module strings required in parallel.

5) Using hourly irradiation and load data (hypothetical, based on the daily means, if no actual figures are available), compute, by re-iteratively tracing the daily charge/discharge cycles, the minimum battery capacity required to meet load requirements throughout the year with the minimum size of array, assuming that the maximum allowable depth of discharge is not exceeded. Check that this capacity is adequate to carry the load over the longest period of consecutive sunless days likely to be experienced, on a probability consistent with the required security level.

6) Estimate the total cost of the array and battery.

7) Repeat Steps 3 to 6 for other array inclinations and for more than the minimum number of modules, in order to determine the cheapest combination of array and battery. (At present, the cheapest option is almost bound to be the one with the smallest array but this situation may change as module prices fall).

8) Check the sensitivity of the design to variations in irradiation within the published limits and variations in module performance within the maker's tolerance. Make any necessary adjustments.

In some applications, there are constraints which may determine the array inclination and thus simplify the sizing procedure. For example, on a buoy free to rotate about its vertical axis, the array would be mounted in or near the horizontal plane. In an installation on a pitched roof, structure costs could be saved by conforming to the roof line.

For systems with a MPPT, the procedure is similar, except that, in Step 2, the number of modules in a series string is chosen to suit the input voltage range of the MPPT and, in Step 3, the daily mean outputs of the module string are based on the maximum power from the module instead of the power at the working voltage.

7.12 Sizing Calculation

7.12.1 The problem

As an exercise in array and battery sizing, we will take the hypothetical case of an off-grid farmhouse just north of Nice (latitude 43°39' N, longitude 7°12' E, altitude 10m). The farmer requires a 240Vac standalone pv system, giving a 5-year security of supply of 0.95 with no back-up.

7.12.2 Input data

Table 7.1 gives the monthly mean daily load figures and the required meteorological data for Nice, taken from the European Solar Radiation Atlas. The maximum number of consecutive sunless days is taken to be ten.

The array design is based on a commercial 50Wp (± 10%) crystalline silicon module, with the I-V characteristics shown in Fig. 7.8. The NOCT of the module is 45 °C. Its temperature coefficient of current, α, is 4×10^{-4} $°C^{-1}$ and its temperature coefficient of voltage, β, is -3.4×10^{-3} $°C^{-1}$.

The selected dc bus voltage is 240V.

The maximum allowable depth of discharge of the battery is taken to be a nominal 80%. It is assumed that, on average, 45% of the

Table 7.1 LOAD, IRRADIATION AND TEMPERATURE DATA

Month	Monthly mean daily load (kWh)	Monthly mean daily irradiation ($kWh.m^{-2}$)				Mean ambient temp. (°C)
		Horiz.	30°	Angle of latitude	60°	
January	7.2	1.72	2.97	3.32	3.54	12
February	7.1	2.46	3.51	3.75	3.81	13
March	6.9	3.91	4.86	4.96	4.79	15
April	6.7	5.36	5.92	5.76	5.25	17
May	6.5	6.09	6.12	5.74	4.98	20
June	6.5	6.79	6.56	6.05	5.11	24
July	6.5	7.13	7.11	6.63	5.68	26
August	6.5	5.92	6.36	6.12	5.49	27
September	6.7	4.39	5.52	5.55	5.26	25
October	6.9	3.27	4.60	4.88	4.91	21
November	7.1	1.99	3.23	3.57	3.76	16
December	7.2	1.64	3.06	3.48	3.76	13
Mean	6.8	4.25	4.99	4.99	4.70	19

load is during the day, leaving 55% to be supplied from the battery. The energy efficiency of the battery is taken to be 70%, so the overall battery loss is 0.55 x 30% or 16.5%.

The energy loss in the inverter is assumed to be 15% and a further 5% is allowed for losses in the array, including the effects of dust and shading.

In the interests of simplicity, the system does not include a maximum power point tracker.

7.12.3 Array sizing

From Table 7.1, it is seen that the maximum ambient temperature T_{AMB} is 27 °C. Therefore, the maximum cell operating temperature, T_M is given by:

$$T_M = T_{AMB} + (NOCT - 20)\ °C$$
$$= 27 + (45 - 20)\ °C$$
$$= 52\ °C$$

From Fig. 7.8, it is seen that, at 1000 $W.m^{-2}$ and 52 °C, the module gives its maximum power at 16V. A suitable working voltage, bearing in mind the ± 10% tolerance on module output, is 13.5V.

At this voltage, the number of modules in a series string required for the dc bus voltage of 240V, after making allowance for the blocking diodes, is 18.

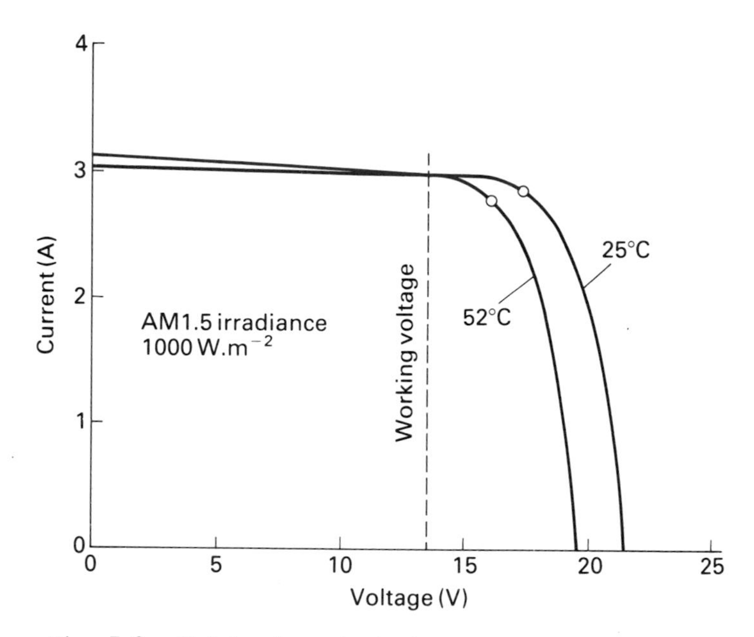

Fig. 7.8. Module characteristics.

Figure 7.8 shows that the nominal power output of each module at the working voltage and an irradiance of 1000 $W.m^{-2}$ = 3.0A x 13.5V

= 40.5W

(Note that this output is not significantly affected by cell temperatures below 52 °C).

The nominal power output of a series string at 1000 $W.m^{-2}$

= 18 x 40.5 W

= 729 W

At a mean daily irradiation of E $kWh.m^{-2}$, each series string will deliver 729 E Wh/day.

This output must be reduced by 45% to allow for the following losses and safety margin:-

	Percent
Battery loss	16.5
Inverter loss	15
Array losses	5
Safety margin appropriate to 0.95 security and ± 10% module tolerance	19
Total allowance	45

Thus, the net mean daily output of each series string of modules

= (729 x 0.55) E Wh/day

= 401 E Wh/day

Table 7.2 lists the monthly mean daily outputs from a module string at three inclinations, calculated from the above expression.

Table 7.2 MODULE STRING OUTPUTS

Month	Monthly mean daily output from a module string (Wh /day)		
	30°	Angle of latitude	60°
January	1191	1331	1420
February	1408	1504	1528
March	1949	1989	1921
April	2374	2310	2105
May	2454	2302	1997
June	2631	2426	2049
July	2851	2659	2278
August	2550	2454	2201
September	2214	2226	2109
October	1845	1957	1969
November	1295	1432	1507
December	1227	1395	1507
Mean	1999	1999	1882

To calculate the minimum number of module strings required, we divide the annual mean daily load by the annual mean daily output from each string. Thus, for an inclination equal to the angle of latitude:

$$\text{Number of module strings} = \frac{6.8 \text{ kWh}}{1.999 \text{ kWh}}$$

$$= 3.4 \text{ - rounded up to } 4$$

This number is also required for the other inclinations.

7.12.4 Battery sizing

A computer is needed to carry out the battery sizing calculations. However, a rough idea may be obtained by calculating the monthly surplus or deficit of energy from the array, as in Table 7.3.

Table 7.3 MONTHLY ARRAY OUTPUT/LOAD BALANCE

Month	Load requirement (kWh)	Angle of latitude		60°	
		Array output (kWh)	Surplus or deficit (kWh)	Array output (kWh)	Surplus or deficit (kWh)
January	223	165	- 58	176	- 47
February	199	168	- 31	171	- 28
March	214	247	+ 33	238	+ 24
April	201	277	+ 76	253	+ 52
May	202	285	+ 83	248	+ 46
June	195	291	+ 96	246	+ 51
July	202	330	+ 128	282	+ 80
August	202	304	+ 102	273	+ 71
September	201	267	+ 66	253	+ 52
October	214	243	+ 29	244	+ 30
November	213	172	- 41	181	- 32
December	223	173	- 50	187	- 36
Total summer surplus			+ 613		+ 406
Total winter deficit			- 180		- 143

The 60° inclination results in a smaller winter deficit than that at the angle of latitude and so is the preferred option from this simple calculation. A more detailed computer study might find a different optimum.

To cater for the winter deficit at 60° inclination, the required battery capacity, with a maximum depth of discharge of 80%, is:

$$\frac{143}{0.8}\ \text{kWh} = 179\ \text{kWh}$$

At 240 Vdc, this is equivalent to $\frac{179000}{240}$ Ah = 746 Ah

Since the average daily load is 6.8 kWh, the number of consecutive sunless days the battery will cater for (allowing 15% inverter loss)

$$= \frac{143 \times 0.85}{6.8}\ \text{days}$$

$$= 17\ \text{days}$$

Thus, the final design, from this rough calculation, consists of a 3600 Wp array of 72 50 Wp modules (18 in series x 4 in parallel) at an inclination of 60°, and a 750Ah lead/acid storage battery of 120 single cells.

7.13 Inverter sizing

The determination of the correct size of inverter depends on the application and involves balancing cost and performance to achieve the best compromise. As inverter efficiency increases with load, it is usually best to select the smallest inverter capable of performing the required function, so that it operates at an average load as near as possible to its full load capacity. In the case of a remote standalone system with storage, the inverter can have a full load rating considerably lower than the rated power of the array. On the other hand, for a system providing peak power to the grid, the inverter may have a full load capability higher than the array rating.

7.14 Conclusion

An attempt has been made in this chapter to lay down general guidelines for the photovoltaic system designer. However, there is no substitute for practical experience and it is important that the lessons learned from pilot and demonstration plants be clearly documented and made available to all concerned. Some progress is being made in this direction by the European Commission but independent publications are also becoming available. Examples are the detailed report on the design, installation and operation of seventeen pv systems in the Gabonese Republic (Kaszeta, 1987) and the Intermediate Technology Handbook on Solar Water Pumping (Kenna and Gillett, 1985).

Chapter 7 References

European Solar Radiation Atlas, Vol. I Horizontal Surfaces (Re vised Edition) (1984), Vol. II Inclined Surfaces (1984). Published by the Commission of the European Communities, Directorate-General Information Market and Innovation, Luxembourg.

Kaszeta, W.J., "Design, development and deployment of public service photovoltaic power/load systems for the Gabonese Republic". NASA/Lewis Research Center Document CR179603, 1987.

Kenna, J.P. and Gillett, W.B., "Solar water pumping - a handbook". Intermediate Technology Publications, London, 1985.

Kondratyev, K.Ya and Fedorova, M.P., "Radiation regime of inclined surfaces". Technical Note No. 152 (WMO No. 467), published by the World Meteorological Organization, Geneva, 1977.

Lof, G.O.D., Duffie, J.A. and Smith, C.O., "World distribution of solar radiation". Report No. 21, Engineering Experiment Station, College of Engineering, University of Wisconsin, Madison, Wisconsin, USA, 1966.

Major, G. et al., "Meteorological aspects of the utilization of solar radiation as an energy source. Annex: World maps of relative global radiation". Technical Note No. 127, WHO No. 577, Meteorological Service of the Hungarian People's Republic, Budapest, Hungary, 1981.

Rodgers, G.G., Page, J.K. and Souster, C.G., "Mathematical models for estimating the irradiance falling on inclined surfaces for clear, overcast and average conditions". Proceedings of UK-ISES Conference C18 on Meteorology for Solar Energy Applications. Jan. 1979.

Chapter 8

POLYCRYSTALLINE SOLAR CELLS

R. HILL
Newcastle Polytechnic, UK

Introduction

The cost of electricity from solar cells depends on many factors, but one of the most important is the cost of the photovoltaic modules. The cost of modules depends on the cost of materials used in their construction and the cost of the manufacturing processes. Both of these costs can be reduced by the use of thin films of the semiconducting materials for the cells instead of the usual slices of crystalline silicon.

A thin film is a layer of material with a thickness of the order of a micron (one thousandth of a millimetre) which is deposited on a substrate which provides mechanical support. Thin films can range in thickness from a hundredth of a micron (about 100 atomic layers) to a few tens of microns (about the thickness of a layer of paint).

There are a wide variety of techniques which can be used to deposit a thin film of material on a substrate. Some techniques produce films which are perfectly crystalline, with electrical properties which are superior to those of slices of single crystal material, and such films are widely used in the electronics industry. Other techniques produce amorphous films, in which the atoms are not in any regular order, and solar cells based on amorphous silicon are the subject of Chapter of this book.

In this chapter, we will consider films which are intermediate between crystalline and amorphous - these are known as polycrystalline films. Polycrystalline films, as the name suggests, are composed of a large number of small crystals. Within these crystallites, the atoms are regularly ordered, and their electrical and optical properties are similar to those of a bulk single crystal of the material. The crystallites in the film are packed together in a purely random fashion so any electric current flowing across the film must cross many boundaries between crystallites, and these grain boundaries exert a strong influence on the characteristics of the films.

Polycrystalline materials seem, at first sight, to be unlikely to be useful for making efficient solar cells, since their electrical properties are so inferior to those of crystalline material. However, the semiconductors which are used in polycrystalline cells absorb light much more strongly than crystalline silicon. About 95% of the light can be absorbed in only a few microns of semiconductor, so the material can be made 100 times thinner than a silicon wafer. Furthermore, if the film is deposited under carefully controlled conditions, each crystallite can be a small column extending through the whole thickness of the film.

The currents then flow up these crystalline columns and the device behaves rather like a large number of tiny crystalline cells connected in parallel. The influence of the grain boundaries is not eliminated, but it is reduced and the larger the diameter of the columnar crystallites or grains, the smaller the effect of the grain boundaries.

There are many ways of depositing polycrystalline films on a substrate, each with advantages and disadvantages. The choice of deposition technique is governed firstly by its proven ability to produce efficient cells from particular materials, but then secondly but equally importantly, by the cost per cell of manufacturing at a given production rate. Some techniques are relatively slow and labour intensive, but have low capital costs. These techniques could produce cells fairly cheaply at quite low production rates. Other techniques are capital intensive, but would produce cells at low cost only at high production rates. As the market for solar cells increases to beyond 100 MW_p per year, these techniques will become viable, but the production rate must always approximately match the rate of sales.

Deposition Techniques for Polycrystalline Thin Films

Physical Vapour Deposition

The most usual method of depositing thin films is by physical vapour deposition (PVD). The deposition takes place in an airtight chamber of metal or glass in which the air pressure has been reduced to about one thousandth of a millionth of one atmosphere (10^{-6} Torr). The material to be deposited is in the form of either solid pieces, powder or pellets. It is heated inside the vacuum chamber until it begins to vapourise and the vapour stream impinges on the substrate on which the film is to be deposited. The substrate is cooler than the hot source material, so the vapour condenses on to the substrate and layers of the film build up at rates typically of one micron per minute to one micron per hour.

The most common method of heating the source material is to place it on a shaped strip of metal, usually tungsten or molybdenum, and heat the metal by passing a large electric current through it. The vapourisation of the source material can be better controlled if it is placed in a closed crucible of quartz or graphite, with a small hole through which the vapour can escape.

The crucible is surrounded by a small furnace which heats it and the source material, to a carefully controlled temperature. The use of crucibles of pure quartz or graphite also avoids contamination of the vapour and gives thin films of higher purity.

Another common method of heating the source material is to focus a beam of electrons on to the material, which is held in a water-cooled crucible. Only the material near the focus of the electron beam is vapourised, whilst all the surrounding material and the crucible remain cool and do not contaminate the vapour stream. The power densities in the electron beam are very high, from tens to thousands of kilowatts per square centimetre, so the rate of vapourisation, and hence the growth of the film, can also be very high.

Many other ways of heating the source material can be used, including beams of atoms or ions, laser beams, radio frequency heating coils and microwave techniques, each of which can have advantages in particular circumstances but are usually more costly and less convenient than thermal or e-beam methods.

There are three stages in PVD - the vapourisation of the source material, the movement of the vapour from source to substrate and the condensation of the vapour on the substrate. Vapourisation of simple materials such as metallic elements is straightforward. As the metal becomes hotter, it first melts and then vapourises and the composition of the vapour is the same as that of the original solid. For compound semiconductor materials, the transition from solid to vapour is more complex. The solid material rarely becomes molten and the vapour is produced directly from the solid surface. It is rare for the vapour to consist of molecules of the compound, instead the compound is usually split into its constituent elements, and, if the vapour pressure of these elements at the source temperature is very different, then the ratio of the atoms in the vapour stream will be quite different from the ratio in the solid. The thin film which condenses from the vapour is then likely to have stoichiometry (atom ratio) which is different from that of the source material. In order to have good control of the stoichiometry of the deposited film, it is necessary for many compounds to vapourise each element independently, from its own source, or to vapourise the compound and one of its constituent elements independently. These multiple source techniques are used very commonly to deposit compound semiconductor thin films for solar cells.

The vapour stream leaving the source spreads out to a greater or lesser extent on its way to the substrate, as shown in Figure 1, and the thickness of the deposited film will not be constant over the surface of the substrate. Increasing the distance between the source and the substrate decreases the angle between the centre and the edge of the substrate and hence improves the thickness uniformity of the film. It also, however, reduces the fraction of the vapour leaving the source which condenses on the substrate. For a single source, it is not uncommon for only 5-10% of the source material to be deposited on the substrate, so it is very wasteful of material.

Both problems, of thickness uniformity and low utilisation, can

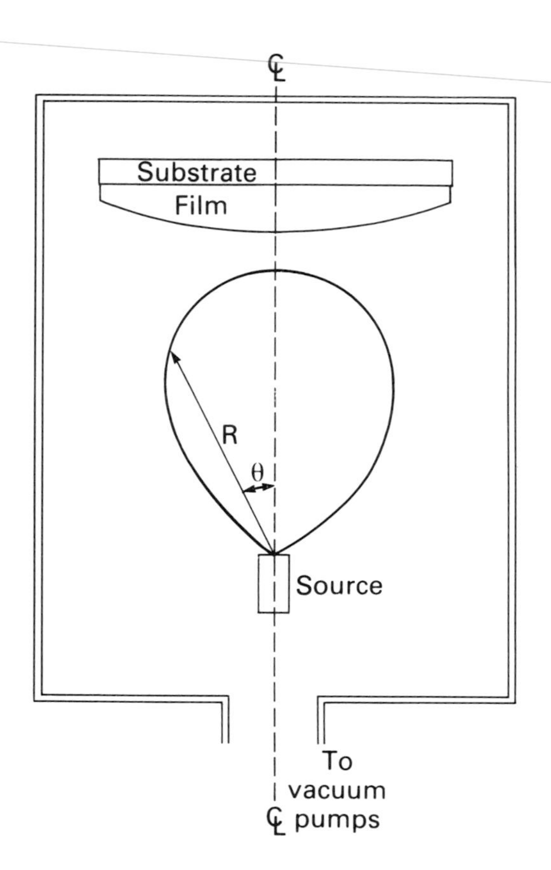

Fig. 8.1. The distribution of the vapour emission from a source is not uniform. The rate of emission R varies with the angle θ about the centre line ($R \alpha \cos^n\theta$ where $2 \leq n \leq 4$). The resulting film thickness therefore varies with distance from the centre line.

be overcome by arranging the sources in a regular array under a large substrate, as shown in Figure 2. If the distance between the sources and the distance from sources to substrate are chosen correctly, then the non-uniformity of the film deposited by one source can be compensated by the nearby sources to yield a total film thickness which is uniform over the whole substrate. Also, since the vapour stream from each source, except those at the edges, will condense somewhere on the substrate, the final film can contain about 50% or even 75% of the original source material, so the material utilisation is improved enormously.

Sputtering is a PVD technique in which the source material takes the form of a large flat plate (the target). The atoms are vapourised not by heating but by bombardment of the target by ions of argon. The substrate is also a flat plate and the target

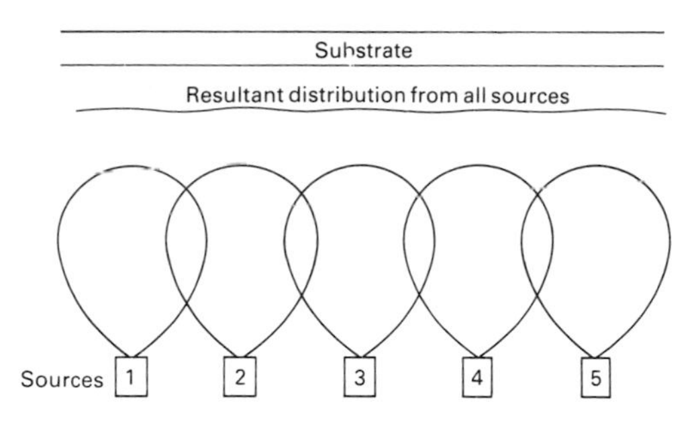

Fig. 8.2. An array of sources, each giving a non-uniform distribution, can be arranged so that the combined emission is uniform over a large area.

and substrate form the plates of a capacitor. A DC or RF discharge is set up in argon gas at low pressure between these plates and the energetic argon ions from the discharge bombard the target, knocking out atoms from its surface, which then condense on the substrate. Both the target and the substrate can be large and close together, so the material utilisation is high. The rate of deposition is usually lower than that in thermal or e-beam evaporation but substrates up to a few square metres in area can be coated uniformly.

Spraying

Semiconductor films can be deposited by spraying a solution of the elements on to a hot substrate. The solvents evaporate and the elements react to leave a film of the semiconductor material. This is, in principle, a very simple technique similar to spraying paint on to a surface and it requires small capital investment to coat large areas. In practice, however, the technique has a number of problems which have prevented it from fulfilling its apparent promise.

Most elements are not soluble in cheap, non-toxic solvents, so they must be combined into soluble compounds that are themselves cheap and non-toxic and which decompose completely at the temperature of the substrate into the element and a gas. This gas must be non-reactive and stable at the substrate temperature, so that it does not contaminate the semiconductor film. The substrate temperature at which the reactions occur must not be too high, so that cheap substrate materials such as glass may be used, but it should be high enough to promote the reaction of the elements into the required semiconductor compound and to promote the growth of crystallites of the materials in a reasonably ordered way.

There are now quite good methods for producing solutions for all the elements in common use in thin film solar cells, and the major problem is in ensuring that the polycrystalline structure is suitable. In particular, it is difficult to ensure that the structure is columnar, i.e. that the crystallites extend through the whole thickness of the film without interruption by grain boundaries.

Electrochemical Methods

Electroplating is a well known technique for producing thin films of metals (e.g. chrome plating) but it is also a powerful method of depositing semiconductor films. The elements to be deposited are first made into ionic compounds and dissolved in an appropriate solution. The compounds ionise in the solution so that the elements exist as electrically charged ions in the solution. If an electric field is applied, the charged ions are forced to either anode or cathode, depending on their charge, and are deposited there when their charge is neutrallised. If the electric field is uniform and the concentrations of the ions in the solution are uniform, then a uniform deposit of the elements or compounds can be produced over large areas.

This method requires a relatively small capital expenditure to deposit films on large area substrates, and as a standard industrial technique, its problems are well understood, with a long history of experience and expertise on which to build.

Electroplating can be used for some or all of the stages of production of solar cells. It can be used to deposit metals for the front and/or back contacts to the cells, as well as for the deposition of the semiconductor materials themselves, so it is possible to envisage the production of a complete module in a series of electroplating baths.

Electroplating is usually done with water as a solvent, so it has to take place well below 100 °C. The structure of the semiconductor films is often one of small granules not very closely packed, and a heat treatment is needed to promote the growth of the larger columnar, close-packed crystallites needed for solar cells. The heat treatment stage can be quite prolonged, and can be the main factor in determining the throughput of the process and the cost of the modules.

An alternative to electroplating is electrophoretic deposition. In this method, a suspension of very fine particles of the semiconductor in a liquid is placed in an electric field. The electric field induces an electric charge on the particles, which then move under the influence of the field and deposit on a conducting substrate which neutralises their charge. The deposit is in the form of a layer of these very fine particles, held loosely together, but a subsequent heat treatment can transform the deposit into a polycrystallime film. If is difficult to produce large columnar grains with good electrical properties, but films with good optical properties can be produced which may be suitable for window layers.

Screen-Printing

Screen-printing is another very widely-used commercial technique which can be adapted to produce layers of semiconductors. An ink is produced which consists of a suspension of very fine particles of the semiconductor or the elements in a carrier fluid. The substrate is held just underneath a fine wire mesh, and the ink is spread over the mesh by a flat blade called a squeegee. The pressure of the squeegee on the wire mesh pushes it in contact with the substrate and squeezes the ink through the mesh on to the substrate. The release of the pressure on the mesh allows it to spring back away from the substrate leaving ink on the substrate surface. The ink is dried and then fired at high temperature to produce the final film.

The II/VI semiconductors such as cadmium sulphide or cadmium telluride have been screen-printed with considerable success. Cadmium sulphide films have been produced from inks containing fine particles on the compound. Cadmium telluride films have been most successful when screen-printed from inks containing the separate elements. The subsequent heat treatment promotes the reaction between the elements to form the compound and also helps to produce large crystallites and good electrical properties.

Polycrystalline Materials

Physical Structure

The structure of a polycrystalline film plays a large part in determining its electrical and optical properties, and hence its suitability for use in solar cells.

With most deposition techniques, a thin film is grown on the substrate, and the growth of the first few atomic layers has an enormous effect on the final structure of the film. As atoms begin to deposit on the bare substrate, they do not remain at the spot where they arrived, but meander around the substrate until they come to a site where the bonding to the substrate is a maximum. These nucleation states trap the wandering atoms, and it is at these sites that the film starts to grow. At first the monolayer of atoms grows outwards until it meets growing monolayers from nearby nucleation sites. The atoms in each of the monolayers are arranged in just about the same way as in a crystal of the semiconductor, but the monolayer from one nucleation site has no orderly relationship with those of neighbouring sites. The boundaries where the monolayers meet therefore remain distinct and there is no merging of the layers. As more atoms arrive, the crystalline layer around each necleation site gets thicker at a rate which depends partly on the crystallographic orientation of the layer. Some crystal directions allow the layers to grow more rapidly than others, and the crystallite layers with these orientations grow more quickly than those with other orientations until crystallites with the "fast-growth" orientation swamp all the others. Films growing in this way produce the large columnar crystallites which are required for solar cells.

The way in which the crystallites meet at their boundaries can strongly influence the properties of the film. The crystallites (grains) do not completely fill all the volume of the film, and there are pores between the grains, which range in size from a few Angstroms to a few tenths of a micron. The small pores fill with liquid water when the film is in a normal atmosphere, so the optical properties are a mixture of those of the semiconductor and of water, weighted according to the relative amounts of each. The pores in the film can occupy from 0% to 20% of the total volume of the film. The proportion decreases as the temperature of the substrate increases, and high substrate temperatures also produce large crystallites, so in general thin films for solar cells are either deposited at high substrate temperatures or are given a subsequent heat treatment. The space around the grain boundaries provides a route for impurities to enter the film and cause shorting paths which reduce the performance of the cells, so it is always desirable to have the minimum possible pore density.

Electrical Properties of Polycrystalline Thin Films

The structure of polycrystalline thin films has a considerable effect on their electrical properties. Electric currents which flow from top to bottom of the film travel through the crystallites. If the film is columnar, as shown in Figure 3, then the current flows through crystallite material, without having to cross a grain boundary. The current flow in this case is similar to that in a single crystal of the material.

Currents which flow across the film must cross many grain boundaries in order to travel from crystallite to crystallite. Within each crystallite the current flow is again similar to that in a single crystal of the material, but each grain boundary restricts the current flow to such an extent that they exert a controlling influence. A grain boundary acts like a diode which restricts the flow of current by an amount which depends on the relative electron concentrations inside and outside the grains. The surface of the grain contains disordered atoms of the material as well as impurities, and the electron concentration in these surface regions is critically dependent on the conditions under which the film was prepared and any subsequent heat treatment. These factors therefore strongly affect the ability of the film to carry current across the crystallite boundaries.

When light is shone onto a semiconductor, extra electrons and holes are created, and it is these excess carriers which give rise to the electrical output of a solar cell. For efficient operation of the cell, the electrons and holes must recombine only slowly, so that they can be separated by the semiconductor junction in the cell and collected at the contacts. Grain boundaries act as efficient sites to promote the recombination of electrons and holes unless the surfaces of the grains can be passivated in some way. Passivation of the surfaces involves the filling of the unfilled bonds of the disordered surface atoms either by a heat treatment to reduce the degree of disorder or by the introduction of atoms such as hydrogen which can react with these "dangling bonds".

The diode-like behaviour of crystallite surfaces has an effect

Fig. 8.3. A scanning electron microscope picture of a polycrystalline thin film. The columnar crystallites can be clearly seen. Film thickness is 10 μm.

on the movement of electrons and holes up and down the crystallite. All diodes have an associated electric field, and this field attracts majority carriers to the surface. The majority carriers, electrons in n-type material or holes in p-type material, can therefore be trapped at the grain boundary. The surface electric field repells the minority carriers, holes in n-type or electrons in p-type material, and forces them away from the surface towards the centre of the crystallite. The output of a solar cell depends on the movement of minority carriers, so a small electric field at the grain boundary can in fact be beneficial in keeping the minority carriers away from the recombination sites at the boundary.

It is frequently found that thin films have a high degree of stress, with magnitudes of 10^8-10^9Nm^{-2} being quite common. A tensile stress, where the film is like a stretched elastic band, can cause the whole film to separate from the substrate. Long before that occurs, the stress causes a deterioration in the electrical performance of the film.

The most obvious cause of stress is a difference in thermal expansion coefficients of the film and the substrate. Films are usually deposited, or heat treated, at temperatures from 200 °C to 500 °C, and any mismatch in expansion coefficients leads to considerable stress on cooling down to room temperature. Care must be taken to match the expansion coefficients of solar cell materials and their substrate, or to use a metal contact to the cell which can take up or relieve this stress.

Stress can also be induced in thin films during their deposition, particularly in those deposited by physical vapour deposition. The magnitude, and the sign of the stress - tensile or compressive - depend on factors such as the substrate temperature, rate of deposition, film thickness, film material and the substrate material and condition. Part of the skill in producing thin film solar cells is to deposit the films in such a way as to reduce this "intrinsic" stress to levels at which it has no significant effect on the cell performance.

Junctions of Polycrystalline Films

Almost all solar cells made from thin polycrystalline films are heterojunctions i.e. they are semiconductor junctions consisting of an n-type material and a different p-type material. To make these cells, an electrically conducting film is first deposited on a substrate, the first semiconductor is deposited on the contact film, the second semiconductor is deposited over the first semiconductor, and the other electrical contact to the cell is then deposited on top of this sandwich.

The process of making a thin film solar cell thus involves the deposition of thin films on other thin films, and in particular the deposition of one semiconductor film on to another to create the junction. The interface between one film and another is not smooth or flat, but occurs on the rough top of the crystallites and is discontinuous because of the pores and grain boundaries of both films. In the time between the end of deposition of the first semiconductor layer and the beginning of the deposition of the second semiconductor, the top surface of the crystallites and the surfaces of the pores can become covered by one or more layers of oxygen atoms or other impurities. In the final solar cell, these layers of impurities are in the centre of the semiconductor junction, and can have a significant influence on the current transport properties of the junction. In PVD deposited cells, it is important to maintain a low pressure in the vacuum system between depositions of the semiconductor films and/or to remove any impurity layers by techniques such as plasma cleaning prior to the deposition of the second layer.

To produce semiconductor layers with good optical and electrical properties, it is necessary to deposit these layers at high substrate temperatures. The optimum temperature is usually different for each semiconductor, and a film will usually be degraded if it is subjected to temperatures higher than this optimum, subsequent to its deposition. The semiconductor layer requiring the highest substrate temperature should thus be deposited first, with each subsequent deposition being at progressively lower temperatures. This then constrains the order in which the layers can be deposited, and therefore constrains the design of the cells.

The first semiconductor layer is deposited on an electrically conducting layer which acts as the contact to the semiconductor. This contact layer must fulfil a number of important functions. It must be a good electrical conductor; it must adhere well to the substrate and to the semiconductor layer and have a reasonable match to the thermal expansion coefficient of both: it must form a good base on which to deposit the semiconductor layer, i.e. it must promote the growth of large columnar crystallites; and it must form a non-rectifying low resistance electrical contact with the semiconductor. It is very rare that any material meets all of these criteria in full, and compromises have to be made. Even so, the possible materials for the bottom contact are very limited for each of the types of thin film solar cell, and when cost and availability have also to be taken into consideration for possible commercial production, the choice is even further limited.

All thin films are liable to have "pin holes". These are holes in the film usually 1-20 microns in size. They are defects in the film and are quite different than the fine pores which are an intrinsic feature of polycrystalline films. Pin-holes arise from such things as dust particles on the substrate during deposition, which subsequently fall out of the film, or from defects on the substrate which prevent nucleation and growth of the film. Dust and other extraneous particles can be avoided by a strict housekeeping regime. The number per unit area of pin-holes from other causes decreases as the film thickness increases, since the nearby crystallites expand into the gap. Film thicknesses need to be above a few microns to ensure pin-hole densities below 1 cm^{-2} or so. The effect of pin-holes is to create pathways by which the cell can be shorted out. A pin-hole in the first semiconductor layer would allow the second semiconductor to be deposited at the bottom of the pin-hole, on the bottom metal contact. A pin-hole in the second semiconductor layer would allow the top contact to be deposited all the way down the hole to the junction region. These cases usually give rise to rectifying contacts which affect the fill-factor. The worst case occurs when pin-holes in both semiconductor layers overlap, allowing the top contact to be deposited down both holes to the bottom contact. The cell can then be completely shorted out and give no output, or at best the shunt resistance is significantly reduced.

The nature of polycrystalline thin films gives rise to a series of problems in the production of efficient solar cells, as discussed above. In practice, polycrystalline thin films can have properties surprisingly close to those of crystalline material and cells made from these films can have good efficiencies. The major potential advantage of polycrystalline thin films comes from the possibility of manufacturing them in complete modules at high rate and low cost using well-established production techniques. There are only two thin film cells which have been proven to have efficiencies over 10% and long term stability and these are described in the next section.

Thin Film Solar Cells

There are a large number of different materials which have been used in thin film form as window or absorber layers in solar cells. The requirements for materials to be used as a window layer are quite different from those of materials to be used as absorbers. A window layer should be transparent to visible radiation, to allow sunlight through to the absorber, it should form an efficient heterojunction with the absorber, it should have a low sheet resistivity so that the photocurrent can be collected with minimum loss and it should be stable, non-toxic and low cost. An absorber layer muct have a high absorption coefficient and an optimum energy-gap (about 1.4 eV for single junctions); it must have a diffusion length many times longer than the inverse absorption coefficient so that all photogenerated carriers can be collected, and agin it must be stable, non-toxic and low cost. The window layer and absorber layer must be of oppostite carrier type so that they form a p-n heterojunction, and they should have electronic properties which can easily and reproducibly controlled.

No real materials match these ideal requirements but two types of polycrystalline thin film cell have been shown to have characteristics which make them attractive for commercial production - the copper indium diselenide/cadmium sulphide cell and the cadmium telluride/cadmium sulphide cell.

Copper Indium Diselenide/Cadmium Sulphide

Copper indium diselenide (CIS) is a semiconducting material with the chemical formula $CuInSe_2$. The material was first produced as single crystals, and p-n homojunctions were made by varying the copper/indium ratio and the metal/selenium ratio. Thin films of CIS were first deposited in the early 1970's and the p-type CIS films were combined with n-type cadmium sulphide (CdS) films to form a solar cell.

CIS has a bandgap of 1 eV, about the same as silicon, but has a very high absorption coefficient of about $10^5 cm^{-1}$. This means that almost all sunlight is absorbed in a layer which is less than 1 micron thick. The mobility of the charge carriers in CIS is quite high, of the order of 100 $cm^2V^{-1}s^{-1}$, and the diffusion length of electrons in the p-type material can be over 1 micron. It is easy to change the resistivity of p-CIS by altering the Cu/In ratio and it forms a good quality heterojunction with n-type CdS. Neither copper, indium, selenium or CIS itself are dangerous chemicals and copper and selenium are readily available. Indium is not available in large quantities and the constraints on large scale production due to materials availability will be discussed later.

CIS layers can be produced by a number of different techniques. The most common method, and the one which has produced cells with the highest efficiencies, is to thermally evaporate each of the elements simultaneously onto a common substrate. The layers can also be produced by electrochemical deposition, by spraying or by sequential deposition of the elements followed by a heat treatment to form the compound. With its excellent optical and electrical properties and the range of deposition techniques, CIS meets most of the requirements of the ideal absorber material.

Cadmium sulphide is a very common window material in many types of thin film solar cell. It can be formed easily into thin films with large diameter, columnar crystallites with dense packing and good optical and electrical properties. CdS has an energy gap of 2.4 eV so it transmits easily all solar radiation of wavelength above 0.5 micron. The material is invariably n-type and its resistivity can be controlled by careful deposition or by extrinsic doping from about 1 ohm.cm to semi-insulating. The low resistivity layers can conduct the photocurrent of a solar cell without serious loss to the fingers of a collecting grid. Ideally a window layer should have a bandgap of over 3 eV to allow all visible light to be transmitted, but apart from this disadvantage, CdS has most of the characteristics required of an ideal window layer.

Production and Performance of CIS/CdS Cells

Physical Vapour Deposition of Cells

Most devices which have been made use a substrate of glass with molybdemun layer sputtered onto the glass. The molybdenum layer acts as a contact to the CIS and also as the nuclcation base for the CIS film.

In order to make low resistance, non-rectifying contacts between the molybdenum and the CIS, the CIS layer must be highly conductive. However, to make a good stable heterojunction with CdS, the CIS should be nearly stoichiometric and of low conductivity. The CIS is therefore deposited in two layers with different Cu/In ratios, the first layer being copper rich to give highly conductive p-type material, whilst the second layer is slightly indium rich, and is nearly intrinsic.

The CIS is deposited from three sources, one for each of the three elements, which are heated independently to control the rate at which the elements evaporate. The evaporation rate from the sources can be controlled in two distinct ways. The rate of evaporation from each source can be measured as a function of the source temperature in a prior experiment and the evaporation rate during deposition is then "feed forward" controlled via control of the source temperature. Alternatively the rate of evaporation can be measured during deposition and feedback control applied to the power to the source heaters to maintain the required rates.

The feedforward technique avoids the need for rate measuring detectors inside the vacuum chamber during deposition. All of the rate detectors employed to date suffer from problems of "cross-talk" - the interference of the mesaurement of the rate of evaporation of one of the elements by the presence of the other two. The feedforward technique does, however, assume that the rate of evaporation from each source has a constant relationship to the source temperature. This relationship does remain constant if the aperture through which the vapour escapes from the source does not change due to the build-up of condensed material and provided that only a small fraction of the total material in the source is evaporated each time, and is topped up again before each run. Both control techniques have advantages and disadvantages, and high efficiency cells have been produced using both techniques. It seem likely that feedforward control would be more easily implemented in large-scale commercial production if thermally evaporated layers were to be used.

The CdS layer is usually thermally evaporated on top of the CIS layer. Again a two layer structure is employed, with a high resistivity layer deposited first to ensure an efficient heterojunction, and a low resistivity layer deposited last to ensure high conductivity current collection paths. The CdS is evaporated from a source containing sintered CdS powder or pellets. The resistivity of the CdS layers can be controlled either by doping with indium, evaporated simultaneously from a separate source, or by changing the stoichiometry of the film by changing the source temperature.

It is very important to control the substrate temperature during deposition of all of these layers, and to ensure that it is uniform over the whole substrate. During the deposition of the copper-rich CIS layer, the substrate is maintained at about 350 °C, which results in a highly conductive film with a sphalerite crystal structure. The layer which will form the junction is deposited at a substrate temperature of 450-500 °C, which, with the higher In/Cu ratiuo, results in the high resistance films with a chalcopyrite crystal structure that are required for efficient junctions. The CdS films are deposited at substrate temperatures of about 200 °C. The high substrate temperatures required for the junction layer of CIS make it impractical to deposit CIS on CdS, since the CdS layer would be badly affected by being subjected to such high temperatures after deposition.

The final step to complete the cell is to provide a top contact to the CdS layer. This top contact can be either a metal grid with lines about 0.01 mm wide spaced a few millimetres apart, or a layer of transparent highly conducting material such as zinc oxide or indium tin oxide. The structure of a cell is shown in Figure 4, which gives the approximate thickness of each of the layers. Both the CIS and CdS layers are much thicker than theoretical calculations suggest because of the need to avoid pin-hole problems and to produce crystallites of reasonable size.

The performance of the cell depends not only on the deposition of the layers but also on subsequent heat treatment of the cell. Heating the cell to 200 °C in air for some hours results first in a great improvement in photocurrent, open-circuit voltage and fill-factor, and then a slow improvement in voltage and fill-factor. The full improvement in output may require as much as two days continuous heat treatment.

A number of research groups in the USA and Europe have produced cells with conversion efficiencies over 10%. Short curcuit currents of up to 38 $mAcm^{-2}$, open circuit voltages of about 450mV

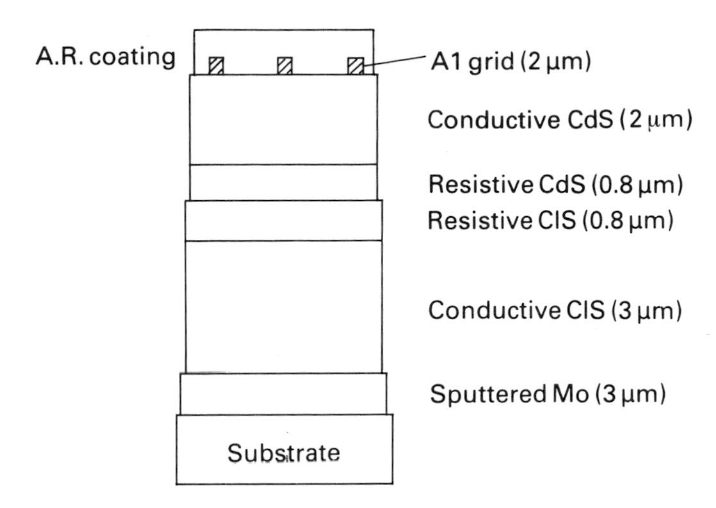

Fig. 8.4. Structure of thin film CdS/$CuInSe_2$.

and fill factors of close to 70% are typical of the best cells achieved by the coevaporation technique.

Other Production Methods

Many other techniques have been used to deposit thin films of CIS. For large scale production, sputtering is a very attractive technique since it is already in commercial use for the deposition of coating on architectural glass up to 3m x 4m in dimensions, at rates up to 1 million m^2 per year. The sputtering of CIS has been studied in detail, and films with good optical and electrical properties can be deposited. The cells made from these layers have so far not shown the same efficiency as those from coevaporated layers. The factors which limit the performance of devices with sputtered CIS are not clear, but it has proved to be very difficult to produce devices with efficiencies more than half or two thirds those with the coevaporated layers.

Spraying of the CIS film is also an attractive option for large scale, low cost production. Solutions of copper chloride, indium chloride and dimethyl selenourea are sprayed onto substrates at temperatures between 150 and 450 °C. The compounds decompose on the hot substrate to leave $CuInSe_2$ if the proportions in the sprayed solutions are chosen correctly. The CIS produced by the spray technique is of lower optical and electrical quality than coevaporated films, and the devices made from sprayed films have not exceeded 6.5% efficiency.

Processes which avoided the use of expensive capital equipment would clearly be beneficial to the economics of solar cell production, at least whilst sales were below 1 MWp per annum or so. Electrodeposition is very attractive from this point of view, since it should be possible to deposit large areas uniformly with relatively simple equipment. It is possible to deposit CIS directly by cathodic electrodeposition from a solution of copper, indium and selenium ions, provided that the solutions are prepared carefully with a complexing agent to prevent precipitation of the metal hydroxides. The results to date show some promise, although the films have very small grain sizes and make poor devices.

It is simpler to deposit only the metals by electrodeposition and then to convert the Cu/In alloy to CIS by selenisation. The alloy is exposed to selenium or hydrogen selenide vapour in a furnace at 400-500°C. This technique produces films with good electrical and optical properties, and devices with "active area" efficiencies of about 10% have been claimed. Selenisation of Cu/In alloy films can be carried out regardless of the method used to deposit the metals, and good results have also been obtained with metal films deposited by PVD techniques. These methods are not as well developed as the coevaporation technique, but they are being actively developed with a view to commercial production.

These ideas can be taken even further, by depositing sequential layers of the three elements, and then causing them to interdiffuse and react together to form CIS. The thickness of the layers is chosen so that the proportions of Cu, In and Se are correct, but the layers must be thin enough to allow the interdiffusion to take place. A laser beam has been used to provide the energy input needed to start the process of diffusion and reaction but since the reaction of the elements to form the compound is exothermic, the heat given out by the reaction plays a significant part in providing the total energy needed. The reaction of single layers of each of the elements is of limited use, in that it cannot produce the conductive region near the contact plus the resistive region near the junction which are needed for efficient devices, nor can it easily produce films with thicknesses over 1 micron or so. The author's laboratory has patented a process which overcomes these limitations. Thin layers of the separate elements are deposited sequentially, but in multiple stacks up to any required total thickness. The ratio of the thicknesses of the Cu and In layers can be varied in each of the stacks, so that Cu-rich CIS is produced near the contact while slightly In-rich CIS is produced at the junction. The reaction can be promoted by a laser beam, but the best results to date have been produced by a short heating step at about 450 °C. The resulting material has good electrical and optical properties suitable for devices.

Cadmium Telluride Solar Cells

Cadmium telluride is a very promising material for thin film solar cells. It can be either n-type or p-type, unlike most other II/IV materials, it has a direct band-gap of 1.4 eV and a high absorption co-efficient, and it can have good electrical transport properties.

Two major problems have retarded the development of CdTe cells:- the existance of deep traps which tend to reduce the carrier lifetime in the material, and the difficulty of making low resistance contacts to p-type CdTe which do not degrade over time. Improvements in material preparation have reduced the density of deep traps to levels where carrier lifetimes are acceptable for solar cell applications, although further improvements are still possible and desirable. The contact problem has proved to be more intractable, although there are a number of very promising routes to the solution of this problem.

One of the attractions of CdTe as a material for thin film solar cells is the variety of methods by which suitable layers can be produced. Methods such as physical vapour deposition, chemical vapour deposition, spraying, electrodeposition and screen-printing/sintering have all been used successfully to deposit layers for solar cells which show good efficiencies.

Thin films of CdTe are easily deposited by the various PVD techniques, but it is less easy to deposit films which have the required electrical and optical properties. Thermal evaporation of CdTe on to hot substrates at 200-500°C can produce p-type films with the active centres being mainly cadmium vacancies. These films also tend to have trapping centres close to the centre of the energy gap due to both anion and cation defects. Films of n-type CdTe can be produced with quite low resistivity, but the p-type films tend to be of high (10^4 ohm cm or more) resistivity.

Closed-space sublimation has been used successfully to deposit p-CdTe films with good carrier transport properties, although again these were of quite high resistivity. Because of its high optical absorption coefficient, the CdTe films need be only a few microns thick to absorb solar radiation effectively, so the high resistivity does not lead to high series resistance in the solar cells. High resistance material is however very difficult to contact, and it is this problem which remains the major obstacle to commercial production.

Electrodeposition of CdTe is a very attractive technique for the commercial production of cells. It is a technique in which it is relatively easy to scale up from centimetre sizes to metre sizes, and the utilisation of cadmium and tellurium is very high. This latter point is important because cadmium is not a very abundant element, and because it reduces the hazard of toxic waste at the solar cell production site. Solar cells with efficiencies of over 11% have been produced using electrodeposition by AMETEC in the USA, and pre-commercial development of electrodeposited CdTe is being undertaken by the BP Research Labs in the UK.

Screen-printing of solar cells offers the possibility of low-cost production at quite low production rates, with a technology which could be used in many developing countries. The key to the successful production of CdTe layers with good electrical and optical properties seems to be the use of cadmium and tellurium powders in the ink, rather than CdTe powder. During the sintering process, the cadmium and tellurium react to produce CdTe, and the high heat of reaction promotes the growth of high quality layers. A similar effect has been found in the laser processing of films with separate cadmium and tellurium layers, where it is possible to produce single crystal films of CdTe.

The Structure and Performance of CdTe Cells

Cadmium telluride cells are in fact heterojunctions with cadmium sulphide or other wide band-gap materials, since CdTe p-n heterojunctions have not proved to be efficient. CdS has a reasonable electron affinity match with CdTe and forms an efficient heterojunction even though there is a significant lattice mismatch.

It is usual to deposit the CdTe layer on to the back contact metal. Molybdenum is the most commonly used contact metal, even though is does not always form a contact which is stable over many years. Deposition of CdTe on to molybdenum sheet metal is rarely satisfactory, as the strip rolling process gives rise to numerous non-uniformities, even if the metal surface is deeply etched. The most satisfactory contact is found with sputtered molybdenum, usually deposited on glass or ceramic. Molybdenum does not have a very high electrical conductivity, and whilst 1-2 micron thickness gives adequately low sheet resistance for small test cells, much greater thicknesses would be required if large area cells are to have a low back-contact series resistance. The sputtering of these thick layers of molybdenum would constitute the rate limiting step in the production of CdTe cells, and would determine the eventual cost of the cells. The alternative would be to produce modules with long thin cells, possibly with an underlying layer of copper to improve the electrical conductivity.

Other metals have been used as contacts with some success, notable a bismuth/antimony alloy on which flash-evaporated CdTe forms films with crystallite sizes measured in millimetres in a narrow range of substrate temperatures. Unfortunately, the cost of these metals would rule them out for commercial production.

Heavily doped layers of other tellurides have recently been used to contact p-CdTe with notable success. Zinc telluride has been used in the AMETEC electrodeposited cell, whilst CdHgTe or HgTe alone have been used in PVD deposited cells. The author's laboratory has recently developed techniques for depositing low resistance p-CdTe, with resistivities controllable between 0.1 and 100 ohm cm. The technique involves the co-evaporation of tellurium with the CdTe, and the resistivity depends on the relative tellurium deposition rate. The electrical properties of these low resistance films make them very suitable for back contacts, but the small crystallite size so far makes these layers unsuitable for use in forming the active heterojunction.

After depositing the p-CdTe layer by any of the techniques discussed earlier, a layer of n-CdS is deposited to form the heterojunction. As with the $CuInSe_2$ cell, the CdS is a window layer and plays little active part in the generation of photocurrent. Other window materials can be used and both zinc oxide and indium tin oxide have been found to give efficient solar cells. These two semiconductors have wider band-gaps than CdS and absorb less of the incident radiation. They also usually have lower resistivities than CdS for the same optical absorption, and thus contribute less to the cell series resistance.

The "back-wall" cell structure described above can easily be turner round to a "front wall" cell. A transparent conducting oxide such as tin oxide, indium tin oxide or zinc oxide is deposited on glass. A thin layer of CdS may be deposited on the TCO, or the junction may be formed directly with the oxide. The CdTe layer is deposited on the window layer, and then a contact made to the CdTe. Molybdenum is not a good contact when deposited on to CdTe, and other materials must be used. Gold forms a good, but expensive, contact to p-CdTe either alone or on an alloy with nickel or copper. Copper plus ITO has also been found to give a good contact, which has the advantage of being optically transparent. Copper-doped graphite also forms a contact with CdTe and can be screen-printed.

A number of structures of CdTe cells, formed by different techniques are shown in Figure 5. All these structures have demonstrated cell efficiencies over 10%.

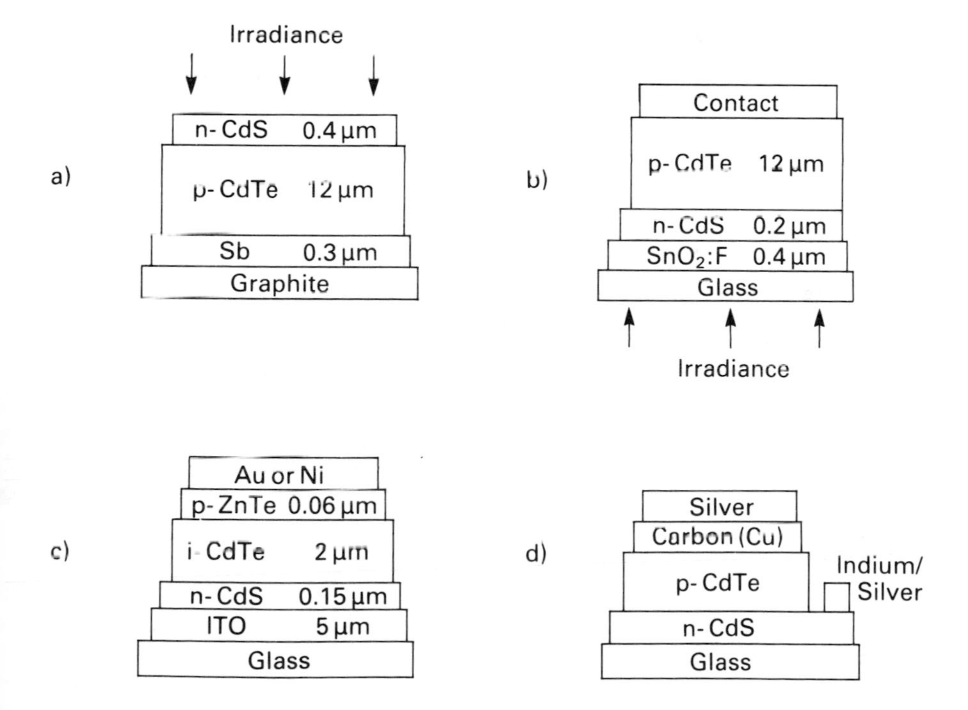

Fig. 8.5. Structure of some CdTe cells
a) Backwall cell produced by Closed-Space Sublimation
b) Frontwall cell produced by CSS
c) p-i-n cell produced by electrodeposition
d) Screen-printed cell (layers all about 25µm thick).

The performance of CdTe cells is similar irrespective of the technique used to deposit the thin films. Table 1 lists the parameters of those thin film cells which have efficiencies over 10%. The open-circuit voltage of CdTe cells is nearly twice that of $CuInSe_2$, and this is clearly advantageous in reducing the number of cells, and hence interconnects, needed in a standard 12 V module. The Fill-Factor of the cells is also poor, reflecting the high series resistance of present cell designs.

Table 1

Structure	Area cm^2	Voc mV	Jsc $mAcm^{-2}$	FF %	Efficiency# %	Preparation technology of CdS
CdS/CdTe	0.78*	754	27.9	61	12.8	Screen-print
	64*	720	26.3	54	10.2	Screen-print
CdS/CdTe	0.1	750	22.7	62	10.5	Closed-space sublimation
CdS/CdTe/ZnTe	1.48	620	27.03	63	10.6	Electro-deposition
SnO_2/CdTe	4	663	28.1	56	10.5	Closed-space deposition

* Active area only - total area ∿ 1.6 x active area.

AM1.5 Global spectrum 100 $mWcm^{-3}$ irradiance.

The short circuit currents, reaching 28 $mAcm^{-2}$, suggest a high collection efficiency for the junctions. For many devices, an efficient junction is found only after an air or oxygen anneal of either the CdTe layer or of the complete device. The precise role of the oxygen anneal is not well understood, but it has the effect of reducing the CdTe resistivity, improving the contact to the CdTe layer, and increasing the diode quality of the junction. However, efficient devices have been made without an oxygen anneal, and it is clear that a good deal of basic materials research remains to be done on CdTe and the window materials.

The highest reported open-circuit voltage on a thin film device is 820 mV, for a dot junction on electrodeposited CdTe. This is to be compared with an OCV of 890 mV reported for ITO/single crystal CdTe, and the theoretical value of 900 mV for a CdS/CdTe junction. A material with the band-gap of CdTe should have fill-Factors well over 70% if the engineering can be optimised. Short-circuit currents of 28 $mAcm^{-2}$ have already been achieved with window layers which have large energy gaps, so one may reasonably look to cells in the future with SCC of 30 $mAcm^{-2}$, OCV of 850 mV and FF of 75% i.e. a conversion efficiency of 19%. The production of integrated interconnected modules with efficiencies over 15% is thus an achievable goal in the 1990's.

Multijunction Solar Cells

The major inefficiency in solar cells arises from the mismatch between the single energy gap of the absorber and the continuous energy distribution of the sunlight. Light of energy less than that of the energy gap is not absorbed, whilst light of energy above that of the energy gap gives rise to a voltage of more than the open-circuit voltage, with the excess energy dissipated as heat.

An ideal solar cell would consist of a large number of thin cells with a range of energy gaps from 0.5 eV to 3 eV i.e. from infra-red to blue. Each cell would absorb light of energy greater than its energy gap and transmit all light of lower energy. A stack of such cells with the high energy gap cell outermost would absorb all energies efficiently, and could in theory convert two thirds of the solar energy into electrical energy. Such a cell would be immensely difficult and expensive to make, and it is at present beyond our technical capabilities. Two cells with different energy gaps operating in tandem can be made and can give a significant improvement in overall efficiency.

There are two ways in which the two-cell tandem structures can be operated. In a Two Terminal Device or Monolithic Tandem, the structure is of the form:

Front Contact/High Gap Cell/TC/Low Gap Cell/Back Contact

Electrical connections are made to the front and back contacts and the voltage is the sum of the voltages from the individual cells whilst the current is the same for both cells. The contact TC between the two cells must provide a low resistance, non-rectifying contact between the p-layer of one cell and the n-layer of the other cell. This can be done if quantum mechanical tunnelling contacts are used, but there are considerable technical difficulties in making reliable tunnelling contacts on large area cells with good process yields in large scale production. The other significant problem with such cells is that each component cell must be designed to generate the same current, since with only two contacts, the same current must flow throughout the cell. Any mis-match in the photogenerated currents of the component cells will result in the operation of both cells away from their maximum power point, and hence a reduction in the overall efficiency. It is not difficult to design the component cells to give the same current to a particular solar spectrum, but in fact the solar spectrum changes quite considerably depending on the amount of scattering in the atmosphere or by clouds etc. The efficiency of a monolithic tandem cell can thus vary considerably on a cloudy day compared to a clear sunny day. Despite these problems, monolithic tandem cells have been made using amorphous silicon as the higher gap cell and amorphous silicon/germanium alloy as the low gap cell. Structures with three component cells have been produced in small areas, with amorphous silicon carbide as the outer cell to the previous two cell structure, and efficiencies over 12% have been measured. The amorphous cells are discussed elsewhere in this book, and the monolithic structure is not well suited to polycrystalline cells.

The simplest form of tandem cell consists of two independent cells, each with their own contacts. The top cell has transparent contacts so that any light not absorbed in the top cell will be transmitted to the bottom cell. In these Four Terminal Devices, each cell can be manufactured independently and the only difference from normal production is the transparent back contact of the top cell. Since each cell is operated independently, they can be operated at their maximum power points regardless of the solar spectrum, so the overall efficiency is less dependent that the monolithic tandems on changes in solar spectrum.

The mix of energy gaps which produce the maximum overall efficiency has been calculated for both monolithic and four-terminal devices. Figure 6 shows the results of these calculations in the form of isoefficiency lines for two cell tendems in both the two-terminal and four-terminal structures. It is clear that the four-terminal structure is less critically dependent than the monolithic on the precise choice of energy gap to achieve optimum efficiency, but in both cases, a high gap of about 1.7 eV and a low gap of about 1 eV could give theoretical efficiencies of over 30%.

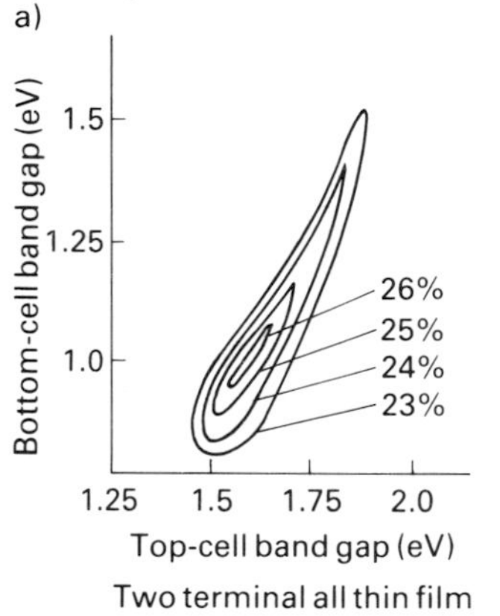

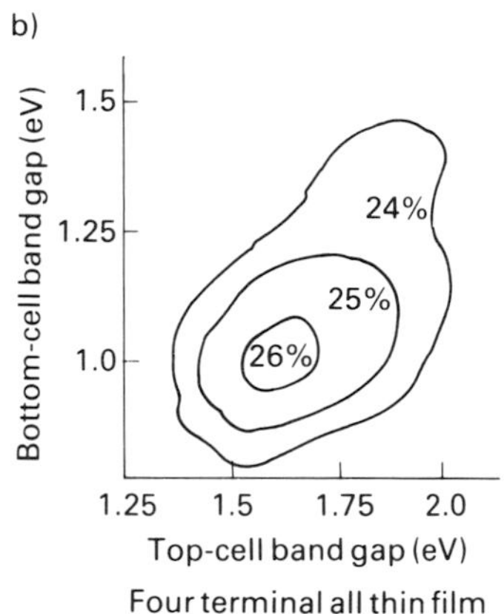

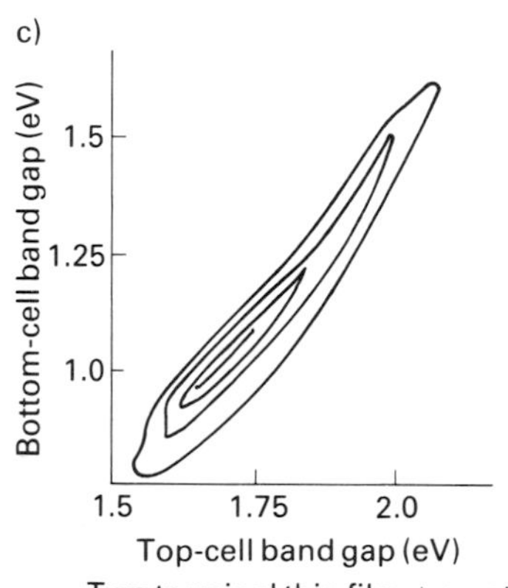

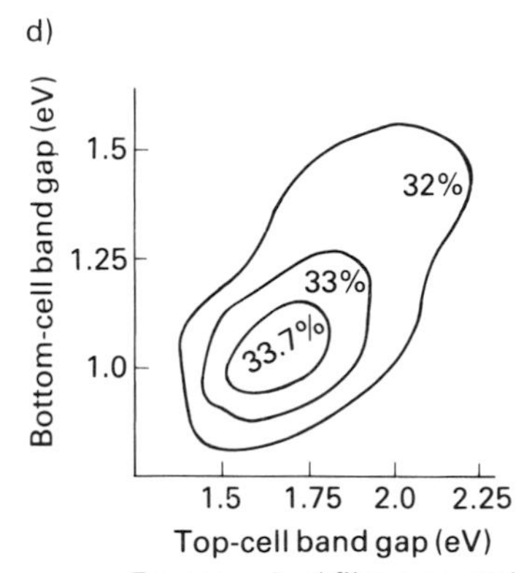

Fig. 8.6. Iso-efficiency curves for tandem cells.
Reproduced from J.C. Fan, Solar Cells.

It must be emphasised that two cells which have efficiencies of say 10% when they are designed as single junction cells will not produce a tandem cell efficiency of 20%. The conditions under which the component cells operate in a tandem structure are quite different from those in which they operate as single junction cells. The top cell must be transparent, so that light, which is not absorbed on its way through the cell, is lost, whereas in single junction operation the back contact is designed to reflect and scatter this light back into the absorber layer. The bottom cell receives only a fraction of the solar spectrum that it would if it were to operate as a single junction cell, so its efficiency is much reduced. Two cells whose efficiency in single junction operation was 10-11%, might have efficiencies of say 9% and 5% (top and bottom cells) as components in a tandem, giving an overall efficiency of 14%.

Polycrystalline thin film cells are very well suited to be used in four terminal tandems. The cells can be made independently on glass substrates and connected optically as the final step in the production process. The wide range of materials which can be used in polycrystalline cells makes it easy to find a cell with the optimum energy gap. For the low gap cell, with an energy gap of 1 eV, copper indium diselenide is the obvious choice. The top cell, with a gap of 1.7 eV is less obvious since no highly efficient cell has yet been produced with polycrystalline materials such as cadmium selenide which have energy gaps of the correct value. The continuing research into such cells is, however, likely to lead to suitably efficient devices.

At present, the only material of the correct energy gap with single junction efficiencies around 10% is amorphous silicon. A four-terminal tandem with an amorphous silicon top cell and a copper indium diselenide bottom cell has been successfully demonstrated. The structure of this cell is shown in Figure 7. The efficiency of the top cell is 9.7% and that of the bottom cell is 4.9%, leading to an overall efficiency of 14.6%. This achievement by ARCO Solar is considerable, but even more significant is their production of foot square tandem modules with efficiencies of 11%.

Economic Considerations

The major reason for considering the production of polycrystalline thin film solar cells is their potential for manufacture at low cost in large areas. The production methods employed in the manufacture of these cells are well-known industrial techniques, used on a scale appropriate to multi-megawatt per annum solar cell production. The adaption of these methods to cell production can be readily envisaged, once the basic scientific and materials engineering studies have defined the requirements for efficient devices.

There are two basic classes of techniques used in polycrystalline cell production:- those which are capital intensive but can produce cells at high rate from single large pieces of equipment, and those which are of relatively low capital cost but also have lower through-put. The PVD techniques are examples of the first

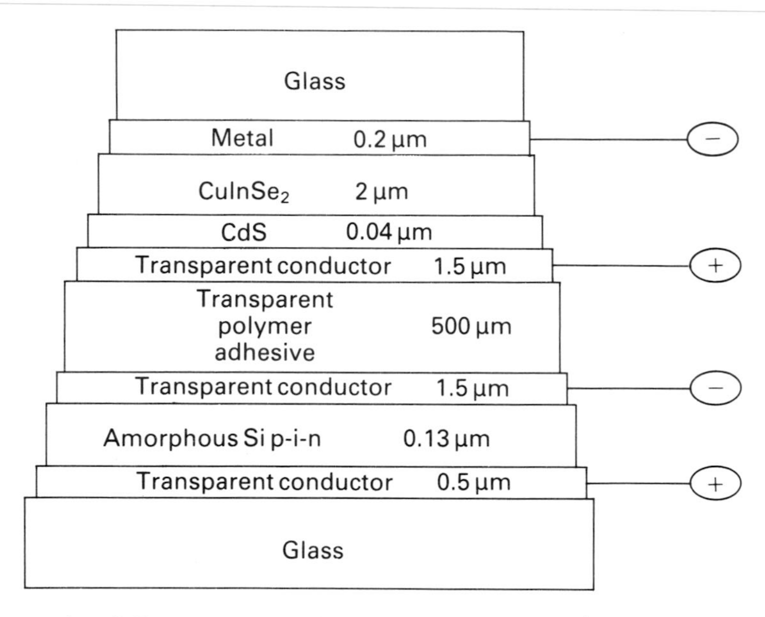

Fig. 8.7. Structure of a Four Terminal a-Si/CIS tandem cell. Fabricated by ARCO Solar.

class whilst spraying, screen-printing and electrodeposition are examples of the second class of techniques.

The cost per unit area of cells produced by these techniques is the sum of the costs of the capital, materials, energy, labour and the indirect costs of management, R&D, marketing etc. The indirect costs are effectively independent of the method of production so we need estimate only the direct manufacturing costs in order to compare the different technologies of different types of cell.

It is essential that whatever production process is chosen for the manufacture of cells, the output of the plant must be equal to the sales of the cells. A considerable amount of time, effort and money must be spent in setting up a cell manufacturing facility, and an estimate has to be made at the planning stage of the likely sales so that the plant can be sized accordingly. If too large a plant is installed, it will be underutilised for a long time until sales increase to match the output, and the onset of profitable operation will be delayed. If too small a plant is installed, the sales will be restricted after a relatively short time by the low production rate, and additional production capacity must be installed. There are always benefits of scale in solar cell production, so a multiplicity of small plant will produce cells at higher cost than one large plant of the same output, so the choice of scale of production of new types of cell is fraught with uncertainties. The strategy with the least risk is to begin with one relatively small machine in single shift operation. As sales increase, the machine is worked more intensively, and then other identical production lines are added. When the sales have increased to a level where multiple small plants are fully utilised, then new large scale production facilities are built. The cost of solar cells under this strategy are shown in Figure 8. These costs were calculated for sputtered CdS/Cu_2S solar cells, but calculations of the manufacturing cost of other types of cell using other capital intensive methods show identical trends, and in fact costs are the same to within the uncertainties of the calculations. The calculations of Figure 8 were first published in 1982, and it is interesting to note that the commercial production of amorphous silicon cells has followed an almost identical strategy. The first plants installed had production rates of about $10^4 m^2$p.a. and the new plants which will come on-line in 1989/90 will have a rate about ten times higher. The manufacturing costs shown in Figure 8 are also very close to those found in the commercial production of amorphous silicon. The usual figure of merit - the cost per peak Watt, depends on the conversion efficiency of the cells (or usually modules) produced, as shown in Figure 8 so there is a considerable commercial advantage to be gained by increasing the cell efficiency.

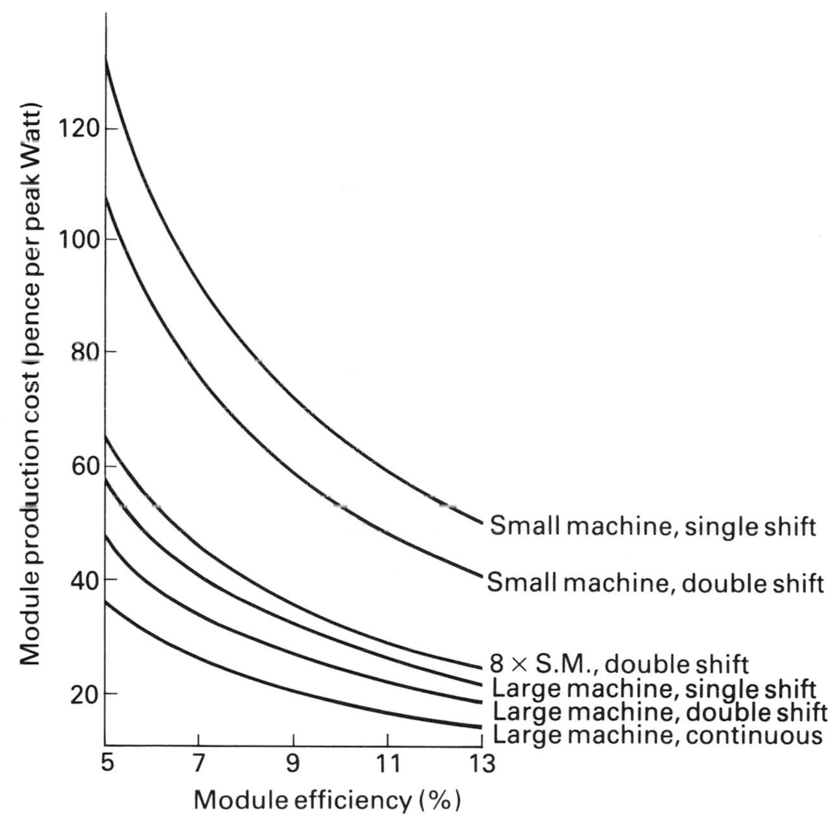

Fig. 8.8. Cost of production of thin film cells by a capital intensive process.

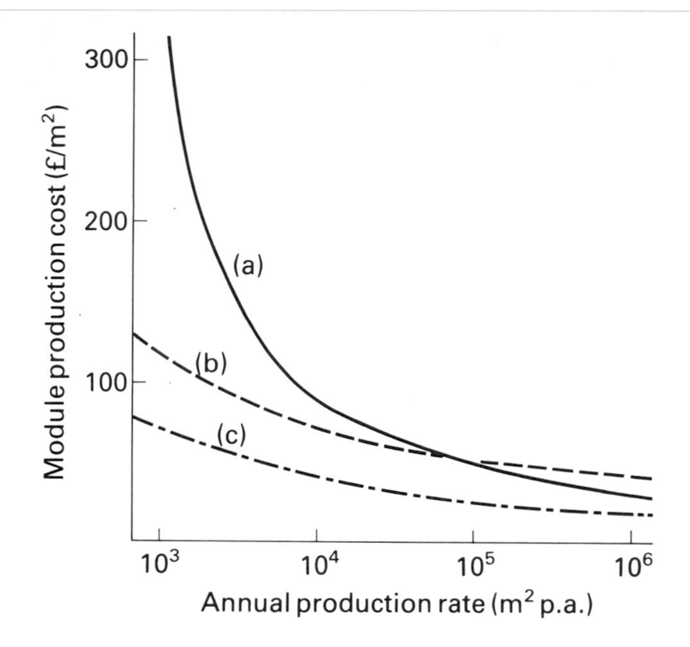

Fig. 8.9. Comparison of cost of production by:
a) Capital intensive process
b) Low capital cost process
c) Low capital cost process with low labour costs.

The cost of capital is a large component of the total manufacturing costs of the cells using capital intensive technologies. The low capital technologies can have significant cost advantages in the early stages of introducing new types of cell on to the market. Figure 9 shows a comparison of cell manufacturing costs as a function of annual production rate for capital intensive and low capital technologies. The comparisons are not meant to be exact since each class contains a number of different technologies with somewhat different costs, but the overall trends are clear. The low capital techniques tend to be labour intensive, and if productivity can be maintained then costs in countries with low labour costs are likely to be significantly lower than for production in industrialised countries. This opens opportunities for some developing countries to consider indigenous production once the technologies have been well-proven. In industrialised countries, it is likely that those technologies would be adapted to use robots rather than labour, increasing the capital required but increasing production rate and lowering manufacturing costs at the higher production rate. The various different production technologies then tend to converge to roughly equal capital intensity and roughly equal manufacturing costs.

Another important factor to consider in comparing the different

different types of cell is the cost and availability of the materials. Table 2 shows the mass of the critical elements required for 1 GWp of 10% solar cells for each micron thickness. The table also shows the estimated annual production capacity, the world reserves and resources of these elements.

Table 2

Cell Type	Rare Element	Tonnes per GWp at 1μm	Production Tonnes p.a.	Reserves 10^3 Tonnes	Resources 10^3 Tonnes
$CuInSe_2/CdS$	Cd	67	20,000	750	18,000
	In	25	65	2	2
	Se	23	23	170	460
[$Cu(InGa)Se_2/CdS$]	Ga	2-10	30	110	55
CdTe/CdS	Cd	108	20,000	750	18,000
	Te	33	406	40	110

It can be seen from Table 2 that there could be some difficulty in the supply of indium, gallium and tellurium if the cells were thicker than a very few microns and/or if the production process does not have a high material utilisation. It must be remembered that production capacity is governed by the level of sales of a material, and can be expanded if the requirement for the material increases, although this takes both time and money. The quoted reserves of a material are the quantities estimated to be available from known sources at present prices and extraction technology, and these are rarely more than the equivalent of 10-20 years production for any material. Resources are the quantities estimated to be available from deposits yet to be explored, but also at present prices and technologies. These two quantities are thus sensitive to both prices and technical improvement, and, of course, to vigorous exploration, and should be taken only as a rough guide to availability. The consequence of a shortage of any of the elements would be a short term price rise, followed in the longer term by an increase in production. The constraints on supply of In, Ga and Te suggested by Table 2 should not be construed as an upper limit on production, but rather as a warning that investment in material supply capacity will be required.

No industrial activity is free of hazard for those in the industry, for the general public and to the environment, and the production of solar cells is no exception. The major hazard for both CIS and CdTe cells is the production and use of cadmium. Cadmium is a heavy metal whose toxicity is well-documented. The most hazardous stages of production would occur during refining, where emissions of CdO need to be kept below an average of 0.5 micrograms/m³ both in the plant and outside. Cadmium dust also originates from the movement and use of cadmium during cell manufacture. To reduce the exposure of workers, the factory must be well ventilated resulting in the dust being emitted into the atmosphere. Unless the ventilated air is very well filtered, this would result in a potential hazard to the population downwind of the production site. Calculations by a number of workers have shown that cadmium exposure would remain well below the permitted limits for both workers and general population in a well designed solar cell production plant up to 100 MWp p.a. capacity. Plant with capacity much larger than this would need to take exceptionally stringent (and expensive) precautions to prevent cadmium hazard. Fortunately almost all of the benefits of scale in production have been realised by a production capacity of 10^6 m^2 p.a. (100 MWp p.a. for 10% cells), so there should be little economic pressure for larger plant.

Conclusion

Solar cells using polycrystalline thin films show great promise for low cost large area production. The two present candidates for commercial production, CIS and CdTe, have excellent properties and are at an advanced stage of pre-production development. Small modules with efficiencies over 10% have already been produced and exhibit stability superior to that of amorphous silicon. There appear to be no outstanding problems which will prevent the large scale commercial production of these cells, although the availability of some rare elements may constrain the production rate below 1 GWp p.a.

There are exciting prospects for the future which are at present in the early research stage. One which has particular promise is the use of the new semiconducting polymers. These can be deposited easily and cheaply in large areas, and their characteristics are such that it should be possible to form efficient p-n homojunction or heterojunction devices from these materials. A good deal of fundamental studies on interfaces and contacts must be done before it will be possible to design devices with confidence, but the promise is considerable.

Polycrystalline thin film solar cells will take their place within the next few years in the range of options available to the designers of photovoltaics systems. Because of the wide range of uses of photovoltaics, it is unlikely that any one type of solar cell will totally dominate the market, and the PV industry will benefit from having a range of cells with different characteristic to fulfill different needs. As the industry begins to fulfill its potential service to mankind, it will find needs for all of the different types of cell, but the mass market will be for efficient, low cost, stable, flat plate cells, and, for this market, polycrystalline thin film cells are likely to be the main option. Module efficiencies of around 15% and purchase prices of £1 Wp^{-1} or less are achievable in the late 1990's, which will open markets of the order of 1 GW p.a. Further developments will continue in improving efficiencies and lowering costs, so that photovoltaics can become one of the major technologies of the 21st century.

Chapter 9

LONG TERM PROSPECTS OF PHOTOVOLTAICS

L. CHAMBOULEYRON

Campinas State University, 13081 Brazil

INTRODUCTION

Energy is a fundamental input in the development of any known human society. However, the amount of energy required per capita to foster or to maintain the development depends largely on the development stage, the local resources, the social and economic model chosen by the country, and several other factors. Most countries of the world today rely on local or imported fossil fuels to supply a large fraction of their energy needs, although it became clear after the oil crisis that supply disruption, either due to political reasons or simply to source exhaustion, was a likely situation to happen during the next decades. Until very recently, many energy planners believed in a successive replacement of fossil energy sources by nuclear, and then nuclear by renewable sources. The accidents of Three Mile Islands and Chernobyl (in particular) showed however, that the environmental risks of the nuclear power are potentially catastrophic for mankind. As a consequence of this, the population of most advanced countries today manifests a strong opposition to nuclear electricity generation. Vast reserves of coal were once considered a viable alternative after oil exhaustion. Nevertheless, in many countries of the world it is now recognized that the negative impacts of massive coal burning on environment (mainly acid rain and liberation of CO_2 into the atmosphere) and human health are important enough to disqualify coal as a viable long-term solution. It appears then that the era of renewable sources of energy has to be hurried-up, even if at the moment the oil prices are not particularly high.

The main favorable aspects of the renewable sources of energy are related to the general recognition that they are environmentally benign. In terms of *direct costs* however, they range between half the price (hydroelectricity) to nearly five times (photovoltaics) the current cost of fuel based systems (Sorensen, 1987). It is important to note that if *indirect costs* (like those derived from the degradation of the environment or from the increased costs of the health care of the population) are taken into consideration the cost differences narrow considerably.

It is clear then that the future of the renewable sources of energy, and particularly of photovoltaics, depends on several variables: technical, economical, social, political, etc. This short introduction intends to be a warning to those believing that in photovoltaic technology there is only a price (or cost) race to be won. A closer look at all aspects intervening in the energy supply problem shows, on the contrary, that the direct costs of an energy option has seldom been the only determinant factor for its massive utilization.

SOCIO-ECONOMIC ASPECTS

It is at present universally recognized that solar electricity is an almost ideal energy source, it's most relevant characteristics being that a high quality form of energy is generated in a modular way, with low maintenance costs and the absence of any negative environmental impact. The question arises then as to why this energy generating technology is not widely used in all places where other conventional energy sources do not exist, or in cases where the energy sources at hand are environmentally risky. The usual answer to the question is the high cost of solar electricity. As already mentioned in the introduction, it might be erroneously considered that the future of photovoltaics depends on a price matter only. It is the present author's opinion that the problem is more complex and involves energy policy matters and energy lobbies interests. As an example, let us simply remember that in most advanced countries, the R&D effort made on solar technologies is ridiculously small as compared to the amounts of money spent on the development of other less benign energy technologies. Since 1975 nearly 113 million sterling pounds have been spent in the United Kingdom on R&D for new (mostly renewable) energy technologies. This has to be compared with over 4 billion pounds on nuclear fission (Sweet, 1987). A similar situation occurred in the USA, France and other highly industrialized countries as well. It is evident that renewable sources of energy, as impressive as their progress might be, have not been sufficiently considered by policy makers, even in the most highly centralized political systems.

It is often argued that the high prices of solar electricity derive partly from the diffuseness of the resource. Rose (1979) made an extensive analysis of the problem and demonstrates that the diffuseness per se of a solar energy farm does not necessarily mean high cost energy. Comparisons are made with well known solar farms that are orders of magnitude more diffuse than a farm of solar cells. The conversion efficiency of forests, for example, is of the order of 10-3. i.e. one hundred times lower than solar cells. Even at this conversion efficiency the cost of energy from forests is comparable with the cost of energy from 10 US dollars/barrel oil.

Let us mention another example, refering this time to hydroelectricity. Itaipu, located on the border between Paraguay and Brazil, with a generating capacity of around 12 GWatts, is the largest hydroelectric dam ever built. The up-stream reservoir however, flooded more than 100,000 hectares of forest, the mean dimensions of the lake being 200 km long by 7 km wide. Half of this area covered with 10 % efficient photovoltaic modules corresponds to an installed peak power capacity of 70 GWatts. It is clear that the water stored by hydroelectric dams may sometimes have other uses (like in Assuan, Egypt). The idea behind the example is not to disqualify hydroelectricity but simply to point out that photovoltaics is much less diffuse than commonly thought. It is useful to make the same kind of calculations of generating capacity vs. total area used in nuclear power plants. Amazing as they may be, the comparison with photovoltaics always indicate the same order of magnitude in generating capacity for both technologies.

These figures show that besides the technical issues, the future of photovoltaics, as well as of other new sources of energy, is strongly linked to the way mankind will organize the post-oil era. It is unfortunate to note however, that as long as oil prices stay at a moderate level, the R&D efforts necessary to make solar electricity a viable alternative are very modestly funded (and planned).

Another relevant question concerning the future of photovoltaics is:- the future where? The question addresses the possible impact of photovoltaics in the developed world as well as in the so called developing countries (Chambouleyron, 1986). In other words, how much can PV modules contribute to the satisfaction of the energy needs of the rich and of the poor. A closer look at the development of solar electricity indicates that different philosophies exist for both kinds of countries. In developed countries the main idea is to obtain electric energy from the sun at the lowest possible price. To that aim, large generating units interconnected, or not, to already existing power grids are planned and built. The complementary approach refers to decentralized systems working at a residential level which will interact with the grid. In both cases the point is to feed the distribution network at the lowest price. In the Third World, and particularly in the rural areas where the energy problem is more severe, such networks do not exist, and, even worse, they are unlikely to exist in a near future. The energy deficiency in these areas not only refers to transportation, lighting and home comfort, but often takes on a calamitous aspect for millions of human beings who suffer from the lack of portable water and food, illiteracy, diseases and isolation. This picture indicates that big differences might appear between countries in the priority list for solar electricity use. The technologies associated to centralized and to decentralized PV applications will not always necessarily coincide. Nor the requirements in performance and reliability. It is unlikely that advanced countries will make the industrial effort to adapt photovoltaic equipment to the local conditions of every Third World country, or to every sought application. The long-term prospects of photovoltaics in the Third World appears then to be somehow associated to the local R&D effort in PV technologies and peripherals. These considerations on problems related to the diffusion of PV technologies and not refering exclusively to the price matter, have not received due attention. It is not intended here to say that the price of the kW-h is not important. The point is that it is not always the only relevant parameter.

TECHNOLOGICAL PROSPECTS

Let us freely imagine what would be the characteristics of an ideal photovoltaic system. It must certainly be cheap, highly efficient, reliable, durable, easy to use and install, without moving parts and maintenance costs and last but not least, it should be nice looking. These characteristics are not all equally important and, depending on the specific application (e.g. large central units or decentralized systems), some of them will be highlighted at the expense of others. In almost all cases however, the cheapness, efficiency, reliability and durability become the most important parameters. A closer look to them shows that all contribute to what we may call the integrated price of photovoltaics. The present state of the art in photovoltaics do not combine them in a single fabrication process or material. In other words, what is today reliable and efficient is not cheap, and what is cheap is not reliable or durable. There are several competing technologies, most of which have been analysed in previous chapters. It is the present author's opinion that the most promising from the point of view of reduced production costs are those refering to polycrystalline and to amorphous silicon.

Single junction devices

Polycrystalline Si solar cells are less expensive and less efficient than single crystal Si devices, they are however equally stable and reliable. It can be roughly stated that large grain polycrystalline Si made out of semiconductor quality powder is as good as single crystal Si. The large price reduction expected however, would derive from the utilization of solar grade silicon, typically four nines purity, and high growth rates. Economic studies conducted in the USA and refering to areas with abundant sunlight indicate that, in order to be competitive with conventional electricity generation, efficiencies of 15% or more would be required in flat plate PV modules (DeMeo, 1987). This appears to be an ambitious target for solar grade Si. A complementary appraoch for cost reduction is the utilization of low cost substrates coated with high quality polycrystalline Si films or the fabrication of self-supported wafers. Both eliminate the problem of expensive material waste resulting from ingot sawing and, in the case of substrate coating, the thickness required for a satisfactory device operation is much smaller (Hamakawa, 1987). R&D efforts in these directions are promising and may eventually combine all the above mentioned benefits. It is important to stress that the figures on efficiency requirements being discussed apply essentially to highly industrialized countries.

A somewhat smaller conversion efficiency would perfectly match the requirements of decentralized systems working in Third World countries, where the land costs are lower and the energy alternatives more expensive or absent. At present prices PV modules are already the most economic energy alternative for electricity generation in most applications in the Third World's rural areas.

Hydrogenated amorphous silicon (a-Si:H) solar cells are becoming a source of electric power generation. The manufacturing processe of a-Si:H PV modules involve a series of well known technologies which appear appropriate for large-scale and low-cost production. Devices can be fabricated over large areas on a variety of different and cheap substrates, like stainless steel, glass, ceramic and special plastics. At present it is possible to coat with high quality material areas of the order of half a square meter. However, the most conventional manufacturing technique is a batch type process which is inherently slow and cannot comply with the high throughput requirements of the solar electricity generating industry. Some of the key factors refer to the quality of the transparent conducting oxide and to the laser patterning techniques that may constitute a serious obstacle to an efficient production (Madan et al., 1987). Continusous processes like the roll to roll approach are also under development. They use either organic polymers (Nakatani et al., 1987), or stainless steel (Nath et al., 1987) substrates. Both require the deposition of a transparent electrode after the p-i-n deposition. Although this is a very interesting approach, the problems associated to cross contamination and high production rates seem not to be completely solved yet.

Besides the purely manufacturing processes but somehow related to them, a-Si:H solar cells present a more basic and still not completely understood problem: the light induced degradation of electronic properties (the so-called Staebler - Wronski effect). This is probably the most challenging issue for amorphous Si modules and systems, and today the main obstacle to their use in power applications. It has been experimentally found that the degradation is less severe for increasing internal electric fields. This fact opens new possibilities for stacked thinner structures that should be more stable, as it has already been shown to be the case (Hanak and Korsum, 1982).

Multiple junction devices

In the conversion of sunlight into electricity, single gap solar cells experience two fundamental and unavoidable losses. On one hand, photons having an energy lower than the band gap are not absorbed. On the other hand, photons with an energy much greater than the forbidden band generate electron - hole pairs deep in the bands. These carriers thermalize to the band extrema and the excess energy is lost as heat in the network. The situation is depicted in Figure 1 where the general shape of the solar spectral density per unit photon energy vs. the photon energy is represented. If the forbidden band of the semiconductor is decreased the left hand losses of Figure 1 decrease while the excess power loss area increases. The opposite occurs if the band gap of the active layer is increased. An optimum band gap exists giving a maximum for the non shaded area.

In order to overcome partially these limitations multiple gap, or multiple junction solar cells were proposed at the very beginning of the photovoltaic activity (Jackson, 1958). The multiple gap devices consists of a series of single gap cells fabricated one on top of the other, each one using a different region of the solar spectrum. The philosophy behind the multiple gap absorption concept is sketched in Figure 2 where the loss

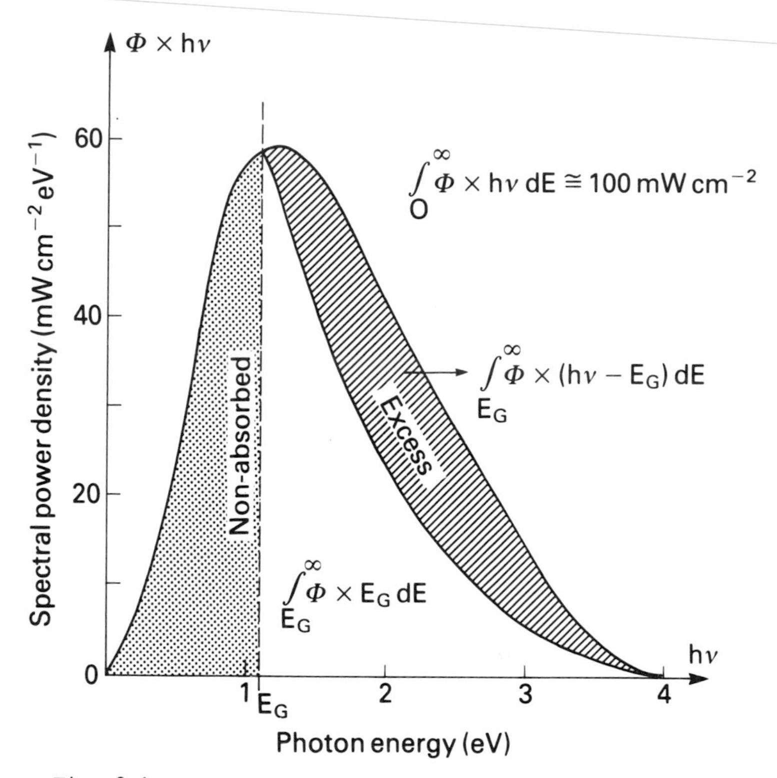

Fig. 9.1.

mechanism analysis of Figure 1 is done for a structure composed of three single gap stacked cells. It can be seen that as the number of individual cells increases the shaded loss area decreases. This is a direct consequence of the better use of sunlight energy.

For a given solar spectrum it is always possible to adjust the individual cell thicknesses and band gaps to optimize the energy conversion efficiency of the structure. The expected benefits are appealing. Under AM1 conditions an ideal single gap solar cell may have a conversion efficiency of roughly 26%. An optimized (and ideal) two band gap cell may increase the conversion efficiency up to 35%, i.e. more than 35% relative increase over the single gap device. Adding a third cell to the structure produces an additional increase in conversion efficiency of approximately 15% over the two gap device. It is clear that as the number of individual cells in the structure increases, the conversion efficiency also increases. The relative gains in conversion efficiency however, become less important as the number of cells raises and three stacked cells is probably the upper practical limit. In principle, single crystal, polycrystalline or amorphous structures can be built. With single crystal struc-

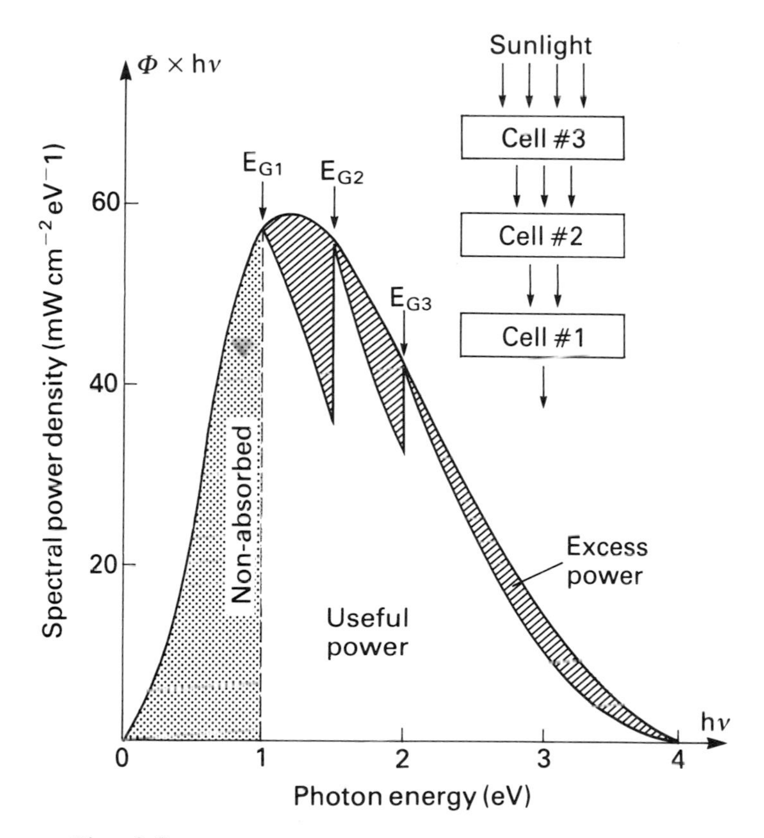

Fig. 9.2.

tures however, the problem of lattice mismatch between different materials appears. A large density of defects producing high recombination losses at the interfaces is normally found in these cases. The problem could be circumvent by mechanically stacking the cells but such stacking would never lend to economical mass production. For technological reasons then, only stacked solar cells made out of amorphous semiconductors or other thin film polycrystalline materials may have any future in solar power applications.

The set can be connected in series or in parallel, each one presenting some advantages and drawbacks. With the manufacturing processes of amorphous semiconductors the series connected cells (monolithic structures) are easier to fabricate and only one current-voltage output has to be conditioned. In the series connected configuration however, the photocurrents of the individual cells must be matched in order to achieve the full potential of the structure. If this condition is not met the overall efficiency will drop. The major drawback of the structure may come then from photocurrent mismatches due to changing spectral distributions of sunlight (Chambouleyron, 1984).

Non-monolithic structures are not constrained to the current matching condition. For practical reasons the individual cells of the structure should be deposited on separate substrates. The final structure can be designed for either parallel operation or series connection, the number of cells in each of the submodules being adjusted in order to provide, either voltage, or current matching conditions. The drawbacks of this approach are essentially constructive. These constraints may increase production costs and, consequently, the price of the solar kW-h.

The question arises then as to whether multiple gap solar cells are a good technological option for a substantial reduction of the price of solar electricity. The answer to this question depends on a number of unknowns referring to the future understanding and eventual mastering of the electronic properties of amorphous semiconductor alloys. In the next section the benefits and limitations of the multiple band gap device concept will be analyzed. Prior to that analysis a rapid survey on the properties of variable band gap-tetrahedrally bonded amorphous semiconductor alloys will be made.

Variable band gap amorphous semiconductors

The two most studied classes of amorphous semiconductors are chalcogenide glasses and tetrahedrally bonded alloys. Only the latter can be prepared as electronic materials in the sense that their electrical conductivity can be modulated by appropriate chemical doping and that they possess a relatively sharp optical edge. It is by now well established that these characteristics are a direct consequence of the role of hydrogen as a dangling bond passivating agent in the netwrok. The deposition technologies of tetrahedrally bonded amorphous semiconductors allow the relatively easy fabrication of stacked multiple gap solar cells, and photovoltaic devices possessing two, three and even four active layers have been fabricated (Yang et al., 1986). The principal methods of preparation of hydrogenated amorphous semiconductors involve either the decomposition of the hydrides of the species composing the material, or the use of a solid source that is bombarded or heated in an atmosphere containing hydrogen. The hydride decomposition may be achieved in an a.c. or in a d.c. electric field (glow discharge) or with the help of high energy photons and/or heat. The r.f. glow discharge of monosilane is the most used decomposition process for a-Si:H film preparation. Among the methods that use solid sources, the r.f. sputtering of a high purity crystalline target is preferred. It is by now well accepted that, referring to a-Si:H, "soft" methods, e.g. methods involving little ionic bombardment, may be beneficial in reducing the density of electronic states in the pseudo gap. These conclusions however are by no means universal in the sense that they might not apply to the preparation of all kind of alloys. In fact, the deposition of high quality films still involves a lot of empiricism, the number of factors affecting the material quality being rather high.

The main characteristics of variable band gap tetrahedrally bonded amorphous semiconductors will be presented. The materials of interest for stacked solar cells are a-Si:H, a-Ge:H, amorphous silicon - germanium alloys, silicon and germanium carbides and nitrides, and compounds incorporating tin in the network of elemental Si and Ge. It will be shown that while the variation of the optical band gap over a wide range of energies is easily obtained, the retention of good electronic properties appears to be a more difficult task.

The elemental amorphous semiconductors: Si and Ge

Early work on the structural, electronic and optical properties of evaporated or sputtered tetrahedrally coordinated a-Ge and a-Si showed that these materials possessed a large density of structural defects ranging from single dangling bonds to microscopic voids. The passivation of a resulting high density of electronic states in the pseudo gap was not possible either by ultra high vacuum deposition nor by high temperature annealing. An important improvement in the understanding of these limitations occurred in 1974 when Lewis and coworkers at Harvard University reported on the properties of a-Ge films deposited by r.f. sputtering in an Ar plasma containing some hydrogen (Lewis et al., 1974). The films showed a reduction of orders of magnitude in their room temperature conductivity and electron spin resonance, a reduced low energy photon absorption and an improved photoconductivity. These changes were attributed to the compensation of dangling bonds and other defects by hydrogen atoms. Simultaneously, Spear and coworkers at Dundee University, prepared high quality a-Si films by the r.f. decomposition of silane. In 1975 a-Si:H films could be doped n-type or p-type and its conductivity increased by orders of magnitude (Spear and Le Comber, 1975).

Today a-Si:H is a structure sensitive and controllable semiconductor. On the basis of the present knowledge in the physics and chemistry of a-Si:H remarkable progress has been made in the device field: high efficiency p-i-n solar cells, thin film transistors, electrophotography and imaging devices. An excellent review on a-Si:H can be found in Pankove, (1984).

a-Ge:H films have been much less studied than a-Si:H, one of the main reasons being that the material produced by the r.f. decomposition of germane does not possess very good electronic properties. Recent efforts to produce high quality films using hydrogen and fluorine atoms as dangling bond terminators resulted in a material having a photoconductivity one order of magnitude higher than the material without fluorine (Nozawa et al., 1983). The role of fluorine however, is not yet perfectly understood. An increased interest also exists for alternative methods to deposit a-Ge:H. This is partly due to the need of producing high quality hydrogenated Si-Ge alloys, the band gap of which would be well matched for solar energy conversion.

Silicon-Germanium alloys

The optical band gap of a-Si:H (e.g. 1.75 eV) is too large to convert sunlight into electricity with high efficiency. Although there is still some controversy concerning the optimum band gap value for solar electricity generation with amorphous p-i-n solar cells, it is generally agreed that it should be lower than 1.5 eV. Amorphous silicon germanium alloys are ideal candidate materials, both elements allowing to prepare perfect solid solutions in any proportion, e.g. an energy band gap adjustable between 1.1 and 1.7 eV. It is important to note that Si-Ge alloys are not only important for single gap solar cells but as low band gap material for tandem cells as well. Different methods have been employed to prepare a-Si:Ge:H films. The dual magnetron deposition system (Rubber et al., 1984) and the glow discharge from germanium hydrides and fluorides (Oda et al., 1984; Paul and MacKenzie, 1987) are reported to produce materials possessing much better transport properties than the films made with the conventional glow discharge decomposition of germane. In all cases however, the mobility lifetime product for carriers decreases monotonically with increasing Ge content, and the films having an appropriate composition for photovoltaic applications show transport properties much inferior to those of a-Si:H. In conclusion, the tailoring of the optical properties of Si-Ge alloys is easily achieved by any method of preparation, but the important problem, that remains unsolved up to now, is the deterioration of the electronic properties of the films with increasing Ge concentration.

Widening the band gap: nitrides and carbides

Nitrogen and carbon atoms widen the forbidden band of both elemental semiconductors. It is worth stressing here than amorphous semiconductor alloys do not have fixed material parameters, so they can be varied by changing the composition or the deposition conditions. This is a particularly useful property for multiple gap photovoltaic devices, where the need appears to adjust the band gaps of the individual cells composing the structure. Large band gap amorphous materials are important in building "window" layers, e.g. transparent eletrodes. With that aim in mind, their conductivity is normally increased by doping. In the case of silicon nitride boron doping is quite effective in reducing the layer resistivity, although a concomitant band gap narrowing is measured. Phosphorus, on the contrary, appears not to be so effective. A likely explanation of the difference has been given in terms of different bonding configurations for the impurities (Alvarez and Chambouleyron, 1984).

In 1977, Anderson and Spear reported on the deposition of a-Si:C:H thin films by the glow discharge of silane and hydrocarbons. Since then many of its structural, optical and electrical properties have been investigated, including the doping with several impurities. The main solar application of silicon carbide films is as a window material for amorphous silicon solar cells. As already mentioned this application requires a material possessing a relatively high band gap and a good conductivity. These somehow conflicting requirements are satisfied partially by boron doped films (Tawada et al., 1982).

Similar band gap variations can be obtained using a-Ge as a host network. Although Ge_3N_4 films have been used in the past to passivate GaAs surfaces, off-stoichiometric a-Ge:N got no attention until recently (Chambouleyron, 1985). In this case the optical band-gap can be adjusted between 1 eV and more than 4 eV. The beneficial action of hydrogen atoms in cleaning the pseudo gap is confirmed with conductivity vs temperature measurements. The dark conductivity becomes activated, an indication of a large decrease in the electronic density of states in the pseudo gap. However, no photoconductivity under AM1 conditions has been measured in hydrogenated films. No reports exist on n or p type doping of amorphous germanium nitrogen films. Nevertheless, the similarities in structure and bonding configuration with amorphous silicon nitrogen alloys, lead us to imagine that the material may be effectively doped.

Very little work has been done on a-Ge:C:H alloys, although the first report on its preparation by the r.f. glow discharge of germane and ethylene is more than ten years old (Anderson and Spear, 1977). The band gap of the material widens as the carbon content increases, the photoelectronic properties however, seem to degrade very rapidly with increasing carbon content.

Narrowing the band gap: a-Si (or Ge):Sn

Amorphous silicon tin alloys have been prepared by the r.f. sputtering of compound targets and by the glow discharge decomposition of silane and tetramethyl tin (Verie et al., 1981; Kuwano et al., 1982). The optical band gap of the unhydrogenated material decreases linearly with increasing Sn content up to 20 at. % at an approximate rate of -0.04 eV/at.% Sn. In the case of a-Si:Sn:H films however, the electronic properties degrade very rapidly with Sn. No Sn-H vibrational modes have been detected in these samples by infrared transmission measurements, an indication that hydrogen is unable to passivate Sn dangling bonds. Mossbauer spectroscopy studies indicate that, for small tine content, Sn atoms go substitutionally into the a-Si network. Moreover, defect structures appear to be strongly dependent on deposition methods and conditions, and at large tin content a clear tendency for metallic segregation appears. More than a decade ago unhydrogenated amorphous germanium tin alloys were prepared by the r.f. sputtering method (Temkin et al., 1972). As in the case of Si-Sn alloys, the optical band gap of a-Ge:Sn recedes linearly with increasing Sn. The beneficial action of hydrogen in the transport properties of these alloys was recently reported, the activation energy of the conductivity in hydrogenated samples being approximately half of the pseudo gap (Chambouleyron and Marques, 1988). Solid solutions of Ge and Sn up to 27 at.% tin have been prepared. Mossbauer spectroscopy studies indicated that in these samples all Sn atoms are in a tetrahedral covalent coordination in the Ge network (Chambouleyron et al., 1988).

In conclusion, enormous progress has been made in the preparation methods as well as in the understanding of the structural, optical and electronic properties of tetrahedrally coordinated amorphous alloys. The tailoring of the optical and compositional properties appears to be relatively well understood and mastered.

Nevertheless, much work remains to be done in the bettering of the electronic properties of the materials that are closely related to defect structures. This is certainly a difficult and demanding task but without doubt a worthwhile challenge. Once these difficulties are overcome, enormous possibilities for amorphous alloys can be foreseen in the photovoltaic industry.

Conversion efficiency of multiple-gap amorphous solar cells

The photovoltaic modules of multiple gap cells can be produced, either on a single substrate passing through successive deposition chambers, or fabricated onto separate substrates and then mechanically stacked. The deposition technologies for amorphous thin films make stacked-series connected (monolithic) configurations more appropriate to continuous mass production. The final structure is limited to a single pair (top and bottom) of electrical contacts and the maximum conversion efficiency is obtained if the photocurrents of the cells composing the structure are matched. In order to avoid collection losses the highly doped n(+)/p(+) junctions should act as perfectly ohmic and low resistance interfaces. The band gaps and the active layer thicknesses of the individual cells must be simultaneously optimized to produce, under some particular irradiation condition, the maximum electric power. The main disadvantage of the configuration may come from the changing spectral distribution of sunlight that might produce photocurrent mismatches which in turn affect the conversion efficiency.

Mechanically stacked modules can have their own set of electrical contacts and, consequently, the photocurrent matching constraint is avoided. The structure however, is of more complex assembly and the shadowing of additional contacts more severe. A double gap cell structure made on separate substrates, having a conversion efficiency of nearly 13% has been demonstrated. The top and bottom solar cells are of amorphous silicon and copper indium diselenide type, respectively (Mitchell et al., 1987). The band gaps of these materials (approximately 1.75 and 1.0 eV) do not correspond to those given by any ideal optimization process. Nevertheless, taking into account the electronic properties of existing alloys they represent one of the best technological options.

It is impossible to give here a full account of all the details, either theoretical or technological, entering into the problem of the proper design and construction of multiple gap solar cells and modules. Instead, reference will be made to a particular aspect that has been somewhat disregarded in the literature; the question of how much the atmospheric conditions and the position of the sun in the sky may affect the performance of these devices. Let us clearly state from the beginning of the discussion that these considerations only apply to high performance devices. If, on the contrary, the main loss mechanism (as with today's technology) come from the interfaces between cells or between layers of the same cell, or derive from a charge collection limited by the low quality of the active layer material, the above mentioned limitations become irrelevant. In other words, the considerations to follow apply to devices having a conversion efficiency near the theoretically predicted maximum, i.e. only to those that matter in the consideration of the long-term prospects of photovoltaics.

As mentioned above the characteristics of the optimum cells are calculated for a particular irradiation condition. It is a common practice to choose AM 1 conditions. Let's consider devices showing no losses, neither optical nor electrical; assume that the forward current is of the diffusive type and that the superposition principle applies, i.e. the I x V characteristic curve of the illuminated device corresponds to the dark transfer curve shifted downwards by the photocurrent under short circuit conditions. Optimized series connected-stacked devices calculated under such conditions possess the band gaps and conversion efficiencies shown in Table I. Table I also shows the relative conversion efficiency increments obtained when going from one to two absorbing layers (32.2%), and from two to three individual cells (13.5%) (Chambouleyron and Alvarez, 1985). These efficiencies will change however if AM 1 conditions are not met.

TABLE I

Number of cells	1	2		3		
Forbidden band energy (eV)	1.41	1.91	1.37	2.03	1.52	1.16
Conversion efficiency (%)	25.8	34.1		38.7		
Relative conv. effy. increase (%)		32.2		13.5		

Table I: Optimized band-gap values, conversion efficiencies and relative conversion efficiency increase as a function of the number of cells in the structure (from Chambouleyron and Alvarez, 1985).

The spectral distribution of the sunlight reaching a planar receiver at the surface of the Earth is influenced to a variable extent by a large number of effects which can be classified as astronomical. geographical, geometrical, physical and meteorological (Robinson, 1966). The last two refer mainly to the various attenuation mechanisms of the pure atmosphere, the absorption by particles and gases, the cloudiness of the sky and the effects of the ground albedo. Figure 3 shows the solar photon flux density per unit photon energy interval versus the photon energy for a clear sky and under AM 1.5 conditions. It is important to note in the figure that for photon energies below

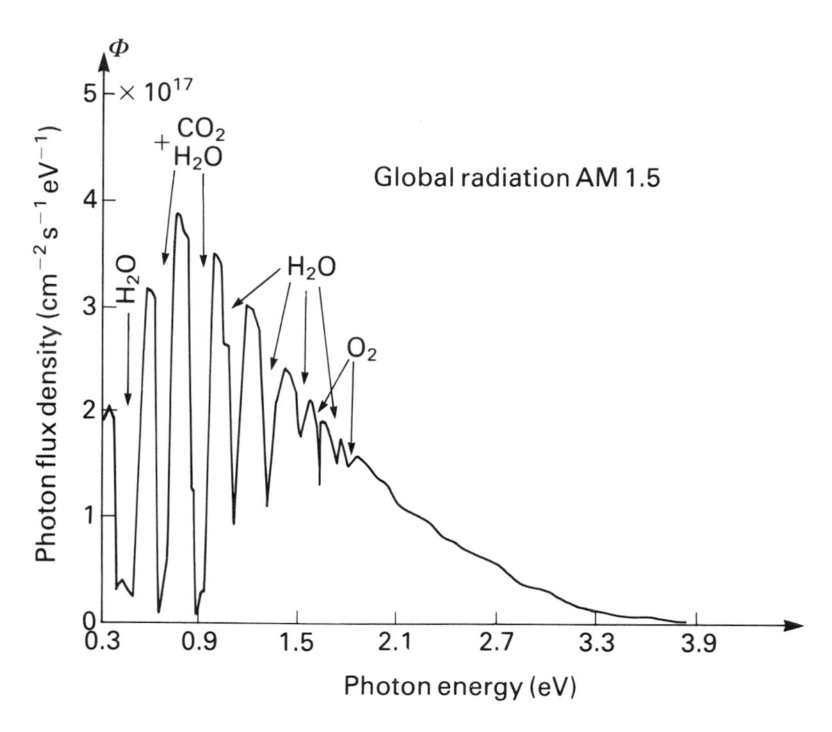

Fig. 9.3.

1.4 eV a series of absorption bands of increasing magnitude begin to appear. They correspond to vibrational frequencies of different gases, mainly water vapor and CO_2, existing in the atmosphere in variable amounts. A rapid look at Figure 3 indicates that the solar spectrum can be roughly divided into two regions. The high photon energy side does not change much its distribution with varying physical and meteorological conditions. The energy region below 1.4 eV on the contrary, is very sensitive to such variations. A detailed calculation is not needed to perceive that multiple gap cells having active layers with forbidden band energies below 1.4 eV will experience larger current mismatches under "real world" operating conditions. These conclusions were tested calculating the conversion efficiency of the structures shown in Table I. To that aim 67 solar spectra measured in the year 1980 at Uccle [51 N], under clear sky (50) and overcast (17) conditions, were used (Inst. Meteor. Belgique, 1981). Together with the spectral irradiances, the reference provides the data on the prevailing meteorological conditions, like the Linke turbidity factor, the percent of diffuse radiation, the atmospheric pressure and the total amount of precipitable water. The air mass and the solar time are also indicated. The results of the calculation is shown in Figure 4 where the relative conversion efficiency increments for a double and a triple cell structure (series connected) versus the total radiation

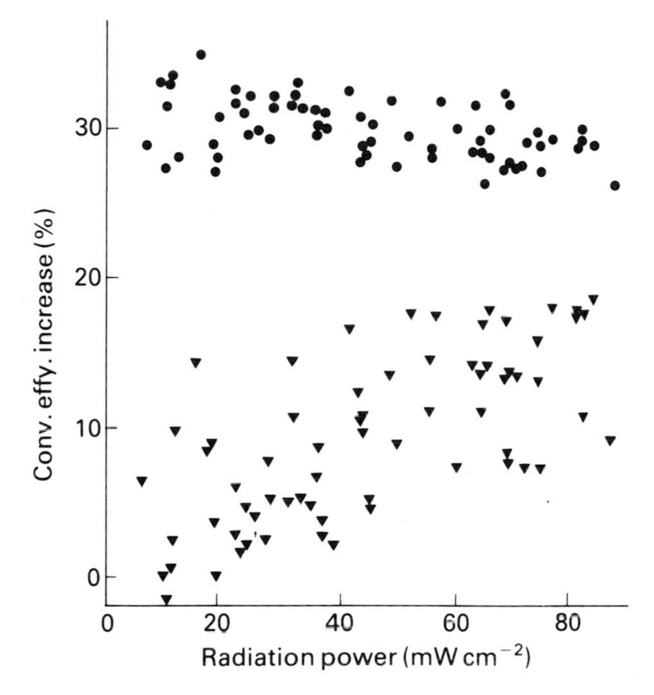

Fig. 9.4.

power are represented (Chambouleyron et al., 1986). It is interesting to note that going from a single gap to a double gap solar cell results in a conversion efficiency increase (approximately 30%) that is almost independent of irradiation level. On the contrary the lower part of Figure 3 shows that a third solar cell in the structure produces a relatively high scattering in the efficiency incresae, going from negative values up to near 20%. A weak correlation appears in this case between efficiency gain and radiation level. A better correlation between conversion efficiency gain and meteorological data appears in Figure 5. Here the same kind of plot has been made but, instead of the total radiation power, the product of the Air Mass and the water content of the atmosphere has been plotted in the horizontal coordinate. In Figure 5 only the spectra corresponding to clear sky conditions have been used. In the triple stacked solar cell structure the strong decrease in conversion efficiency as the water content, the Air Mass, or both increase, is due to mismatches in the photocurrent of the individual cells, which in turn come from photo losses in the absorption bands of water vapor. On the contrary, the active layers of the double gap solar cell possess forbidden band energies above the main absorption energies of water vapor. Consequently, the structure becomes almost insensitive to atmospheric humidity or solar altitude variations. No photocurrent mismatch between the cells occurs and consequently, no limitations to the benefits of the multiple gap absorption concept are detected. The above example indicates the optimized multiple junction monolithic devices may, for practical, economical and meteorological reasons, have an upper limit of two solar cells. Of course, more active layers can be added, but the potential benefits might not compensate the increased technological complications.

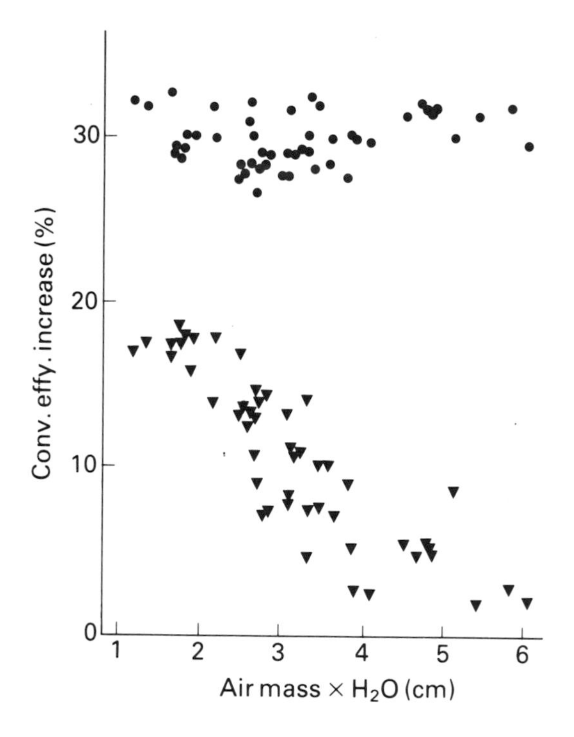

Fig. 9.5.

For non-monolithic modules two active materials is also a practical limit because the glass substrate is the single most expensive item of the manufacturing process. If one of the submodules is made with solar cells having a band gap less than around 1.4 eV, it would be interesting to connect the two subsystems in parallel, because output voltages are far less sensitive than photocurrents to solar spectral diversity. These predictions have been experimentally verified in the case of the photovoltaic module using amorphous silicon and copper indium diselenide active layers (Gay, 1986).

CONCLUSIONS

The long-term prospects of photovoltaics appear to depend on several issues of both, technical and non technical nature. As it was clearly shown when considering the socio-economic aspects of the diffusion of new energy technologies, besides the cost of the energy being produced, there are other constraints that should be taken into account in any forecasting assessment. Among them, it is worth mentioning the reluctancy shown by energy planners to consider the new and renewable sources of energy as a valuable contribution to the future energy needs of mankind. Last decade has been rich in examples of a vicious circle kind of reasoning stating that solar technologies are not cost competitive, the consequences of this thought being the allocation of very modest funds for R&D. The very poor planning and funding slow down the pace of technical progress, giving new arguments to those disregarding solar technologies as viable alternatives. On the other hand the diffuseness of the solar resource does not constitute by itself a serious obstacle to the production of cheap solar electricity. The differences in PV generation philosophy, as well as in system applications, between the advanced countries and the poor nations of the Third World are important issues to be considered. They are closely related to technical issues, in the sense that some PV technologies that prove to be too costly for power generation in the rich nations, may be the correct answer for low power descentralized applications in the Third World. The importance of an independent R&D effort in the developing countries derive from such considerations.

From the technical point of view, the future of photovoltaics appears at present to be linked to the development of good quality solar grade polycrystalline Si and of stable amorphous silicon films. The performance of a-Si:H solar cells has been steadily improving since 1975, but it seems to attain its practical limits. Figure 6 shows the impressive advances in conversion efficiency experienced by a-Si:H solar cell technology. The dotted lines at the right hand side of Figure 6 indicate what is believed to be the efficiency saturation range of a-Si:H. The fabrication technologies of amorphous semiconductor alloys allow the manufacturing of multiple gap solar cells, in which the use of the solar energy is optimized. At present however, the transport properties of all known amorphous semiconductor alloys are inferior to those of a-Si:H. As a consequence the potential benefits of the multiple gap stacked solar cells are severely limited by the poor collection of carriers and by interface losses. Nowadays, the overall conversion efficiency of multiple gap amorphous solar cells is hardly higher than that of a good a-Si:H solar cell. The question of whether the poor quality of the alloys is the consequence of the preparation methods, or derives from more fundamental reasons, has not received yet a definite answer. Figure 6 also shows the upper practical limit of the conversion efficiency of a couple of optimized stacked amorphous solar cells of good quality. These limits are impressive and without doubt a technological worthwhile challenge. Finally, an important point to consider is the dependence of the conversion efficiency of stacked solar cells on the meteorological conditions. Efficiency calculations using measured solar spectra on ideal cells, as well as technological considerations, indicate that two stacked cells is the upper practical limit for multijunction structures. Nevertheless, if with low cost-high throughput manufacturing processes, a couple of amorphous solar cells attain the conversion efficiency limits shown in Figure 6, the technical conditions required to begin the post-oil era would have been met.

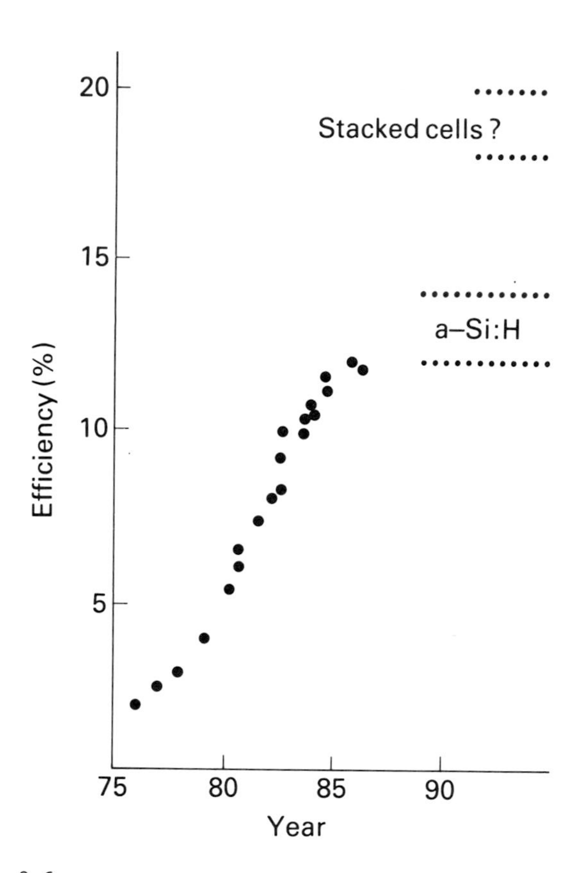

Fig. 9.6.

References

Alvarez, F. and Chambouleyron, I. (1984). Solar Energy Materials, 10, 151.

Anderson, D. and Spear, W. (1977). Phil. Mag. 35, 1.

Chambouleyron, I. (1984). Solar Cells, 12, 393.

Chambouleyron, I. and Alvarez, F. (1985). Conf. Rec. IEEE Photovoltaic Spec. Conf. 18, 533.

Chambouleyron, I. (1985). Appl. Phys. Letters, 47, 117.

Chambouleyron, I. (1986). Solar Energy, 36, 381.

Chambouleyron, I., Chello, D. and Baruch, P. (1986). Proc. 2nd Photovoltaic Science and Engn. Conf., Beijing, China, Adfield Advertising Co., Hong Kong, 361.

Chambouleyron, I., Marques, F.C., de Souza, J.P. and Baumvol, I. (1988). J. Appl. Phys. To be published.

Chambouleyron, I. and Marques, F.C. (1988). MRS Meeting on Amorphous Silicon Technology, Reno, NV, USA, April 5-9.
DeMeo, E.A. (1987). Technical Digest of the 3rd International Photovoltaic Science and Engn. Conf., Tokyo, Japan, 777.
Gay, C. (1986). Private communication.
Hamakawa, Y. (1987). Technical Digest of the International 3rd PVSEC, Tokyo, Japan, 147, for an up-to-date review of solar cell fabrication technologies.
Hanak, J.J. and Korsum, V. (1982). Conf. Rec. IEEE Photovoltaic Spec. Conf., 16, 1381.
Institut Royal Meteorologique de Belgique (1981). Distribution Spectrale du Rayonnement Solaire a Uccle, Miscellanea Serie B, 52-53, Bruxelles, Belgium.
Jackson, E.D. (1958). Trans. Conf. on the Use of Solar Energy, Tucson, AZ, 1955, University of Arizona Press, Tuscon, AZ, 22.
Kuwano, Y., Ohnishi, M., Mishiwaki, H., Tsuda, S., Fukatsu, T., Nakashima, Y. and Tarui, H. (1982). Conf. Rec. IEEE Photovoltaic Spec. Conf., 16, 1338.
Lewis, A.J., Connell, G.A.N., Paul, W., Pawlik, J.R. and Temkin, R.J. (1974). International Conference on Tetrahedrally Bonded Amorphous Semiconductors, A.I.P. Conference Proceedings # 20, 27.
Madan, A., Von Roedern, B. and Muhl, S. (1987), 7th Int. Conf. on Thin Films, New Delhi, India, December 7-11.
Mitchell, K., Eberspacher, C., Wieting, R., Ermer, J., Willet, D., Knapp, K., Morel, D. and Gay, R. (1987). Tech. Digest 3rd Photovoltaic Science and Engn. Conf., Tokyo, Japan, 443.
Nakatani, K., Suzuki, K. and Okaniwa, H. (1987). Tech. Digest 3rd Photovoltaic Science and Engn. Conf., Tokyo, Japan, 391.
Nath, P., Hoffman, K., Call, J., Vogeli, C., Izu, M. and Ovshinsky, S.R. (1987). Tech, Digest 3rd Photovoltaic Science and Engn. Conf., Tokyo, Japan, 395.
Nozawa, K., Yamaguchi, Y, Hanna, J. and Shimizu, I. (1983). J. Non-Cryst. Solids. 59&60, 533.
Oda, S., Yamaguchi, Y., Hanna, J., Ishibara, S., Fujiwara, R., Kawate, S. and Shimizu, I. (1984). Tech. Digest 1st Photovoltaic Science and Engn. Conf. Kobe, Japan, 429.
Pankove, J., Ed. Hydrogenated Amorphous Silicon, Semiconductors and Semimetals Series, Vol. 21 (A-D), Academic Press, Inc., (1984).
Paul, W. and Mackenzie, K.D. (1987). SERI Report STR-211-3107, Golden, Co., USA.
Robinson, N. (1966). Solar Radiation, Elsevier, N.Y.
Rose, A. (1979). Phys. Stat. Solidi (a), 56, 11.
Rubber, R.A., Cook Jr., J.W. and Lukovsky, G. (1984). Appl. Phys. Letters, 45, 887.
Sorensen, B. (1987). Science and Public Policy, 14, 252.
Spear, W. and Le Comber, P.G. (1975). Solid State Comm., 17, 1193.
Sweet, C. (1987). Science and Public Policy, 14, 246.
Tawada, Y., Kondo, M., Okamoto, H. and Hamakawa, Y. (1982). Jap. J. Appl. Phys., 21, 297.
Temkin, R.J., Connell, G.A.N. and Paul, W. (1972). Solid State Comm., 11, 1591.
Verie, C., Rochette, J.F. and Rebuillat, J.P. (1981). J. Phys. C, 10, 667.
Yang, J., Glatfelter, T., Burdich, J., Fournier, J., Boman, L., Ross, R. and Mohr, R. (1986). Proc. 2nd Photovoltaic Science

and Engn. Conf., Beijing, China, Adfield Advertising Co., Hong Kong, 361.

Chapter 10

BATTERY STORAGE FOR PHOTOVOLTAIC APPLICATIONS

A. O. NILSSON

SAB NIFE AB, S-26124 Landskrona, Sweden

1. INTRODUCTION

Terrestrial photovoltaic (PV) systems generate clean, dependable electricity directly from sunlight. Commercially available for the past fifteen years. Today these systems are being used in a variety of stand-alone applications, serving locations where reliable power, if it was formerly available at all, required the use of high cost remote diesel generating sets.

Typical requirements for PV installations are ruggedness, environmental flexibility, unattended operation, ease of installation, and reliability. In applications such as telecommunications, navigational aids, railroad signaling, and vaccine refrigeration, reliability and system availability are paramount. In less critical installations, such as those for cathodic protection, off-grid residential, and rural electrification, system availability is less a factor.

To meet these demands, photovoltaic modules have become increasingly dependable. Typical failure rates have diminished from 0.5% per year to 0.1% per year in the present generation of modules [1]. Obviously, for optimum reliability their electrical subsystems, designed in much the same way as any other electric power source, must be equally dependable.

The common denominator for all stand-alone PV systems is that they rely on a battery bank to supply continuous power at night and during hours of limited sunshine. Two basic types of batteries, the lead-acid and the nickel cadmium pocket plate, are generally found in PV installations. The type of battery used affects other design considerations in the system.

For example, there are often electronic components, used as blocking diodes and logic circuits in power conditioners and as controllers or voltage regulators. These are the most vulnerable parts of the system. When the PV system is exposed to surges, resulting from a direct lighning path to its conductors, induction of surges for other reasons, experience has shown that the electronic devices fail [2]. In a system using lead-acid batteries as its storage medium, voltage regulator failure results in battery overcharging or undercharging - depending on site location and seasonal weather conditions.

When a PV system is designed with such controls, (as is generally the case when lead acid batteries are used), the battery abuse which results from control failures is a product of the special electrical characteristics of PV power supply systems. A PV generator behaves like a current source with limited current. It is non-interruptible when the sun is shining and it is environmentally sensitive with respect to temperatures and irradiance.

The PV system is a dispersed energy source that is exposed to ground and lightning, and it is difficult to control. Because of this, electronic devices ought to be avoided wherever possible. Further, the system's battery should be of industrial design and of a type that can take abuse in harsh environments, maintaining the same long operational life as the PV modules.

Stand-alone PV systems are generally installed in remote areas, at sites accessible only by foot, helicopter, or boat in good weather conditions. Normally, skilled labor to service the sites is limited. The ideal remote PV power system, therefore, requires an extremely reliable, fuel free installation with infrequent maintenance calls.

Modern PV modules can meet these requirements, and at the same time can supply a 20 year service life. The weak link in the past has been the battery bank used for energy storage. The demands made on batteries in PV applications differ widely from those used for standby and emergency power. In PV installations, the most important battery qualities are:

* Ability to withstand cycling daily and seasonally;
* Ability to withstand high and low environmental temperature;
* Ability to withstand deep discharge;
* Ability to operate reliably unattended and with minimal maintenance;
* Resilience to damage during transportation to remote sites;
* High charging efficiency;
* Low self discharge;
* Easily installed with limited handling equipment and unskilled labor;
* Reliability and availability during the 20 years service life of the PV modules;

Not all types of batteries can meet these requirements. Batteries have different characteristics depending on their electrochemistry and design. Selecting the right battery for a PV installation requires specialized knowledge and the consideration of many factors, including the technical aspects of battery sizing. There is a large variance in cost per energy unit among different battery types, and it is more important - and more economical - to buy a battery which is properly designed for the application than one that has simply the right capacity.

2. STORAGE BATTERIES

The storage battery [3] is the heart of the PV system. When the battery fails, the whole system is inoperable. This means that the selection of battery type is vital to the system's design.

Typically, batteries used in PV systems have been of the lead-acid type. These are avialable in a wide variety of configurations. More recently, nickel cadmium pocket plate batteries have been introduced for PV systems. They have a long-time reputation for reliability in harsh environments and offer many valuable characteristics in PV applications.

2.1 Lead Acid Batteries

A lead-acid battery consists of one or more battery cells, each of which is rated at 2 volts. A 12V battery system is assembled from six cells, and a 24V system from twelve cells. The cell has five basic components.

* Positive plates with lead dioxide;
* Negative pates with sponge lead;
* Separators;
* Electrolyte of sulphuric acid;
* A container.

The plates can be of different configurations, the most commonly used being Planté, pasted, and tubular. The plate type generally refers to the positive plate. In most lead-acid batteries the negative plate is of the pasted type.

The type of plate affects the characteristics and performance of the battery. All types have lead oxide (PbO) applied during manufacture. The plates are formed by a process similar to normal charging, in which the lead oxide is converted to lead dioxide (PbO_2) in the positive plates and porous sponge lead (Pb) in the negative plates. The positive plates are welded to form a plate group and the negative plates are assembled in the same manner. The plates of different polarity are insulated from each other by separators. The complete plate group assembly is immersed in sulphuric acid in the container.

These components - an electrochemical couple of two different plates or electrodes (PbO_2 and Pb) an an acid electrolyte (H_2SO_4) - are all that is required to make a rechargeable, or storage, battery using lead and acid.

The formula for the chemical reaction in the battery is [4]

$$\underset{\text{Positive Plate}}{PbO_2} + \underset{\text{Electrolyte}}{2H_2SO_4} + \underset{\text{Negative Plate}}{Pb} \underset{\text{Charge}}{\overset{\text{Discharge}}{\rightleftharpoons}} \underset{\text{Positive Plate}}{PbSO_4} + \underset{\text{Water}}{2H_2O} + \underset{\text{Negative Plate}}{PbSO_4}$$

Because this chemical reaction is reversible, the battery can be both discharged and charged. The discharge / charge process is illustrated in Figure 1.

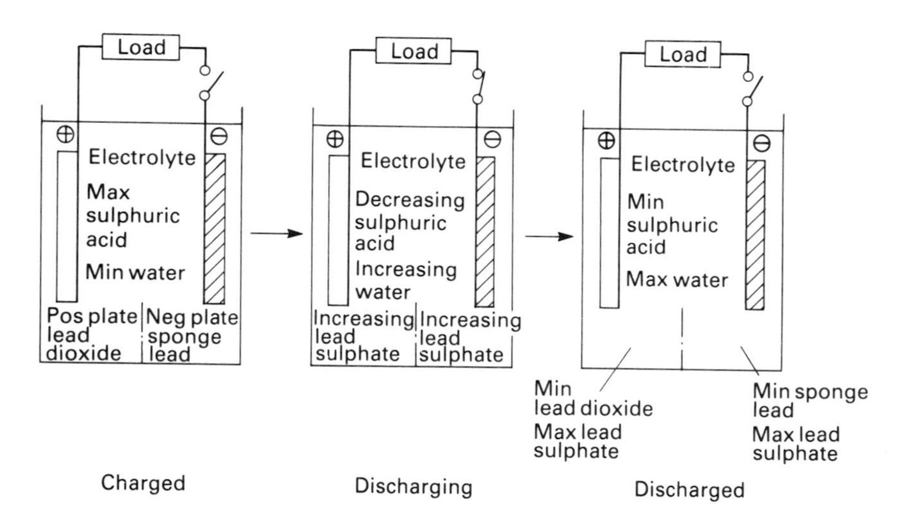

Fig. 10.1. Lead acid battery discharge.

When a load is connected to a charged battery, a current is supplied by the battery. The positive and negative plates react with the sulphuric acid. The lead dioxide and sponge lead are transformed into lead sulphate ($PbSO_4$). The sulphuric acid is consumed in the production of lead sulphate. This reduces the density of the electrolyte, and in a fully discharged battery the sulphuric acid is transferred to water (H_2O).

To reverse this reaction, a charging instrument (PV array, rectifier, or generator) is connected to the discharged battery. The plates will revert to their original lead dioxide (PbO_2) and sponge lead (Pb), while the diluted acid is converted to the original sulphuric acid from the lead sulphate that covered the plates when being discharged (Figure 2).

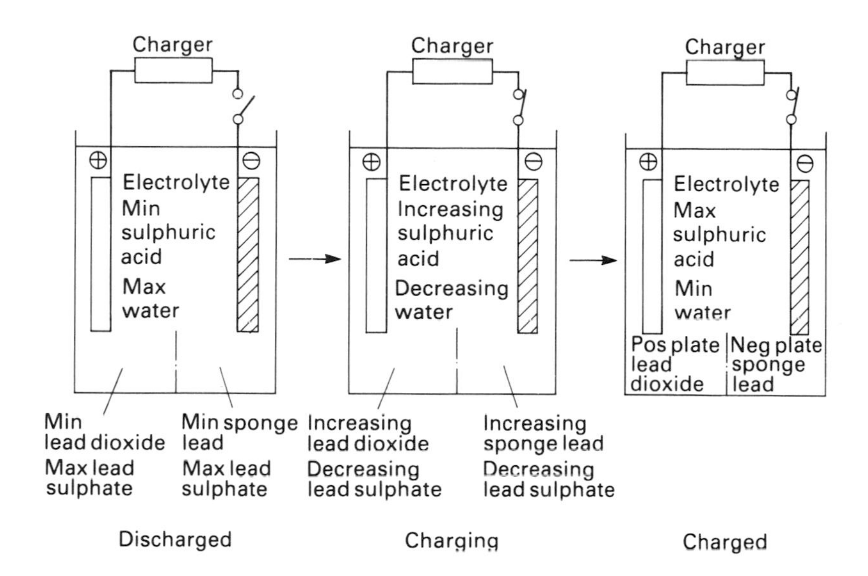

Fig. 10.2. Lead acid battery charging.

In the lead-acid battery the positive active material (lead dioxide) negative active material (sponge lead), and the sulphuric acid electrolyte are all components in the electrochemical reaction which takes place during discharge and charge.

All of the electrolyte (that is, the electrolyte contained within plates and separators, between plates and separators, and that both above and below the plate group) is utilized in the electrochemical reaction. The sulphuric acid reacts with all lead parts: the current carrying grid, the mechanical structure, and the active materials. This means that the electrochemical system functions on the basis of corrosion and, consequently, has a finite life.

2.2 Grid Alloy

As mentioned there are three basic types of plates used in most lead-acid batteries. Because lead is a soft metal, it has poor mechanical strength. Originally, batteries manufactured with pure lead had very thick plates to overcome this problem. Another way of solving the problem is to use an alloy to harden the grids as well as to improve the casting properties.

Batteries can be divided into categories based on how the grids for the plates are manufactured.

* Planté types: Positive plates are made from pure lead, negative plates are pasted with lead-antimony alloy grids.

* Tubular type: Positive plate grids are cast from a high lead-antimony alloy or low lead-antimony alloy; negative plates are pasted with lead-antimony grids.

* Pasted type: Positive plates have lead-calcium alloy grids and pasted negative plates with similar alloy grids.

* Pasted type: Positive and negative plates have grids of various alloys. The alloys may contain antimony arsenic, tin, silver, selenium, sulphur and copper.

The design of the positive grids is the most critical parameter with respect to the battery's service life. The positive grid is where one of the basic irreversible processes occurs, with anodic corrosion as a result. A corrosion layer of lead dioxide is formed, and this has an initial beneficial effect of increasing the capacity in the early part of a lead-acid battery's service life. However, the structural, electrical, and the physical properties of the grid changes in a manner that is ultimately detrimental to the battery's performance. The grid cross-section decreases during corrosion, and this both reduces mechanical strength and increases electrical resistance. The process continues until, eventually, the grid cross-section reaches a critical value at which the grid breaks and the battery is suddenly dead. To combat the corrosion, different grid designs are used and that is where various alloys play a significant role. The only other way to reduce the impact of corrosion is to control the charging conditions very carefully.

2.3 Valve Regulated Batteries

The chemical reactions previously described apply to all lead-acid batteries, no matter what design or alloy is used for the plates. Many types of lead-acid batteries are available - starter batteries, SLI (Starting Lighting Ignition) batteries, industrial batteries, traction batteries, stationary batteries, etc. More recently introduced have been maintenance free batteries, sealed batteries, and gel cell batteries, to mention just a few. Each of these is a lead-acid battery with the same basic chemical reaction, but each has been designed to meet the requirements of a special application.

Today all types of the lead-acid batteries can be divided into two main categories:

1) Flooded or vented

2) Valve regulated or sealed

Within these categories are many varieties which will be discussed more in detail under "3. BATTERY DESIGN."

Originally, all lead-acid batteries were flooded or vented. In this type the liquid sulphuric acid covers the entire plate group and the gases generated during charging are passed into the open air through a vent. However, the last two decades have seen the development of a lead-acid battery with immobilized electrolyte, and these are found in two types:

1) Gelled electrolyte;

2) Starved electrolyte.

In the gelled electrolyte battery the sulphuric acid is immobilized between the plates by the addition of silica. In the starved electrolyte battery, the acid is immobilized in an absorbent separator made of glass fibers and retained between the plates.

The principal characteristic of both these batteries is that they can be manufactured with no free electrolyte. The gases generated during charging are combined in the plate group and not vented off to the air. These batteries are often referred to as "maintenance free" or "sealed". More accurately, they are now referred to as "valve regulated" lead-acid batteries, both in British Standards (BSI) and International Standards (IEC). Each cell is equipped with a pressure release valve which opens in case the internal gas pressure reaches a preset level.

A lead-acid battery can only operate as valve regulated (sealed) under certain conditions. The quantity of active materials in the plates, and alloys used in the manufacture of the grids supporting the active material, ensure oxygen is first produced on the positive plates. The internal design of the cell must allow for the diffusion of the oxygen (O_2) towards the negative plates, where it will react chemically with the sponge lead (Pb) of the negative active material to produce lead oxide (PbO). The sulphuric acid will react with the lead oxide to produce lead sulphate ($PbSO_4$), which is transformed electrochemically into lead and gives back some sulphuric acid [5].

$$2Pb + O_2 \rightarrow 2PbO$$

$$2PbO + 2H_2SO_4 \rightarrow 2PbSO_4 + 2H_2O$$

represents the chemical reaction at the negative plate and explains the oxygen reduction process in the valve regulated cell.

2.4 Nickel Cadmium Pocket Plate Batteries

A nickel cadmium battery [6] consists of one or more battery cells. Each cell has a rated voltage of 1.25 volts. A 12V battery system is normally assembled from 9 cells and a 24V system from 19 cells.

A nickel cadmium pocket plate cell has five (5) basic components.

* Positive plates with nickel hydroxide;

* Negative plates with cadmium hydroxide;

* Insulators;

* Electrolyte or potassium hydroxide;

* Container.

The active materials nickel hydroxide for the positive plates and cadmium hydroxide for the negative plates are in powder form packaged into perforated nickel plated steel strips that are formed into pockets, hence the name pocket plates (Figure 3).

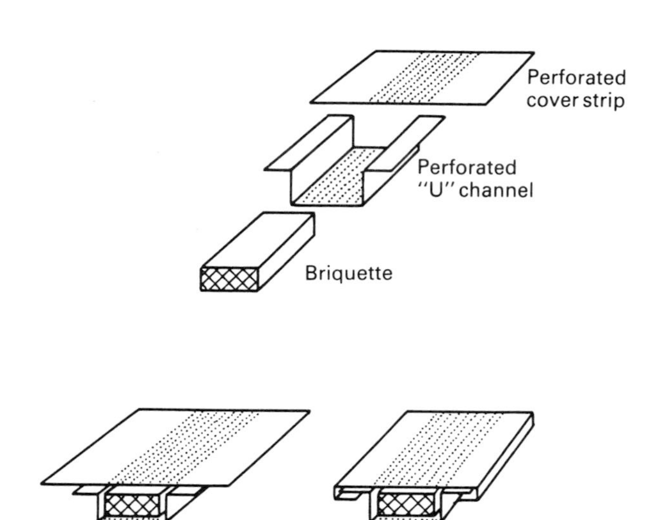

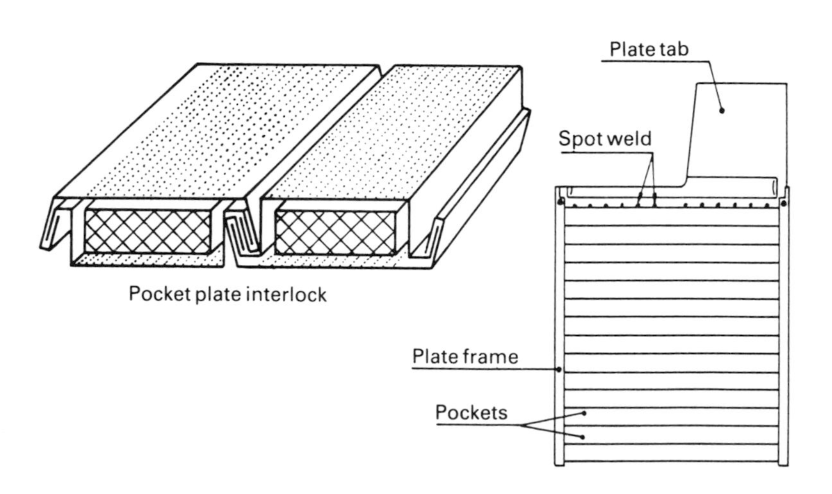

Fig. 10.3. Pocket plate construction.

The steel strips are folded so that one pocket locks into the next when they are placed side by side, with every second pocket placed upside down. The active materials for both positive and negative plates are totally enclosed in the steel pockets. The electrolyte has access to the active materials through the perforation of the steel strips, but the holes are too small for the active material particles to escape. As a result, the surface of the plates remains inactive. The perforated steel enclosure also applies physical pressure to the active materials, promoting conductivity and minimizing plate swelling.

The positive plates are welded or bolted together to form a plate group. The negative plates are assembled in the same manner. The current collectors and all internal mechanical parts are made of steel. The plates are insulated from each other with plastic rods or grids.

The alkaline electrolyte (potassium hydroxide) in the cell does not react with steel, which means that the supporting structure and current carrying parts remain intact and unchanged for the life of the battery. Not only is there no corrosion in a nickel cadmium cell, the alkaline electrolyte actually acts as a preservative to the steel components in the cell mechanical structure.

The formula for the chemical reaction in the cell is:

$$\underset{\text{Positive Plate}}{2NiOOH} + \underset{\text{Negative Plate}}{Cd} + \underset{\text{Water}}{2H_2O} \underset{\text{Charge}}{\overset{\text{Discharge}}{\rightleftharpoons}} \underset{\text{Positive Plate}}{2\ Ni\ (OH)_2} + \underset{\text{Negative Plate}}{Cd(OH)_2}$$

As can be seen from the formula, the alkaline electrolyte does not take part in the electrochemical reaction for the discharge-charge process, acting instead merely as an ion conductor. The potassium (K) in the potassium hydroxide (KOH) electrolyte is not active in the reaction. The formula shows that water is produced during charging, but the quantity is too small to have any noticeable impact on the electrolyte density. The electrolyte density remains practically the same at any state of charge.

The discharge/charge process is illustrated in Figure 4.

When a load is connected to a charged battery, current is supplied by the battery. A reduction reaction takes place at the positive plate and an oxidation reaction at the negative plate.

When a discharged battery is connected to a charging instrument (PV array, rectifier or generator), the reactions in the battery are reversed: oxidation occurs at the nickel electrode and reduction on the cadmium electrode. This is a simple way of demonstrating the reactions in a nickel cadmium battery, showing that there is no corrosion process and that the electrolyte is not active in the reaction.

Since there is no corrosion, and plate surface is inactive in the nickel cadmium battery, the aging of the battery is related

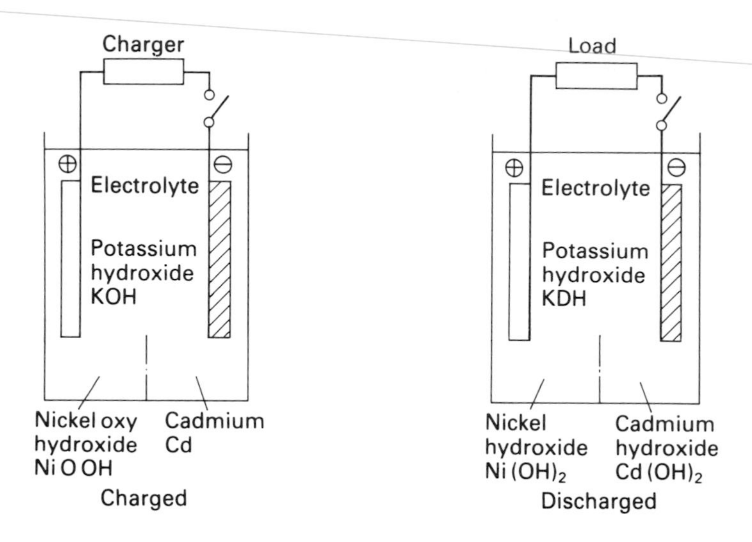

Fig. 10.4. Nickel cadmium battery.
Charge/discharge process.

only to changes in the active materials. Because this process is a linear degradation in the steady state, the life of a nickel cadmium pocket battery is predictable.

3. BATTERY DESIGN

Batteries are generally not well understood and there are many misconceptions about them among PV system design engineers. It is obvious from the above that the characteristics and features differ significantly between lead-acid batteries and nickel cadmium pocket plate batteries. But the PV system designer has also a large variety of lead-acid batteries to select from, and must make the important decision as to which characteristics and features are most appropriate to the specific application and environment where the battery is to be operated.

3.1 Flooded Lead-acid Batteries

Flooded or vented lead-acid batteries, those with their plates structures immersed in sulphuric acid and the gases generated vented off, are described by their plate construction and the metal alloys used to strengthen the plate grid.

3.1.1 Planté Plate Battery

The positive plate is a thick, pure lead plate with grooves or gaps to increase the active surface area of the plate. The negative plate is normally a pasted plate. The separator allows for a large volume of acid between the plates. The Planté plate battery is a long-lived battery and has traditionally been used in telecommunication applications, where it has a life expectancy of twenty five years or more. Modern Plante plate batteries have low internal resistance and can be used for short duration high rate discharge. However, they require more floor space than other designs and are generally very heavy.

3.1.2 Tubular Plate Battery

The tubular positive plate has a grid made of spines connected at the upper end with a common current collector. The spines are surrounded with active material in the center of the tubes manufactured from braided, felted, or woven synthetic fibers or glass wool. The negative plate is of the pasted type. The positive plate design, in combination with separators, allows for free flow of the electrolyte.

The plate design is capable of many relatively deep discharge cycles. This battery type is commonly used in cycling applications, such as fork-lift trucks and other electric vehicles. The battery can be used for short duration high rate discharge but requires more floor space than other designs.

3.1.3 Pasted Plate Battery

The pasted flat plate design is the most common of the flooded lead-acid batteries. The active material is formulated into a paste mixture that is applied to a lead alloy grid. This battery design is predominantly used in automotive batteries, but is also common in industrial applications, particularly where high rate discharge currents are specified. It is the least expensive plate to manufacture. Without special retention, the plate design does not have good cycling ability. Different plate thicknesses can provide batteries with different energy and power density characteristics.

3.1.4 High Antimony Battery

Batteries with both tubular and pasted plate can have their grids strengthened with high content of antimony. The amount of antimony is generally 5-11% and includes also other additives such as arsenic, tin, selenium, copper, and sulphur. High antimony grids have good mechanical strength and are easy to cast. During charging, the lead alloy grid is subjected to corrosion. The composition of the alloy affects not only the corrosion rate but also the rate of water decomposition during charging. During corrosion of the positive grid, antimony is transported to the negative plate where it is deposited. This is referred to as antimony poisoning, and will result, eventually, in higher charging current and increased water consumption during charging. This makes maintenance more costly. High antimony batteries have good cycle life, but to a certain extent this is offset by high maintenance costs.

3.1.5 Low Antimony Battery

In batteries with tubular and pasted plates, grids with low antimony have been introduced. The amount of antimony is normally 1-3% and other additives include arsenic, tin, selenium, copper, and sulphur. Grids with an antimony content less than 5% are more difficult to cast and cracks can easily form in the grids. For this reason other additives are included in the alloy.

The low antimony content reduces the water consumption and self discharge in the battery. Low antimony batteries are, therefore, often a better choice for PV applications, as they have fairly good life and low maintenance cost.

3.1.6 Lead Calcium Battery

In order to overcome the high maintenance costs in a battery with antimony grids, the lead calcium battery was developed for stationary application in standby power service such as telecommunications

The calcium content in the alloy is normally 0.07%, with an additive of tin. Lead calcium batteries require very low currents to keep them fully charged. They have a low rate of self discharge. This results in a low gassing rate and low water consumption. Lead calcium batteries serve well under float charge conditions, but have a limited cycle life.

Operating in areas where the lead calcium battery will be exposed to high temperatures for prolonged periods will drastically shorten the life of the battery.

3.2 Valve Regulated Lead-Acid Battery

In principle, all valve regulated batteries function on the process of oxygen reduction explained previously. The oxygen tends to slightly discharge the negative plate and suppress the evolution of hydrogen at the negative plate.

As the cell is under pressure from the pressure release valve, oxygen is kept in the cell, not released from the battery. Hydrogen produced at the negative plate cannot react to the positive plate. When the battery is overcharged or abused, it vents off hydrogen through the valve (hence the term "valve regulated").

This battery type is very sensitive to changes in charging voltage and ambient temperature, with even small variations resulting in dramatically reduced life. These batteries must be used cautiously in complete discharge applications [7]. The electrochemical and physical characteristics currently limit the use of valve regulated batteries to applications requiring only shallow discharge cycles.

3.3 Nickel Cadmium Battery

There are several characteristics inherent in the nickel cadmium batteries which make them versatile for a variety of applications. There is no shedding or plate growth, and the charge/discharge process does not cause any corrosion of the internal parts. The electrolyte maintains its high conductivity in a discharged battery as well. It does not change in density, which means that the battery has excellent performance at low temperatures. The chemistry in the active material is also stable at high temperatures. The cycle life is good in deep discharge applications. There is a large electrolyte reserve, which means that maintenance costs are low. However, the cost of the materials in a nickel cadmium battery is high, resulting in a high capital cost per energy unit. A PV system must, therefore, be sized properly to ensure that an unnecessarily large nickel cadmium battery is not purchased. When the number of cells in the nickel cadmium battery is matched correctly with the voltage of the PV array, the battery can operate satisfactorily without a voltage regulator.

4. BATTERY CHARACTERISTIC

4.1 Depth of Discharge

In a stand alone PV installation, the battery is exposed to daily and seasonal discharges at various depth of discharge (DOD). The DOD is the percentage of the rated capacity that is drawn from the battery. Depending on the application and the number of days of autonomy, the battery may be exposed to shallow or deep discharge cycles. The daily discharge can be shallow while the seasonal can be a deep discharge. Service experience indicates that, in general, batteries are more deeply discharged than the amount they were originally sized for. This results from lower-than-expected efficiency of the PV arrays, or shading and dust collection on the arrays.

The DOD must be considered when the system is designed. The least expensive battery available from a capital cost point of view is the automotive starter (SLI) battery, either vented or valve regulated. However, this battery is designed to deliver high power for a few seconds during engine cranking and is not intended for cycling operation as in PV installations. Therefore it will not last long in any PV application. It is bad economy to use the SLI car battery in a stand-alone PV system - an industrial type battery is always a better and more economical solution.

The lead-acid battery industry classify its batteries as shallow-cycle or deep-cycle. Shallow-cycle lead-acid batteries are lighter, with thinner plates, and will not last if the manufacturer' recommended DOD levels (25-50%) are exceeded. Deep-cycle lead-acid batteries have thicker plates and are more often used for stand-alone PV systems. The maximum DOD recommended by manufacturers is 80%. If this level is exceeded regularly, the battery life will be drastically reduced.

Deep discharges have no harmful effects on nickel cadmium pocket plate batteries. The manufacturers accept 100% DOD for this type, meaning that, generally, a smaller nickel cadmium battery can be used compared to a lead-acid battery. All nickel cadmium batteries are deep-cycle batteries.

Lead-acid batteries should not be left in a discharged condition for an extended period of time. Manufacturers recommend that the batteries be recharged immediately after a discharge. If not, there is a risk that the lead sulphate formed during the discharged cannot be transformed into lead during the charging of the negative plates, a phenomenon called sulphation. In nickel cadmium batteries, this cannot occur because the potassium hydroxide (KOH) electrolyte does not take an active part in the chemical reaction.

4.2 Charge

A PV array charges the battery with variable current depending on weather conditions up to the voltage limit of the array, from then on with almost constant potential related to the array IV-curve characteristic and regulated by the battery voltage or the setting of the voltage regulator. As the charging current varies with the light intensity on the array, the battery is actually exposed to a kind of pulse charging.

All lead-acid batteries are sensitive to overcharging. When a lead-acid battery is overcharged, its life is reduced due to corrosion, plate growth, and active material shedding from the plates. In order to eliminate the risk of overcharging, all PV systems with lead-acid batteries must be equipped with a voltage regulator. It is important that the voltage regulator be adjusted to the charging voltage recommended by the battery manufacturer and that it be of good quality, preferably temperature compensated. Voltage regulators are very sensitive to lightning, and it is recommended that they have lightning protection. Cheaper voltage regulators often use mechanical relays to switch the current. These relays will not react quickly enough to protect against lightning-induced surge currents.

When a voltage regulator fails, the battery will be overcharged or deep discharged. With lead-acid batteries this eventually results in reduced life and failure. A good voltage regulator must have the following features:

- Temperature compensation;
- Adjustable set points;
  - High voltage disconnect
  - Low voltage disconnect
- Reverse current protection;
- Low voltage warning.

Valve regulated lead-acid batteries require an even more accurate voltage setting during charging. It is imperative to adjust the charge voltage to temperature variations. The manufacturer's recommended voltage setting [8] for a gelled electrolyte battery is indicated in the graph.

Overcharge or deep discharge is not harmful to nickel cadmium

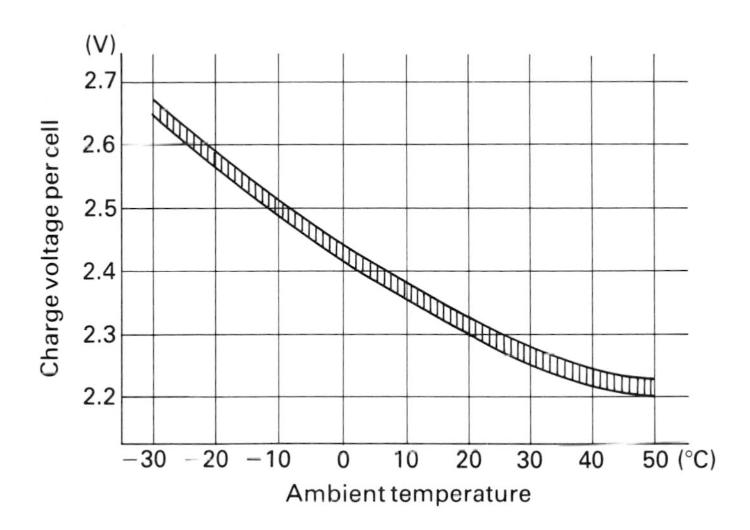

Fig. 10.5. Gelled electrolyte lead-acid battery. Charge voltage variation with temperature.

batteries. As a result, voltage regulators are not required for PV systems with nickel cadmium batteries, providing that the system is sized to match array voltage with battery voltage. In some cases, PV system designers have shown a tendency to overestimate the load requirements or to include too high a safety margin for the load. This will always result in battery overcharging. A nickel cadmium battery can tolerate this condition, but it will result in higher water losses, meaning increased maintenance. Overcharge should be eliminated, as far as possible, through best possible system sizing and design.

4.3 High Temperatures

Grid corrosion in the lead acid battery, which primarily limits the life of the positive grid, is the result of electrolyte and oxygen corrosion [9]. Corroded grids lead to loss of conductivity and, ultimately, grid breakage. Corrosion increases with high temperature and decreasing acid density, conditions which occur when batteries operate in warm climates during days of limited insolation.

When lead acid batteries operate at elevated temperature, life is affected seriously. For each rise in temperature of approximately 10°C above ±25°C, the battery life is reduced by 50% [10]. The nickel cadmium pocket plate battery is also affected by high temperatures, but not to the same extent as lead-acid batteries. The difference in life expectancy at high temperatures for the two battery types can be seen in the graph of Figure 6.

4.4 Low Temperatures

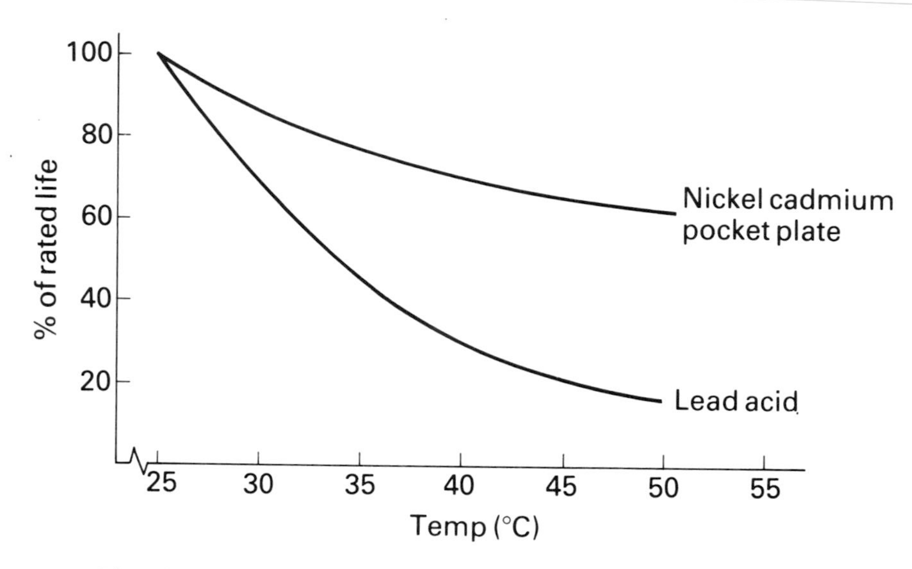

Fig. 10.6. Battery life expectancy at high temperatures.

4.4 Low Temperatures

When batteries operate at low temperatures they must be oversized to deliver the same performance as at room temperature. Lead acid batteries have to be protected against even a single day of low temperatures as they will freeze solid in a discharged condition, causing cracks in the electrode structure and cell containers. Flooded lead-acid cells which discharge to an electrolyte specific gravity of 1.205 will freeze at -32°C, thereby stopping all operations while only 14% discharged [11].

In contrast, a nickel cadmium pocket plate battery can sustain long periods of low temperatures at any state of charge. It can deliver 60% of rated capacity at a temperature as low as -40°C (Fig. 7). The temperature coefficient of the battery has the same characteristic as the PV module temperature coefficient, which is another reason why a voltage regulator is not required in a nickel cadmium battery PV system. By eliminating the voltage regulator, the reliability and MTBF (Mean Time Between Failure) of the system is improved significantly and the cost is reduced, particularly for small systems.

4.5 Efficiency

The charging efficiency of a battery is normally referred to as the ampere-hour efficiency or coulombic efficiency. This is the ratio between the quantity of electricity passed during discharge and charge. For lead-acid batteries, the coulombic efficiency is around 90% at +25°C, depending on charging current and acid

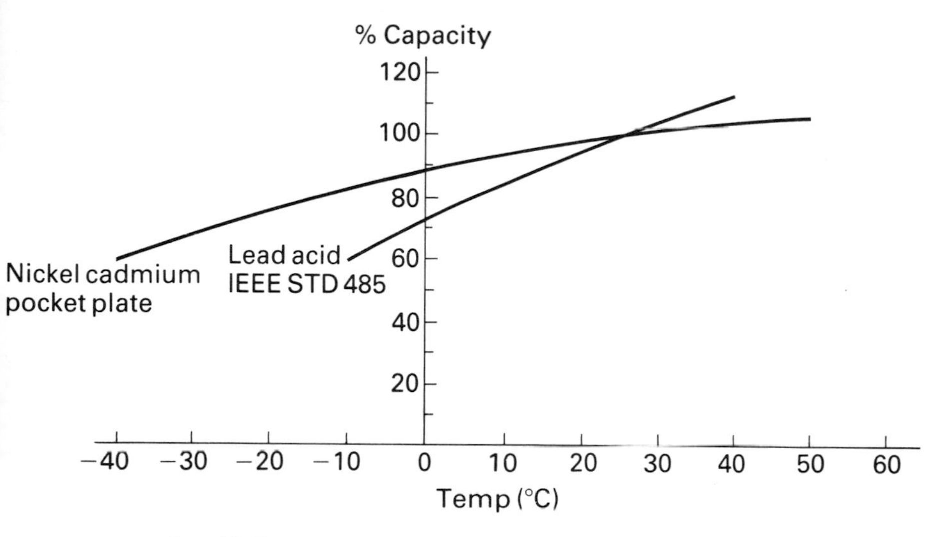

Fig. 10.7. Available capacity at various temperatures.

used. For a nickel cadmium battery it is 90-95% at +25°C, depending on state of charge [12].

The voltage efficiency is the ratio between discharge and charge voltage. This coefficient depends on the value of the discharge and charge current density. In PV systems, the battery is charged and discharged with low currents because the systems are normally sized for a minimum of 4-5 days of autonomy. For longer autonomy periods, the currents are correspondingly lower. The mean voltage for a lead-acid battery is 1.95V during discharge and 2.25V during charge, giving a voltage efficiency of 87%. For a nickel cadmium battery under the same conditions, the mean discharge voltage is 1.25V and 1.47V during charge, an efficiency of 86%.

Watt-hour efficiency, or energy efficiency, is the ratio between energy discharged and energy consumed during the charge. This is calculated as the product of coulombic and voltage efficiency. Continuing the above example, for a lead-acid battery it is .90 x 0.87 = 0.78, or 78%. For a nickel cadmium battery it is 0.92 x 0.86 = 0.79 or 79%. The efficiency of the two types of batteries is approximately the same.

4.6 Failure Mode

During the life of a battery there are reversible processes that occur during discharge and charge cycles, and irreversible processes that limit the battery life. Because of the difference in chemistry between lead acid and nickel cadmium batteries these processes are inherently also different. The irreversible processes in a lead-acid battery are related to corrosion of the positive plate and plate growth; shedding and softening of the positive active material; leadification and bulging of the negative plate; sulphation of the negative plate; and shorts between plates, through or around the separator, in the form of treeing and mossing. In the nickel cadmium battery the irreversible processes are fewer, and are related to recrystallization of active material, carbonization of the electrolyte, and iron poisoning of the positive plates. These harmful processes in the batteries are influenced by internal parameters which affect the battery cycle life [13].

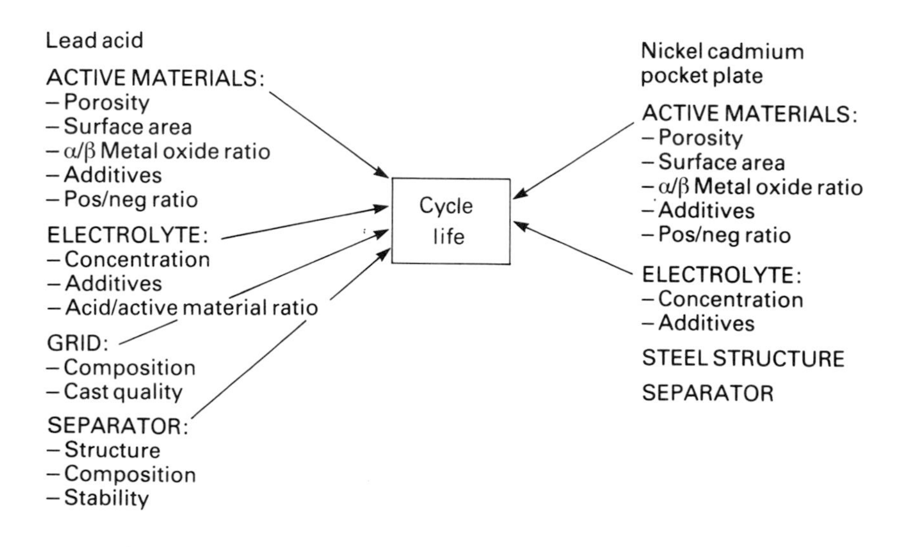

Fig. 10.8. Internal parameters affecting battery cycle life.

Lead acid	Failure modes		Nickel cadmium pocket plate
ACTIVE MATERIALS: – Porosity – α/β Metal oxide ratio – Additives ELECTROLYTE: Concentration – Additives GRID: – Composition SEPARATOR:	Pos A.M. shedding Pos A.M. sulphation Neg A.M. leadification Neg A.M. sulphation Mossing Treeing Corrosion	Recrystallization Carbonization Iron poisoning High resistant shorts	ACTIVE MATERIALS: – Porosity – α/β Metal oxide ratio – Additives ELECTROLYTE: – Concentration – Additives STEEL STRUCTURE: – Ni – Plating SEPARATOR:

Fig. 10.9. Battery failure modes.

However, there are also external parameters affecting the battery life.

The aging of a nickel cadmium pocket plate battery is distinctly different from that of a lead-acid battery. As shown in Figure 2,

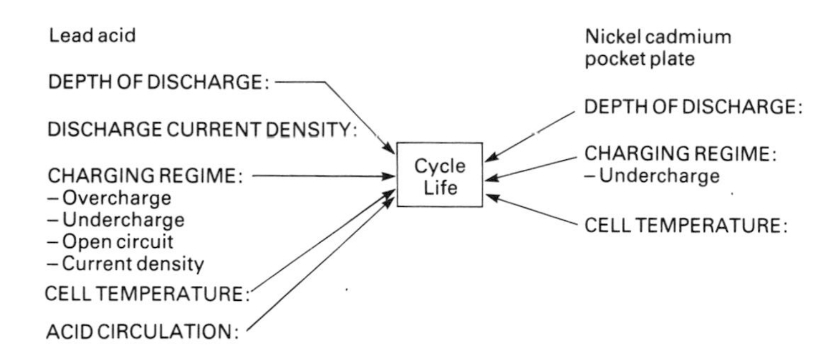

Fig. 10.10. External parameters affecting battery cycle life.

the corrosion process in the lead-acid battery starts as soon as the battery is filled with electrolyte or connected to the charging equipment. At approximately 60% of the expected service life, the degradation of capacity and performance begins to accelerate, and the life becomes unpredictable. At this stage, the corrosion in lead-acid batteries can result in a sudden total loss of capability to supply current, a state known as "sudden death" [14]. Because the steel structure in a nickel cadmium pocket plate battery is inactive, only the gradual aging of the active materials influences the service life of the battery. This aging process is linear in the steady state, and there is no sudden degradation. Degradation is mainly due to the change in the graphite component and recrystallization of the nickel hydroxide in the positive plates.

This characteristic, inherent in the nickel cadmium pocket plate batteries, means that there is always capacity available, even if it has been reduced.

5 RECOMMENDATIONS

Before selecting a battery for a PV system, it is imperative to make a careful analysis of the load requirements for the system and the environmental conditions under which is has to operate.

The capacity of the battery has tradiationally been related to the number of days of autonomy for the system. More recently, this approach has been increasingly abandoned, Battery capacity, or the number of days of storage is now based on the required system availability. Many factors may influence system availability but the must important is local weather patterns. A system designer with both PV array and battery experience should always be consulted in determining the size of the complete system. Remember that it is more important to buy a quality battery designed for PV applications than to calculate exactly the number of hours of battery storage. There are too many variables in the weather pattern to meet an exact figure.

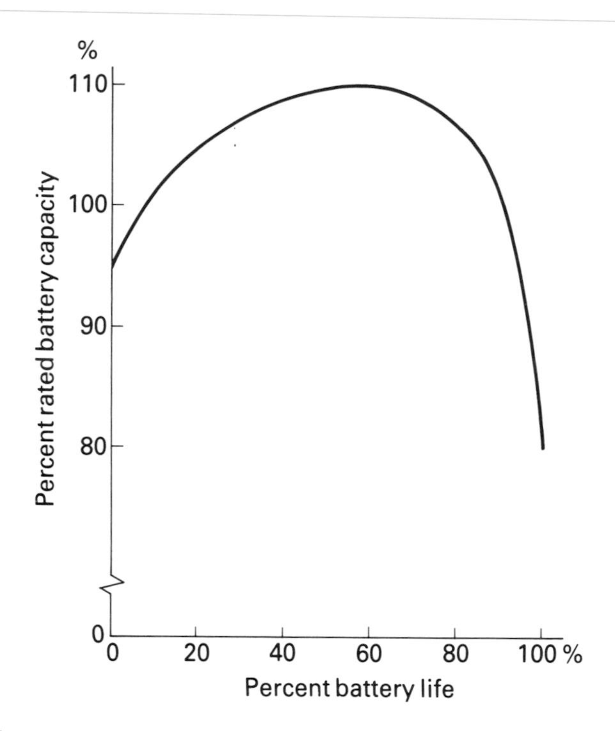

Fig. 10.11. Lead acid battery. Change of capacity over useful lifespan.

As mentioned previously, the nickel cadmium pocket plate battery is often considered too expensive when compared with lead-acid batteries. If the comparison is with a low quality, low cost lead-acid battery, this is doubtless the case. But a low quality battery will compromise the reliability and availability of the equipment that the PV power system is installed to protect. It is always poor economy to install anything but a high quality industrial battery in a PV power system, and here the price difference between lead-acid and nickel cadmium is much closer.

A correctly sized battery is the key to a reliable battery system at a competitive price. As shown above, the characteristics and performance are quite different for lead-acid and nickel cadmium batteries. Do not fall into the trap of sizing a nickel cadmium battery to the same amper-hour capacity as the lead-acid battery.

Good engineering practice dictates that, in most cases, a smaller nickel cadmium can support the same critical load as a larger lead-acid battery. This results in less floor space used and lower installation costs. Because the nickel cadmium battery can be installed next to electronic equipment, it does not require a special battery room - a significant cost saving in larger installations.

In order to arrive at an accurate system cost estimate for a PV power system, an engineering analysis which determines true cost over the expected service life of the system is recommended. This kind of analysis is called life-cycle costing, and allows a direct comparison of the costs of various alternatives.

There are several reasons to use life-cycle cost (LCC) analysis rather than a simple comparison of the initial cost of each alternative. Installation cost is only one of many components. Power systems also require varying amounts of maintenance, repair etc. LCC analysis allows the designer to evaluate all the costs associated with installing and using a power system over its lifespan. Because LCC analysis breaks out specific cost components (such as battery replacement), the designer can analyze the economic effect of using different components with different reliability factors.

The nickel cadmium pocket plate battery has a large reserve of free electrolyte. The water dissociation into oxygen and hydrogen is lower than for most other battery types (Figure 12). Therefore, the water topping up intervals are generally more than 2-4 years, depending on battery type. This means that the service cycle for nickel cadmium batteries is the same as the replacement cycle for low quality batteries of other types.

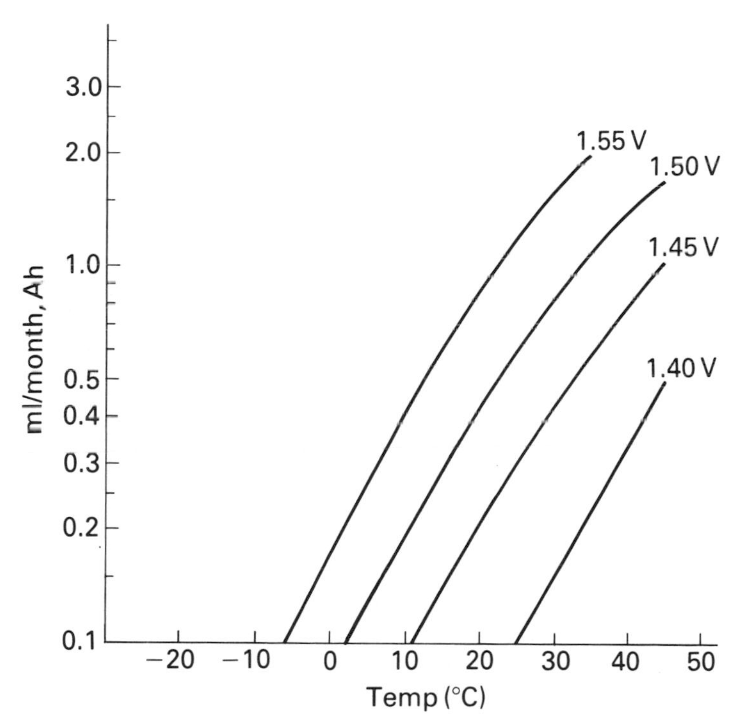

Fig. 10.12. Nickel cadmium pocket plate battery. Water dissassociation in relation to temperature and charging voltage.

As there is no corrosion or plate growth in the nickel cadmium pocket plate battery, the inspection periods can be extended and should not be as frequent as for lead-acid batteries. Generally once per year is sufficient, whereas lead-acid battery manufacturers normally require quarterly inspections.

The life-cycle cost can be calculated from the formula as recommended by Sandia National Labs, USA [15].

$$LCC = C + M + R - S$$

C = capital cost of project including initial capital expense

M = the sum of all yearly scheduled maintenance and operation costs

R = the sum of all repair and equipment replacement anticipated over the life of the system

S = the salvage value of a system in the final year of the life-cycle period.

All factors in this formula must be presented in terms of their "present worth" or actual value today.

In most cases, due to their long, predictable service life, nickel cadmium pocket plate batteries have a very favorable life-cycle cost.

REFERENCES

1. David F. Menicucci and Anne V. Poore, Today's Photovoltaic Systems: An evaluation of their performance, Sandia National Laboratories, SAND 87-2585.
2. David F. Menicucci and T. Key, The Fundamentals of Photovoltaic System Electrical Design, ASES Annual Meeting, Portland, Oregon, July 1987.
3. G.W. Vinal, Storage Batteries, 4th Ed., John Wiley & Sons, Inc., New York, 1955.
4. B.D. McNicol and D.A.J. Rand, Studies in Electrical and Electronic Engineering II, Power Sources for Electric Vehicles, Elsevier Science Publishers B.V. Amsterdam 1984.
5. D. Berndt, Maintenance and Reliability of Stationary Batteries VARTA Special Report, Hannover W. Germany, 1985.
6. S.U. Falk and A.J. Salkind, Alkaline Storage Batteries, John Wiley and Sons, Inc., New York 1969.
7. B.L. McKinney, T.J. Dougherty and M. Geibl, The Comparison of Flooded, Gelled and Immobilized Lead-Acid Batteries, Proceedings INTELEC/IEEE 1984.
8. Accumulatorenfabrik Sonnenschein Gmbh, Dryfit A600 Solar, 10 1561 01/3/2/486 Hell. West Germany.
9. H. Bode, Lead Acid Batteries, The Electrochemical Society Series, 1977.
10. Eltra Corp. C & D Batteries Division, Stationary Batteries Section 12-610, 1979.
11. W. Brooks, Cold temperature testing of Absolyte, AAR-C&S

Division Western Regional Meeting, Ft Worth, Texas, Sept 5-7, 1986.

12. Bruce A. Mork, Optimizing Photovoltaic Project Design on a Mountain Top, Electrical Systems Design, May 1988.
13. Arne, O. Nilsson, Nickel Cadmium Pocket Plate Batteries in Photovoltaic Installations. Progress in Batteries and Solar Cells, Vol. 7, 1988, JEC Press Inc., Cleveland, Ohio, USA.
14. Alber Engineering INC., The Battery Test Notebook, Boco Raton Florida, USA.
15. Stand-alone Photovolatic Systems, A Handbook of Recommended Design Practices, Sandia National Laboratories, Albuquerque, New Mexico and New Mexico Solar Energy Institute, Las Cruces, New Mexico USA, SAND 87-7023, April 1988.

Chapter 11

NICKEL CADMIUM POCKET PLATE BATTERIES IN PHOTOVOLTAIC APPLICATIONS

A. O. NILSSON

SAB NIFE AB, S-26124 Landskrona, Sweden

1. INTRODUCTION

The terrestrial market for photovoltaic (PV) power supply systems is today the largest proportion of the total market for photovltaic technology and is forecasted to grow at a rate of 25-30% annually through 1990.

Among the largest applications for photovoltaics in 1984 (Starr (1)) are telecommunications, cathodic protection, navigational aids, off-grid residential, and rural electrification accounting for close to 40% of the total 25 MWp shipped in photovoltaic modules. Lately railroad signalling has, particularly in North America, become another large photovoltaic application. The common denominator for all these applications is that they all require a battery bank to supply continuous power during hours of limited sunshine and nighthours.

2. THE PROBLEM

In many remote photovoltaic installations if has been found that the weak link in the system is the lead acid battery conventionally used and the voltage regulator installed to control the charge and discharge conditions of the battery. This adds substantially to the cost of the system in terms of maintenance costs and the cost of replacing worn out components (Newham (2)).

As a result many photovoltaic installations have lost credibility because the systems have in many cases not met the expectations of the users. The designers and system builders have because of ignorance not paid due regard to the battery characteristics and performance in the equipment. The battery, which is the heart of the system must be selected with great care.

3. REQUIREMENTS

Photovoltaic applications are different from battery installations for stand-by and emergency power as the demands are different. In photovoltaic installations the most important battery qualities are:

- Must withstand cycling, daily and seasonal;
- Able to withstand high and low environmental temperatures;
- Operate reliably with minimal maintenance in unattended operation;
- Ruggedness for transportation to remote sites;
- Ease of installation with limited handling equipment and unskilled labour;
- Reliability and availability during the 20 years service life of the PV modules.

We must realize that the most appropriate locations for PV installations do not have existing electric power within relative proximity. Therefore the difficulties are more logistical than technical.

4. LEAD-ACID BATTERIES

The most frequently used batteries in terrestrial self-contained photovoltaic systems are lead acid batteries of various design.

In the early days the so called maintenance free lead acid battery of the SLI (Starting, Lighting, Ignition) type was often used. In home electric systems standard SLI batteries are installed. The whole range of industrial lead-acid battery types with pasted, Plante, and tubular plates having grids of low antimony, high antimony, pure lead or calcium alloys are frequently used in photovoltaics. To make the battery selection even more confusing there are vented, gelled and recombination types of all these varieties.

Because of the competitive situation many system builders make their battery choice based on lowest price to generate sale rather than customer benefits in system reliability and durability.

As indicated above the lead-acid battery is a very complicated storage power source with a large number of internal parameters that effect cycle life. This fact has also been demonstrated by Dr E. Voss (3), see Figure 1. In the four internal design elements in a lead-acid cell there are 13 parameters impacting the cell life, whereof 5 in the active materials. However it must also be noticed that the electrolyte and the mechanical parts, grid and separator play an important role in the designed life of a cell. This is in contrast to a nickel cadmium pocket plate battery which will be dealt with more in detail later. The combination of these parameters will result in different

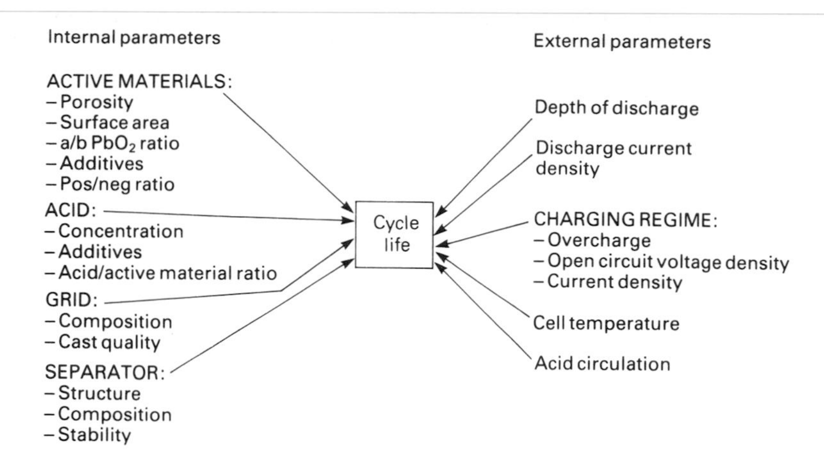

Fig. 11.1. Parameters affecting the endurance of lead-acid cells.

failure modes in the cell from active material sulphation, positive active material shedding, negative active material leadification, mossing and treeing to grid corrosion, when combined with the external parameters the cell is exposed to in operation. In photovoltaic applications the external parameters mostly affecting the cycle life are depth of discharge (DOD), charging regime, cell temperature, and acid circulation.

The immediate effect of deepdischarging a lead-acid battery is positive active material shedding and negative active material leadification. Therefore battery manufacturers recommend that shallow cycle batteries be discharged only to 50% DOD and deep cycle batteries to 75-80% DOD.

The charging regime with overcharge and charging current density influence the active material sulphation. It is a well-known fact that a lead-acid battery is not to be left in a discharged condition for an extended period of time. However, this cannot be avoided in a photovoltaic application where the system is sized for a number of days of autonomy. In many applications 15-20 days of autonomy is specified and in certain geographical areas such as arctic climates up to 90 days of autonomy is required. During these long periods the battery is discharged with a very low current.

When better weather returns, only the excess of what the PV module delivers compared with the load demand is available to recharge the battery, and to fully charge the battery may take several weeks. During long charge and discharge periods the sulphation occurs in the electrodes reducing the battery available capacity.

Probably the most detrimental external parameter affecting the lead-acid battery is the cell temperature. At low temperatures the battery delivers only fractions of its rated capacity and when discharged the electrolyte freezes to a solid block causing cracks in electrode structures and cell containers. At -32°C flooded lead acid cells which discharge to an electrolyte density of 1.205 will freeze, stopping all operations while only 14% discharged (Brooks (4)).

When lead-acid batteries operate at elevated temperatures life is affected seriously. For approximately each 10°C rise in temperature above 25°C the battery's life is cut be 50%. At a constant temperature of only 42°C the battery life is reduced by 75% (Eltra (5)). The reason for this reduced operating life is to be found in positive active material shedding and sulphation, negative active material leadification and grid corrosion. The grids of the electrodes which serve as support to the active material and conductor for the electric current consist of lead and its' alloys. Grid corrosion which primarily limits the life of the positive grid is the result of electrolyte and oxygen corrosion (Bode (6)). Corroded grids break very easily and are extremely brittle, which leads to loss of conductivity and ultimately grid breakage. Corrosion increases with rising temperature and decreasing acid density, which occurs when batteries operate in warm climates during days of limited insolation.

Another phenomenon occurring in lead-acid batteries in PV applications is acid stratification. When the battery is frequently deepdischarged and recharged with a very low charge current, which is actually the case in PV systems, the increased acid density at the bottom of the cell results in increased corrosion of the positive grid and sulphation of the negative active material (Berndt (7)).

5. THE NICKEL CADMIUM BATTERY

One battery known for its high reliability and durability in industrial applications is the nickel cadmium pocket plate battery (Eurobat (8)). SAB NIFE has now developed a new nickel cadmium pocket plate battery, SUNICA, especially suited for the storage of photovoltaic energy. Its design is based upon the well established nickel cadmium pocket plate battery design which has been in use in high reliability railroad and industrial applications for more than 75 years.

The nickel cadmium battery is not such an extremely complex battery as the lead-acid battery, that has been indicated before. The reason for this is that in the nickel cadmium battery there are not thirteen internal parameters that affect cycle life. There are no grids to support the active materials and the mechanical parts used in the battery do not corrode in the internal cell environment.

The active material in the nickel cadmium pocket plate battery is enclosed in pockets of perforated nickelplated steel strips which are joined to the plate materials. The positive plates contain nickel hydroxide as the active material and the negative plates contain cadmium oxide.

Like the plates, the current collectors and mechanical connections are made entirely of steel. The alkaline electrolyte does not react with steel, which means that the supporting structure of a nickel cadmium pocket plate battery stays intact and unchanged for the entire lifetime of the battery.

In contrast to lead-acid batteries the electrolyte of the nickel cadmium pocket plate does not change during charging and discharging. It retains its ability to transfer ions between the cell plates irrespective of the charge level. The nickel cadmium pocket plate battery can operate at any state of charge so there is no need to rush to charge up the battery again after it has been partially or totally run down. The battery can be left uncharged for long periods because there is no sulphating process.

A nickel cadmium pocket plate battery contains plenty of electrlyte. Positive and negative plates are kept apart by separator rods whose only purpose is to keep the plates electrically separated from each other. The separator has no impact on internal battery parameters unlike lead-acid batteries.

It is only the active materials and the electrolyte that are the important internal parameters in a nickel cadmium pocket plate battery affecting the cycle life. The amount and composition of active materials in the SUNICA battery is balanced to match the charge-discharge characteristics of PV applications. The new battery has a special electrolyte that makes it possible to charge the battery with low currents at temperatures from -50°C to +55°C. A large volume of free electrolyte between the plates prevents the cells from drying out. An ample electrolyte reserve allows periods of as long as 10 years before topping up. These features in combination with spill-proof vents, make the battery maintenance free and there is no electrolyte stratification to be concerned about.

6. BATTERY CHARACTERISTICS

A photovoltaic module charges the battery with variable current up to the voltage limit of the module and from then on with almost constant potential. This charging characteristic is ideal for a nickel cadmium pocket plate battery. As the charging current varies with the light on the module the battery is actually exposed to a kind of pulse charging. As mentioned before the SUNICA battery can be charged with a low current. The battery can easily be fully charged with 0.01 x C ampere which corresponds to the 100 hour rate. This magnitude of current is what is normally delivered from the modules for battery charging. At this current density the end of charge voltage at +25°C is 1.53 V per cell, which is well within the voltage limits of the voltage supplied by a PV module.

As a nickel cadmium pocket plate battery is not damaged by over- and undercharging and the temperature coefficient of the battery has the same characteristic as the PV module temperature coefficient it is not necessary to include a voltage regulator in a nickel cadmium battery PV system. By eliminating the voltage regulator the reliability and MTBF of the system is improved significantly and the cost is reduced particularly for small systems.

In a PV system the battery is as a rule of thumb sized with 5 days of autonomy. This means that the battery is exposed to shallow daily cycling of not more than 15% DOD. This cycling regime in combination with variable current charging of low density has a beneficial effect on the ampere-hour charging efficiency of the nickel cadmium pocket plate battery. The efficiency is 90-95% at +25°C and at a temperature as low as -20°C the charging efficiency is as high as 85-90% when the battery is operating at the state of charge typical for PV systems.

A standard vented stationary lead acid battery has an efficiency between 83 and 95% at +25°C depending on the charging current and the capacity used (Gutzeit (9)).

The performance of all batteries is affected by temperature extremes. However nickel cadmium pocket plate batteries are affected to a much lesser extent than lead-acid batteries.

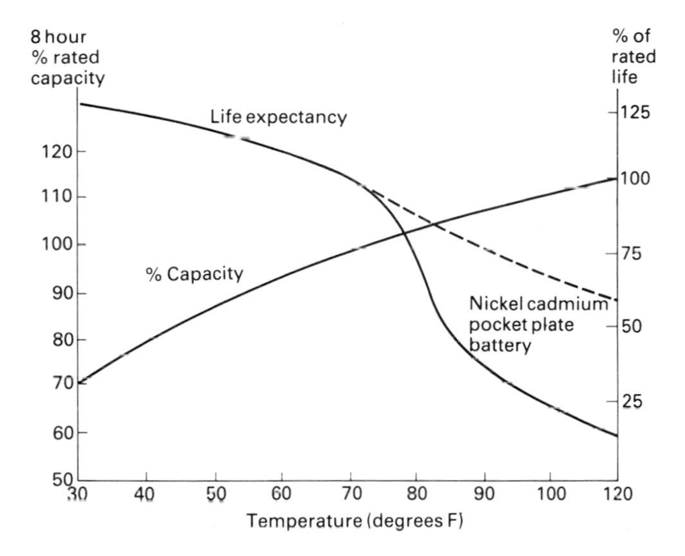

Fig. 11.2. Stationary lead acid battery. Performance and life expectancy at various temperatures.

From the graph, see Figure 2, presented by R. Blohm (10) it can be seen that the life of a lead-acid battery is reduced by 68% at 38°C (100°F). The plotted graph in dotted line indicates that the life is only reduced by 25% for a nickel cadmium pocket plate battery at the same temperature. While a given lead-acid battery in PV installations may have an expected life of 8-10 years at 25°C.

That life is only 2.5-3 years at 38°C. The expected service life of a nickel cadmium pocket plate battery is 20 years at +25°C which is reduced to 15 years at +38°C. At this relatively modest temperature the life of the nickel cadmium battery is 6 times the life of a lead acid battery. This means that a lead acid battery has to be replaced 6 times in a remote installation in a tempered climate for the life of one nickel cadmium battery.

When batteries operate at low temperature thet must be oversized in relation to their performance at normal temperature. Lead-acid batteries have to be protected against a single day of low temperature as the battery will freeze solid in a discharged condition. In contrast a nickel cadmium pocket plate battery can sustain long periods of low temperatures at any state of charge and has 60% of rated capacity left at temperatures as low as -40°C.

7. SIZING

Nickel cadmium pocket plate batteries have been looked upon as expensive in photovoltaics. It is a fact that the materials used in nickel cadmium pocket plate batteries are more costly and durable than the materials used in lead-acid batteries. Consequently the cost per kilowatthour which is in direct relation to the amount of active materials is higher for a nickel cadmium battery than for a lead-acid battery.

The imporatnt thing is however how the active materials are utilized or how the battery is sized to match the requirements of the PV installation. This is where the characteristics of the different battery types must be considered. It is here that most system designers make their fatal errors because of ignorance as referred to in the beginning of this presentation. As explained lead-acid batteries have to be oversized due to restriction in DOD, low temperatures, state of charge (sulphation), aging etc. A nickel cadmium pocket plate battery can be utilized to 100% DOD, is not affected by over- and undercharging and has better performance at low and high temperatures.

There has not been available in the industry a computer sizing model for PV systems including the salient characteristics for batteries and including state of charge, charging efficiency, temperature derating etc. SAB NIFE has developed a new computer model for sizing stand alone PV systems.

The program covers the following areas.

The metrological station data:

- Location name;
- Latitude;
- Longitude;
- Altitude;
- Insolation;
- Maximum and Minimum temperature.

The battery data consists of:

- Electrolyte temperature (-40°C to +50°C);

- Capacity derating relative to temperature;
- State of charge (0-100%);
- Internal Resistance relative to temperature and state of charge;
- Charging efficiency relative to temperature and state of charge;
- Charging voltage relative to temperature and state of charge;
- Discharge time and available capacity;
- Cell type and rated capacity (100 h to 1,2 V per cell).

The panel data consists of:

- IV-curve;
- Maximum power point;
- Number of cells in parallel and series;
- Reference temperature;
- Normal operating cell temperature;
- Cell area;
- Voltage coefficient;
- Current coefficient;
- Curve shape factor;
- Series resistance.

The SAB NIFE sizing program differs from other avialable programs on the following points:

- Capacity, temperature derating.
  The program derates the available capacity at every calculated value. This is sepcially important in cold climates. Some programs include a derating with regard to the expected absolute minimum temperature. This is done mainly to avoid freezing of discharged lead-acid batteries. They cannot take into consideration that the load can be smaller when it is cold.

- Charging efficiency.
  The program uses charging efficiency values related to electrolyte temperature. This is important when the temperature goes above 30°C when the charging efficiency decreases mostly due to "self discharge" at state of charge 60 - 100%. This is also important below 10°C when the charge acceptance for lead-acid batteries is reduced (the reduction depending on the magnitude of the charging current).

- Charging voltage.
  The program uses a charging voltage that is both temperature and state of charge correlated which is important when the system is used in temperatures below 15°C and places with low insolation. There are some programs that use a temperature coefficient one for lead-acid and one for nickel cadmium. But this does not include the dependence of state of charge.

8. CONCLUSION

Probably the worst application for a lead-acid battery is in a

PV installation because it is exposed to:

- Low charge and discharge currents;
- Temperature extremes;
- Harsh environment;
- Operating at low state of charge;
- Daily cycling.

All of which are eternal parameters affecting failure mode.

Industry awareness of the benefits of using nickel cadmium pocket plate batteries in photovoltaic installations has been very poor. The nickel cadmium battery can cope with the characteristics of PV installations and it is a viable and less-costly alternative in a properly sized total photovoltaic power system.

REFERENCES

1. Starr, M.R. Proceed. UK ISES Meeting C41, "Applications of PVs" 1985.
2. Newham, M. "Photovoltaics, The Sunrise Industry", Financial Times Business Information Ltd., October 1986.
3. Voss, E. Dr. "Testing of Batteries in Germany", Electric Vehicle Progress, Nov. 1, 1985.
4. Brooks, W. "Cold temperature testing of Absolyte". AAR - C & S Division Western Regional Meeting. Fort Worth, Texas, Sept. 5-7, 1986.
5. Eltra Corp. C & D Batteries Division "Stationary Batteries Section 12-610" 1979.
6. Bode, H. "Lead-acid batteries" The Electrochemical Society Series, 1977.
7. Berndt, D. Dr. "Using batteries in solar power" Middle East Electricity, Vol. 7, No. 9, October 1983.
8. Eurobat, "Guide to pocket plate nickel cadmium batteries". Association of European Accumulator Manufacturers, Alkaline Accumulator section, Bern, Switzerland.
9. Gutzeit, K. Dr. "Batteries for telecommunication systems powered by solar energy". Proceed. INTELEC. -86 Toronto, Canada, October 19-22, 1986.
10. Blohm, R. "Selecting and sizing stationary batteries", Proceed. The Power Sources Users Conference,October 15-17, 1985.

Chapter 12

PHOTOVOLTAIC CONCENTRATION

A. LUQUE
Universidad Politechnica De Madrid, Spain

1. INTRODUCTION

While solar cells constitute today the best electricity source for scattered applications including many solutions for rural electrification, present technologies do not permit cost competitive production of centralized electricity. After some fifteen years of intensive research contemplating several solutions today the conviction grows among an increasing number of experts (among which ourselves), that concentration will be the first solution to permit centralized cost-competitive electricity in countries of high direct insolation.

The reasons for this is that while efficiencies of 20% with silicon solar cells were forecasted (in 1976 by the US administration) as the achievable goals for 1986, actual values of 28.3% have been effectively measured, unlocking the way for high efficiency photovoltaic generation. As high efficiencies (at least of 14% even with zero module cost) have been recognized as a necessary condition for centralized electricity production, that no other approach can give today (at least at low cost) it is concluded that concentration is the only viable solution for PV. electricity production in the medium term.

From the beginning concentration has been an option for the use of photovoltaics: the first photovoltaic device of which we have notice is a Pt/Se Shottky barrier device (1) that operated under concentrated sunlight in 1876. More recently concentration has been a preferred way of reducing the cost of photovoltaic generators based in the lower cost of the optical surfaces used as collectors with respect to the solar cell area.

However photovoltaic concentrating approaches have not been successful up to now, if we measure this by its market share, nevertheless they experienced a promising growth in 1983-1984 (2) when based on a favourable tax policy in some states in the sunbelt of the USA where a photovoltaic conversion business started to develop.

The analysis of the past experience shows, that an inadequately selected level of concentration has been one of the causes of the failure of this business. The level of concentration usually employed in the past has been around 50 suns and the cost of the concentrating silicon cell has been at around \$ 0.5-1 per cm^2, some ten times the cost of a flat panel cell. In consequence a concentration of 50 only reduced by five the cell cost and the complexity of the concentrator has to be paid with this margin. The result has been that only modest price reduction has been offered by the companies involved in concentrating photovolatics.

Additional drawbacks associated with the concentrating solution: panel size of some few kW_p is, too big for the more common applications of several hundreds of W_p; inadequacy of a system with moving parts for remote operation and waste of the diffuse light, which is a sizeable portion of the available light in most climates. Yet we have to add the concurrence of the solar thermal approaches based on similar collecting concentrators but in - apparently - cheaper thermodynamical converters.

As stated before the present cost of concentrating solar cells is too high and can only allow its competitive electricity generation cost, if mass production of the cells in undertaken, or if very high concentration is used, to permit the higher cell cost to be reduced by this concentration. The physical limit of high concentration operation are described in this chapter as a framework to guide future experimental work.

It will be seen that while top efficiency cannot be achieved beyond concentrations of 500 suns for present silicon solar cells due to unavoidable physical limitations this top efficiency concentration rises to perhaps up to 3000 suns for silicon cells of novel design. For the operation at these high concentrations special attention is to be paid to the optics of the concentrators and to the thermal problems. Both factors seem to show that operation up to the value above mentioned is perhaps possible. At least concentrations of around 1000 suns are almost surely achievable.

The preceding approach is the most suitable for central production of electricity but it requires intensive research and is less adequate for applications where small amount of energy is required in scattered uses. For this case a simpler approach is to take profit from the substantial cost reduction achieved in flat panel cells, mainly of single or polycrystalline silicon. For it the low cost technology of these cells must be applied to the manufacture of concentrating cells by a procedure of optimization of the technological process without renouncing to any of the characteristics that render it cheap, as, for instance, the use of screen printed contacts or polycrystalline substrates. The level of concentration at which reasonable efficiency is to be obtained is not too high, with all probability nearly below 40 suns; neither the efficiency can be expected to be very high. In consequence the concentrator to use with this approach must be very cheap, and again this fact makes appeal to a careful optical design that allows for a simplified manufacturing.

Yet the preceding application is suitable for cases in which a strong direct insolation is to be expected. For situations in which the solar resource contains a strong diffuse component or when moving parts are not acceptable, stationary concentrating panels collecting most of this diffuse light is a solution that can also be offered. The moderate value of the concentration that can be achieved, that is reduced to some 6 suns and therefore the reduction of cost achievable with this approach is moderate.

2. CONCENTRATING SOLAR CELLS

2.1 Different aspects of concentrating solar cells

Conceptually, concentrating solar cells are very much similar to flat panel crystalline cells. However, its lower weight in the total cost of the generator justifies, to a certain extent, the refinement of its fabrication techniques so that a higher efficiency is usually sought. This approach is economically advised by the convenience of increasing the overalll generator efficiency. As a consequence concentrating cells often present a behaviour closer to the ideal behaviour that results from the application of simplified models. Nevertheless some aspects of the concentrating cell behaviour merit a special mention in this chapter. These are mainly the high injection behaviour in base, the series resistance and the thermal behaviour. Here we devote some lines to them. Other aspects of special interest in concentrating cells are those related to the achievement of very high efficiencies. These will be treated in more detail later.

Variable injection in base

In the cell base, under concentration, high injection conditions are often found as well as low injection ones, depending on the base doping on the level of concentration and on the operating point in the J-V curve. Although the study of semiconductor devices in variable injection cannot be considered a novelty it is often presented in complicated form. At our Institute we have developed a quasi-analytical variable injection model that allows for an intuitive comprehension of the mechanisms involved. It is included in the book by Luque and collaborators (3). Previously a very high injection model was developed (4) and was combined with the low injection models in a comprehensive review of the behaviour of bifacial solar cells (5).

In all cases the continuity equation, jointly with the description of the minority currents as a function of the minority carrier density and of the electrical field in a one-dimensional base is the origin a second order differential equation for the profile. The value of the electrical field, when relevant, is obtained from the total current equation, as a function of the majority and minority carriers densities, and the latter is related to the former by the assumption of charge neutrality in the whole base, which is quite accurately fulfilled in the operational conditions of interest. The boundary conditions for this differential equation are the minority carrier densities at the base extremes.

The differential equation so obtained is linear in very low or very high injection and not at intermediate injection levels, but it can still be approximately solved using superposition arguments as in a linear differential equation in which the equation coefficients, now concentration dependent, are taken as independent with values that are only correctly assigned once the solution is known using an iterative procedure.

The voltage is separated into four terms: voltage at the p-n junction, voltage at the homopolar junction (or high-low junction, either nn+ in pnn+ cells or pp+ in npp+ cells) Dember voltage and Ohmic voltage (both in the base volume, due to electric field there) which are added together. The expressions of these terms are (in a npp+ cell) the following functions of the base-end minority carrier densities n_u (at the pn junction) and n_H (at the homopolar junction):

$$V_U = \frac{kT}{q} \ln \frac{n_u N_s}{n_i^2} \tag{1a}$$

$$V_H = \frac{kT}{q} \ln \left(1 + \frac{n_H}{N_s}\right) \tag{1b}$$

$$V_D = \frac{kT}{q} \frac{\mu_n - \mu_p}{\mu_n + \mu_p} \ln \frac{n_H + \mu_p/(\mu_n + \mu_p)}{n_u + \mu_p/(\mu_n + \mu_p)} \tag{1c}$$

$$V_R = - \int^{W_s} \frac{J\, dx}{q(\mu_n n + u_p p)} = - J\, r_s \tag{1d}$$

It is to be observed that in high injection a substantial part of the cell voltage is built-in at the homopolar junction, contrarily as what occurs in low injection. This makes the cell voltage more insensitive to the base doping than if the low injection model is used, in which case a strong decrease is predicted for high resistivity bases; we shall discuss this point later again.

The total current at the base end by the p-n junction is the base minority current, calculated from the base density profile plus the emitter minority current at the emitter end if the net generation at the space charge zone is not considered. The emitter minority current is calculated taking into account the complexities of heavily doped zones, but these complexities do not affect to the J-V law that is a diode equation with an illumination term J_{EUL} and a dark saturation current J_{EUO}. As it is usual the saturation current multiplies an exponential function of the quasi-Fermi level splitting at the neighbouring base-end divided by the thermal voltage. The terms J_{EUL} and J_{EUO} include all the complexities of the heavily doped zone.

At the homopolar junction the continuity of minority currents is required. For it the minority current at the corresponding base end is equated to the minority current at the heavily doped region forming the homopolar junction, the latter given by an illuminated diode law with constants J_{EHL} and J_{EHO}.

As the quasi-Fermi levels splitting $_F$ at the base-ends are related to the minority carrier density (in n+pp+ cells) n_u and n_H at these base extremes by

$$n\,(n+N_S) = n_i^2 \exp \frac{q\ _F}{kT} \tag{2}$$

the cell generation-recombination currents at the emitters can be written as a function of n_u and n_H. Indeed the same can be done for the current base terms and the two relationships described above, at the zone ends, gives the following couple of equations

$$J = (J_{EUL}+J_{BUL}) - [J_{EUO}(\frac{n_u}{N_S}+1) + (J_{EFO}-J_{BEU})]\frac{n_u N_S}{n_i^2} + J_{CRO}\frac{n_H N_S}{n_i^2} \tag{3a}$$

$$O = (J_{EHL}+J_{BHL}) - [J_{EHO}(\frac{n_H}{N_S}+1) + (J_{ERO}+J_{BEH})]\frac{n_H N_S}{n_i^2} + J_{CFO}\frac{n_u N_S}{n_i^2} \tag{3b}$$

where the constants are given in Table 1 for the case of non strongly variable density profiles (a piece-wise calculation is required in the opposite case). An illustrative drawing of the minority carrier density profiles and of the current components appearing in the preceding formulas are given in Figure 1.

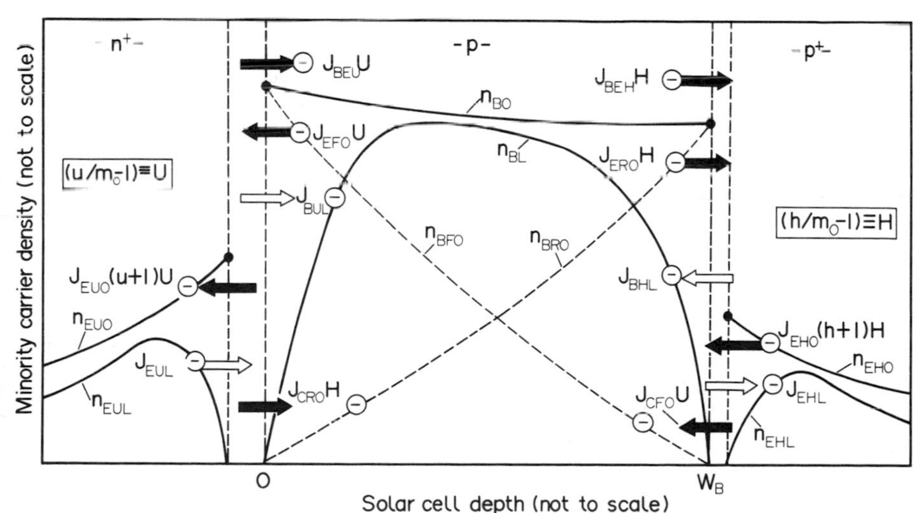

Fig. 12.1. Minority carrier densities and currents in a solar cell.

Table 1. Expressions for the current constants in the base

$$J_{EFO} = q\frac{D}{L_S}\frac{n_i^2}{N_S}\left\{\frac{\cosh(W_S/L_S)}{\sinh(W_S/L_S)} + \beta L_S\right\} \quad ; \quad J_{ERO} = q\frac{D}{L_e}\frac{n_i^2}{N_S}\left\{\frac{\cosh(W_S/L_S)}{\sinh(W_S/L_S)} - \beta L_S\right\}$$

$$J_{CRO} = q\frac{D}{L_S}\frac{n_i^2}{N_S}\frac{\exp(B\,W_S)}{\sinh(W_S/L_S)} \quad ; \quad J_{CFO} = q\frac{D}{L_S}\frac{n_i^2}{N_S}\frac{\exp(-\beta W_S)}{\sinh(W_S/L_S)}$$

$$J_{BEU} = q\frac{n_i^2}{N_S}\mu_m E_O \quad ; \quad J_{BEH} = q\frac{n_i^2}{N_S}\mu_m E_O$$

$$J_{BUL} = q\,D\,\Sigma\lambda\,\{N^F ph\lambda\ \emptyset(\alpha\lambda,\beta) + N^B ph\lambda\ \exp(-\alpha\lambda\ W_S)\ \emptyset(-\alpha\lambda,\beta)\}$$

$$J_{BHL} = q\,D\,\Sigma\lambda\,\{N^F ph\lambda\ \exp(-\alpha\lambda\ W_S)\ \emptyset(-\alpha\lambda,-\beta) + N^B ph\lambda\ \emptyset(\alpha\lambda,-\beta)$$

$$\emptyset(\alpha,\beta) = \frac{\alpha\,\tau}{\alpha^2 L^2 - 2\alpha\beta L^2 - 1}\ \frac{(\alpha - L_S)\sinh(W_S/L_S) + \exp[(\beta-\alpha)W_S] - \cosh(W_S/L_S)}{L_S\sinh(W_S/L_S)}$$

$$D = \frac{(p+n)\,D_n\,D_p}{D_n n + D_p p} \qquad \mu = \frac{(p-n)\,\mu_n\,\mu_p}{\mu_n n + \mu_p p}$$

$$E_o = \frac{J}{q(\mu_n n + \mu_p p)} \qquad E_o = -\frac{kT}{q}\frac{\mu_n - \mu_p}{q(\mu_n n + \mu_p p)}\frac{dn}{dx}$$

$$\tau = \frac{\tau_p n + \tau_n p}{p_o + n_o + n'} \qquad n' = n - n_o,\ n_o \text{ is the equil. dens.}$$

$$\beta = \mu E/2D \quad ; \quad L^2 = D\tau \quad ; \quad L_S = (\beta^2 + 1/L^2)^{-\frac{1}{2}}$$

The normal procedure to obtain the J-V curve is to give a certain value of J and obtain n_u and n_H from Eqs. (3) that are subsequently introduced in Eqs. (1) to obtain the different voltages; these are then added together to obtain V.

However in good cells it is usually found (3) $n_u/n_H \cong 1$ and thus Eqs. (3a) and (3b) can be added together to obtain, for the very high injection case, an equation of the type

$$J = J_L - J_{01} \exp \frac{(V+J r_s)}{kT} - J_{02} \exp \frac{q(V+J r_s)}{2\ kT} \qquad (4)$$

where J_{01} and J_{02} are nearly constant. In general the term J_{01} is due to the recombination in emitters while the term J_{02} is based in the recombination in base; typical values for J_{01} and J_{02} in a good cell are 10^{-12} A/cm² and 10^{-7} A/cm² respectively; the term in J_{02} becomes negligible for V high enough but which term dominates in practical operation is a matter to be discussed further in this chapter (J_{02} may increase with the voltage due to Auger recombination).

In low injection cells a term in J_{01} is the only one appearing but in this case it includes also the recombination in base and therefore is always higher than J_{01} in high injection. The J-V equation is now

$$J = J_L - J_0 \exp \frac{q(V+J r_s)}{kT} \qquad (5)$$

The series resistance

The behaviour of a concentrating solar cell is strongly affected by series resistance. While a solar cell following a superposition of the illumination-proportional current $J_L = J_{LS}\ X\ F$ (J_{LS} is the short circuit current at one sun, in a cell without any metal grid, F is transparency factor or the fraction of cell area uncovered by the metal and X the concentration factor) and a voltage-increasing dark current $J_d(V)$ presents an efficiency that increases mononically with the concentration, the effect of a series resistance (that makes $J_d(V)$ to become $J_d(V+J r_s)$) produces a maximum of efficiency at a concentration X such that, approximately, the ohmic drop equals the thermal voltage (6), i.e. $J_{LS}\ X\ F\ r_s \cong kT/q$. This expression shows very clearly the importance of the series resistance in the cell behaviour.

The series resistance is strongly dependent on the transparency of the metallisation grid on the front of the cell. It is also dependant on its shape and on the cell size. Grids transparent produce strong series resistance but grids very dense produce excessive shadowing so reducing the current. In consequence an optimal grid density can be found for a given solar cell (7).

The specific series resistance, which is the series resistance times the cell area is the magnitude that multiplied by the cell current denisty gives the voltage drop. This specific series resistance, in cells of classical structure, contains three terms. One of them, r_E, is due to the lateral flow of majority carriers along the top layer, to find the metal fingers. This term takes the form

$$r_E = \frac{r_e}{(1-F)^2} \qquad (6a)$$

It is clear that the bigger this transparency factor the bigger the series resistance. The constant $r_e = (1/12)\ R_{se}\ w^2$ for all the cells having a rectangular basic grid (not for circular grids with a trapezoidal shape where the factor 1/12 is modified); R_{se} is the sheet resistance of the top semiconductor layer and w is the width of the metal finger. It is very clear that reducing the metal finger width can produce an unlimited reduction of series resistance. Also the dependence on the sheet resistance is shown in this formula.

The metal grid also gives a series resistance term expressed by

$$r_M = \frac{r_m}{(1-F)} \tag{6b}$$

where $r_m = (1/24)\ R_{sm}\ L^2$ with L the length of the side of the cell. The factor (1/24) is valid for a crossed grid structure as shown in Fig. 2 and other grids may have different factors. In the appropriate literature of r_M is given for different grid shapes (8). In all cases this term is reduced for the smaller areas.

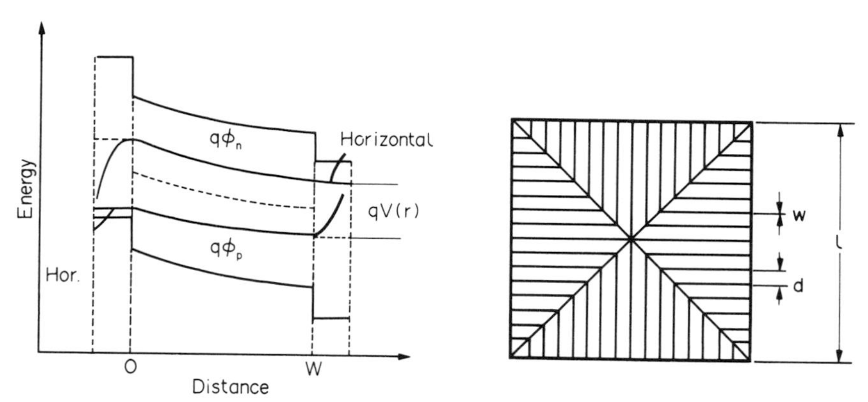

Fig. 12.2. Crossed metallization grid.

Finally a grid independent term r_s is caused by the ohmic drop when the current traverses the cell thickness. This specific series resistance, for a cell of constant base resistivity p_s, is

$$r_s = p_s\ W \tag{6c}$$

As mentioned before

$$r_s = r_E + r_H + r_s \tag{7}$$

is dependent on F and increases with it. Since the cell current can be written as

$$J = J_{LS} \times F - J_d\ (V+Jr_s) \tag{8}$$

an optimization can take place such that $\Gamma = J\ V$ is maximum with respect to V and with respect to F at a given X. If Γ is maximized also with respect to X then we obtain the highest efficiency achievable for a given cell family where its basic technology is defined but the grid is to be optimized for each concentration.

An expression, not very accurate but illustrative, for the solar cell efficiency, roughly valid at not very high concentrations is given by

$$n \cong n_o\ F[1 - \frac{r_s\ J_{LS}\ X}{V_{oc}}] \tag{9a}$$

with

$$n_o \cong \frac{J_{LS}\ V_{oc}}{E_{LS}}\ [1 - \frac{kT}{qV_{oc}}] \tag{9b}$$

where E_{LS} is the irradiance on the cell at one sun and

$$V_{oc} = \frac{kT}{q}\ \ln \frac{J_{LS}\ X}{J_o} \tag{9c}$$

that reveals the increase of the cell intrinsic efficiency n_o with X (due to the increase of V_{oc}) and the decrease when the series resistance and the transparency factor are considered. In Eq. (9a) it also appears the compromise between transparency and series resistance.

Silicon cells operating at very high concentration are limited by r_s while with GaAs cells this is usually not a limitation. However in the latter cells the series resistance tends to be limited by r_E due to the lower doping of the emitter top layer required for effective collection of the carriers generated near the surface.

A detailed treatment of the series resistance effects on solar cells can be found in the already mentioned book (8).

Thermal behaviour

In a solar cell whose equation is given by Eq. (5) the main non-explicit dependence on the temperature is to be found in J_{o1} that contains n_i^2 A exp-(Eg/kT) and therefore is strongly increasing with the temperature. A small increase of J_L is also found due to bandgap reduction but this effect is almost negligible (compared to the increase of J_{o1}). At constant J we find that

$$\frac{dV}{dT} \approx - \frac{Eg/q - V - Jr_s}{T} \tag{10}$$

Thus the cell voltage, at constant current, decreases with the temperature and so does the efficiency. This decrease is stronger when the voltage at the junction $V_J = V + Jr_s$ is farther from E_g/q.

This usually happens more in Si cells than in GaAs cells and because of it the latter tend to be less affected by the temperature. However the better a solar cell, so that V_j approaches to Eg/q, the less dependent it is on the temperature. The same can be said for operation under concentration that makes V_j to increase. The last is a fortunate circumstance because it is in concentrating cells where more heating usually occurs.

In addition Eq. (10) clearly shows that the points of lower voltage in the J-V curve are more affected by the temperature than those of higher voltage. In particular the maximum power point voltage is more affected than the open circuit point; accordingly it is customarily to affect this dependence to the Fill Factor. However we prefer to use Eq. (10) as it is because we find it clearer and more physically meaningful.

2.2 Theoretical limits of efficiency

An idealized solar cell is a device that absorbs all the photons with energy above the material bandgap Eg and convert them into electricity at a rate of one electron every one photon. The energy (voltage) at which this electron can be extracted is the bandgap energy: the same for the electrons generated by every absorbed photon no matter its energy. The photons with energy below Eg are not absorbed and therefore lost.

Using detailed balance arguments for the mechanism of radiative recombination, so linked to those of generation Henry (9) finds the J-V characteristic of this ideal solar cell as

$$J = J_{ph} - J_{th} \exp \frac{qV}{kT} \tag{11}$$

with J_{ph} the total flux of photons of the appropriate energy (above Eg) times q and J_{th} given by

$$J_{th} = \frac{2\pi q(n^2+1)Eg^2 kT}{h^3 c^2} \exp - \frac{Eg}{kT} = 5693\, Eg(ev)^2 \exp - \frac{Eg}{kT} \ (A/cm^2) \tag{12}$$

where n is the index of refraction of the concerned semiconductor. The numeric expression at the right is referred to T = 300° K. The open circuit voltage is given in Eq. (11) by

$$V_{oc} = \frac{Eg}{g} - \frac{kT}{q} \ln \frac{2\pi q(n^2+1)\ Eg^2\ kT}{h^2 c^2\ J_{ph}} \tag{13}$$

that for T 0, V_{oc} tends to Eg/q. This is the highest voltage at which we can extract current from the cell.

Equation (11) allows for establishing an upper limit of efficiency for solar cells as a function of Eg which is given in Fig. 3 (in accordance with a more recent discussion by Araujo and Marti (10). When this is done it is found that the best efficiency, around 44% for $E_g \cong 1.1$ ev is the top for any solar cell operating at around 46000 suns which is also the maximum concentration achievable in Earth, as we shall discuss later.

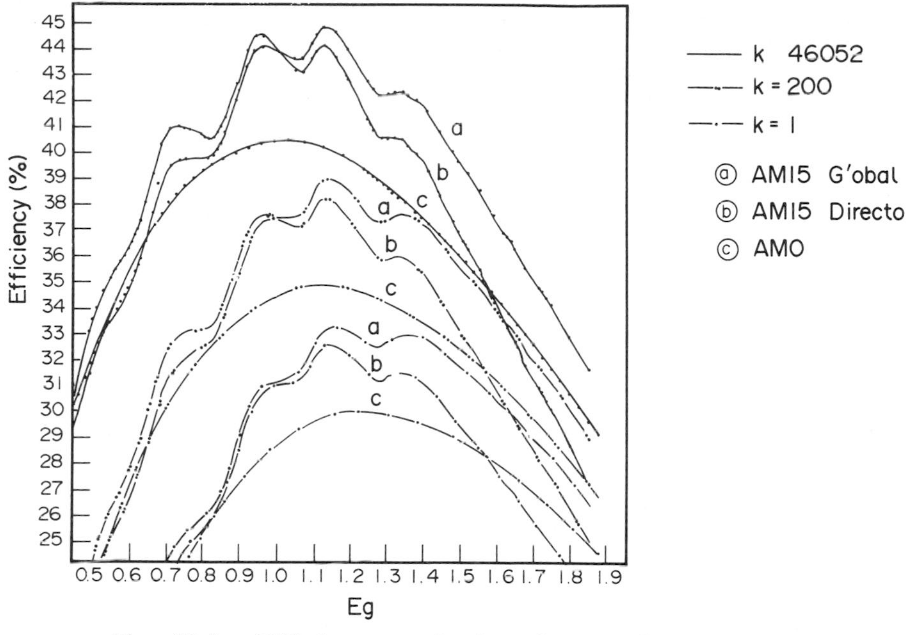

Fig. 12.3. Efficiency vs. bandgap for a cell with radiative recombination for AM0 and AM1.5 direct spectra and concentrations of 1, 200 and 46050 suns.

This bandgap is closer to that of Si (1.12 ev) than to that of GaAs (1.42 ev). The opposite is often affirmed and this is more correct at one sun but not under high concentration. Present practical efficiency for silicon (28.3% at 100 suns) is higher than that of GaAs (12) (26.1% at 500 suns) that seems to confirm the preceding statement, but it still requires a more critical discussion.

Another argument often used is that the higher the semiconductor bandgap the better cell temperature stability. We know, by Eq. (10), that dV/dT depends on $(V_{oc} - Eg/q)/T$ and thus

$$\frac{dV_{oc}}{dT} = \frac{k}{q} \ln \frac{2\, q(n^2+1)\, Eg^2\, kT}{h^2 c^2\, J_{ph}} \tag{14}$$

that shows that the smaller Eg the better the thermal stability (which is improved by the fact that J_{ph} also increases when Eg decreases).

In Si the condition is opposite: the semiconductor is of indirect gap and radiative lifetime is very long. Nonradiative mechanisms dominate. Finally the indirect nature of the gap permits an Auger recombination that becomes the fundamental mechanism, which is non-radiative. Thus the voltages are reduced with respect to those predicted by radiative recombination models and efficiencies and thermal stabilities are reduced. The reason of Si to have a better practical behaviour today is, in my opinion, the today's better understanding of its device physics and technology. Actually for 1000 suns the Si cell efficiency upper limit is to be considered (11) close to 31-32% due to limitations by Auger recombination instead of the thermodynamic (9) 37-38% to be considered for GaAs.

To insist more in the preceding arguments which are, I think, very important conceptually, Si cells, because its indirect bandgap, combine this long radiative lifetime with a weak absorption: both features are linked by the detailed balance principle. Therefore thick Si cells are commonly used as a consequence of the last characteristic and this enhances the role of nonradiative recombination which is proportional to the cell volume. On the other hand GaAs cells links the strong absorption with a short radiative lifetime, but this is not a hindrance to the cell efficiency because it merely implies that the cell can be done thin and the nonradiative recombinations can be reduced greatly. If we make the cell thicker Eq. (12) which is thickness independent implies that the radiative lifetime increases: the detailed way in which this occurs is through self-absorption of the photons generated in the radiative recombination and thus the effective lifetime is enlarged (keeping J_{th} constant) but in this case it starts to be overreached by the shorter radiative lifetime.

Best cells for concentration are thus cells with the bandgap value found in silicon but with direct bandgap: some III-V alloys present this property and if they are made thin enough as to avoid nonradiative recombination while keeping total light absorption then they can present the best efficiency for a single bandgap cell.

Tandem - Cascade Cells

One important drawback of the idealized solar cell is that it extracts effectively its energy only to the photons with energy near the bandgap. Thus important efficiency improvements must be expected from system in which more than one semiconductor is used. In particular the use of two or more bandgaps can be implemented by using a series of filters in the optical path that transmit certain wavelengths and deflect the remaining ones towards the cell of appropriate gap as presented schematically in Fig. 4 (beam splitting scheme). Another possible approach is to stack the cells from the one with higher bandgap to the one of lower bandgap so that the light is partly absorbed in the top semiconductor, permitting the lower wavelengths to go towards the remaining layers below as shown also in Fig. 4 (stacking scheme). A monolithic stacking can be done in the last case but a more realistic implementation of this concept, at least for the shorter term is presented in Fig. 5 with an adhesive to stick the cells in the stacking.

If every cell is connected in series with the next one as in Fig. 4b (two terminals approach) then the total voltage is the sum of the voltages. To avoid losses in such device the current extracte

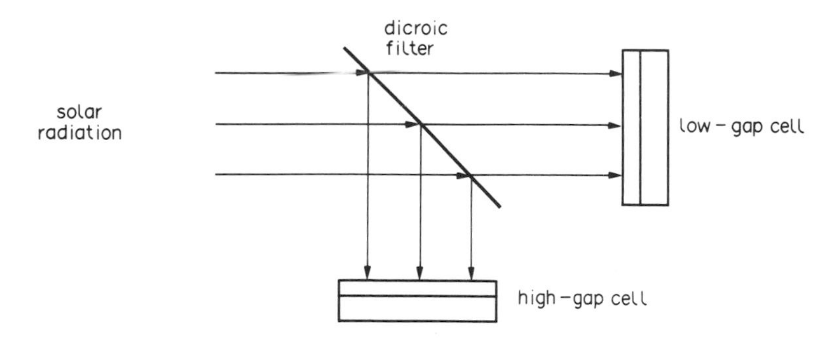

(a): Spectrum splitting

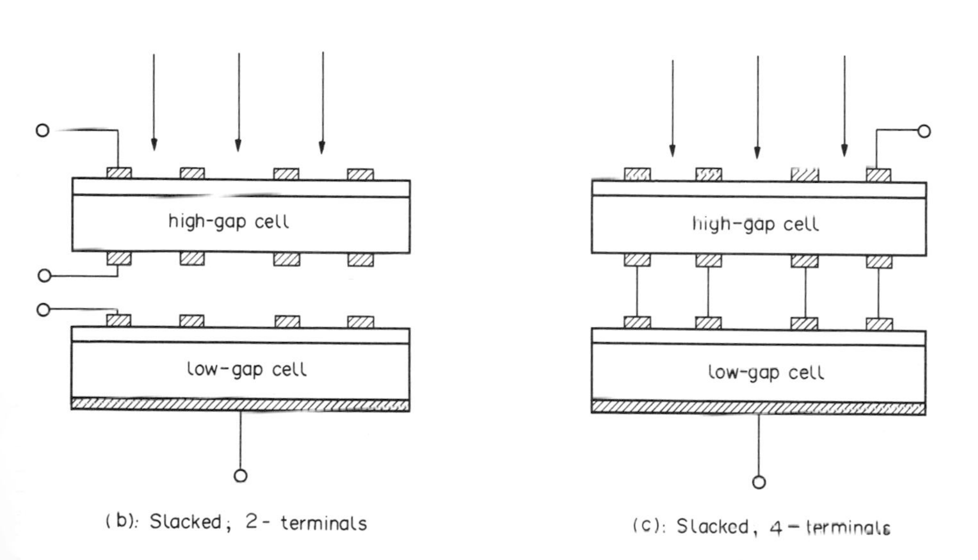

(b): Stacked; 2- terminals

(c): Stacked, 4 - terminals

Fig. 12.4. Different configurations for multiple bandgap cells: (a) beam splitting, (b) four terminal stacked, (c) two terminal stacked.

from each cell must be exactly the same and this places a restriction in the choice of the bandgap of each cell. In the opposite approach every cell has its two terminals and is matched independently to a load as in Fig. 4c. In theory this approach allows for a slightly better efficiency. The thermodynamic limit for 2, 3, and 36 bandgaps at 1000 suns (9) is respectively 50%, 56% and 72%. Because of its great complexity they do not give today higher efficiencies than simpler schemes. Best efficiency to date with two cells is 28.3% but the achievement of efficiencies around 40% seems a totally realistic goal for practical devices in the next century.

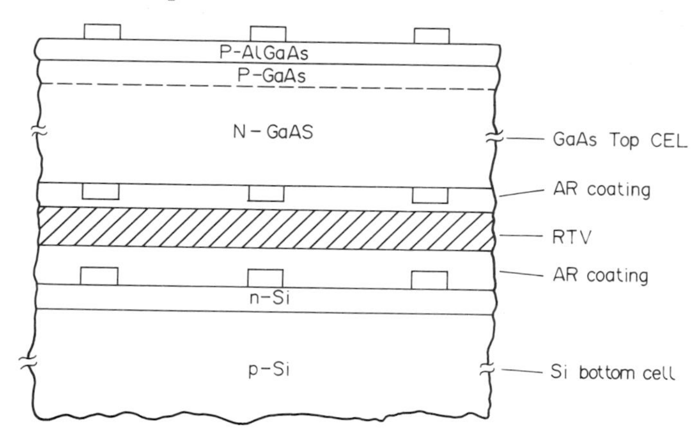

Fig. 12.5. Stacked configuration of GaAs and Si cells with an adhesive in between.

2.3 Light confinement

As we have stressed the reduction of the cell thickness is of paramount importance but when making cells thin light tends to be lost by transmission. Means to avoid it and the ultimate limits of these techniques are described in this section.

A first obvious idea is to place a mirrored surface in the back and this will produce a second passage of the light through the cell so doubting its effective thickness. However additional strategies are yet possible. If the front surface is roughened so as to produce the incident ray to be refracted into a random direction (and how to fully achieve this is not clear) then the rays travel obliquously through the cell thickness and its effective path is increased. For weakly absorbed rays, in average, it is doubled. Then these rays are reflected back and travel in average twice more the cell thickness, i.e. in total a path of 4 W (W is the cell thickness) and reach the front surface with a random direction. At any element of surface only those rays within the cone bounded by the critical angle $\sin^{-1}(1/n)$ will be able to escape the cell, the rest will be sent back by total internal reflection. The fraction of rays lost in this reflection is only $1/n^2$: in silicon 7%; all the rest of the rays make an additional trip back to the rear mirror. This process is repeated indefinitely.

The consequence is that the light becomes confined between the back mirror of reflectivity close to unity and the front textured surface of reflectivity $(1-1/n^2)$. Because of it the absorption, for weakly absorbed photons is enhanced up to $4n^2$ (i.e. 55) times the absorption produced in a single passage of the light and so is the generation of electron-hole pairs (13). The cell thickness can thus be greatly reduced for almost total capture of this weakly absorbed light.

Note that when placing the back mirror, the simple deposition of a metal layer result in a quite poor mirror due to the high index of refraction of the semiconductor but if placing a thin layer (of precise thickness) of a dielectric in between the reflectivity can be excellent.

Some geometrical engravings (14) as well as the random texture often made on flat panel cells can produce almost full randomization. This randomization is not incompatible with the double incidence produced by the texturization.

Actually the enhancement of the absorption caused by the light confinement has a similar effect than the one caused by concentration and has the same physical limit. If the light comes from a restricted cone of rays, as it happens if it comes from the sun without any concentrator, and we are able to build a preferential coating (we do not know how) that collects all the rays from this cone and none from the outside then the light absorption can be still enhanced in a factor $1/\sin^2\phi$, ϕ being the semiangle of this cone (15,16). For the solar angular diameter this factor is around 46747. As a general rule the more is achieved in light confinement (beyond the factor $4n^2$) the less can be achieved concentration; both effects cannot be achieved simultaneously beyond a certain limit.

In the case of the thermodynamically limited cell studied above the term (n^2+1) appearing in Eq. (12) is changed into $\sin^2\phi$ (10), so reducing the recombination current but then the maximum concentration that can be included in the term J_{ph} is also reduced in $\sin^2\phi$. Thus it is found here that the same limit of efficiency is obtained with an ideal concentrator or with an ideal light confiner.

A practical approach for restricting the angle of interchange of photons with the exterior is to locate the cell inside a non-absorbent cavity, that keeps the radiation confined at the cell neighbourhood (17) and communicated with the exterior through an aperture smaller than the cell. Again in this case enhancing confinement reduces concentration and presents similar efficiency limits (16).

Therefore optical treatments can greatly enhance the light confinement, permit very thin cells with almost total capture of radiation and thus increase the cell efficiency towards its theoretical limits. Obviously in practice a balance between performances and cost incurring complexity must be sought.

The University of New South Wales cell

Among the cells of classical structure the cell developed by the Martin Green's team, at the University of New South Wales has achieved the best efficiencies. Its schematic is presented in Fig. 6a.

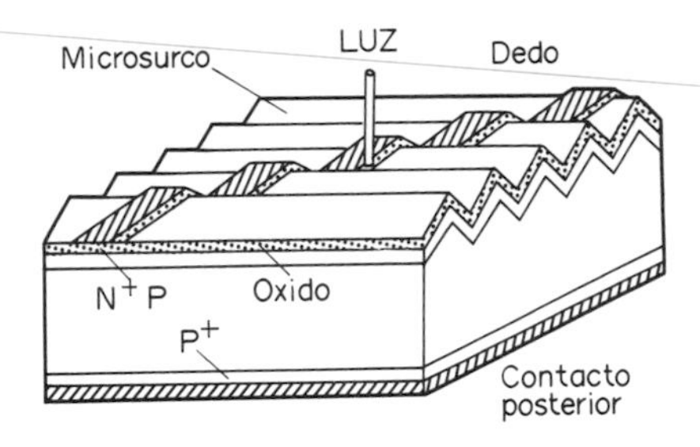

Caracteristica A I sol

	0.1 Ω^{-cm}	0.2 Ω^{-cm}
A (cm^2)	0.09	0.09
Jsc (MA/cm^2)	38.8	40.2
Voc (v)	0.663	0.651
FF	0.836	0.825
η(%)	21.5	21.6

Fig. 12.6(a). University of New South Wales UNSW cell showing the wedged surface and metal connections through small holes.

The emitter has been fabricated following, probably, the philosophy described above. High quality WASO Wacker wakers are used for its manufacture and thinned to about 100 μm to further reduce recombination and a high quality Al doped high low junction is formed at the cell back and a good reflecting layer is deposited there.

On the top are grc ves as those represented on the figure provide a second chance to the reflected light to enter into the cell. This fact, jointly with the use of an antireflecting layer provides almost total (98%) light transmission into the cell. The grooves also provide an almost total randomization of the light so that most of it is captured in spite of the small base thickness.

The metal fingers on the cell top are delineated with a 45° angle with respect to the grove direction so that the light reflected in these metal fingers is cast on the grove wall and has a chance to be absorbed. In this way the shadow created by the metal fingers is almost totally avoided and the trade off between shadowing and series resitance is no more the same permitting a much denser grid and so reducing the series resistance. In addition the lateral path of the current along the emitter to reach the metal fingers is enlarged (as the surface is enlarged by the groves) and the emitter resistance reduced.

Finally the metal is contacted to the silicon through small holes so that most of the surface, including a sizeable part of the area under metal, is passivated with a dry silicon dioxide layer to allow for small surface recombination, Fig. 6b.

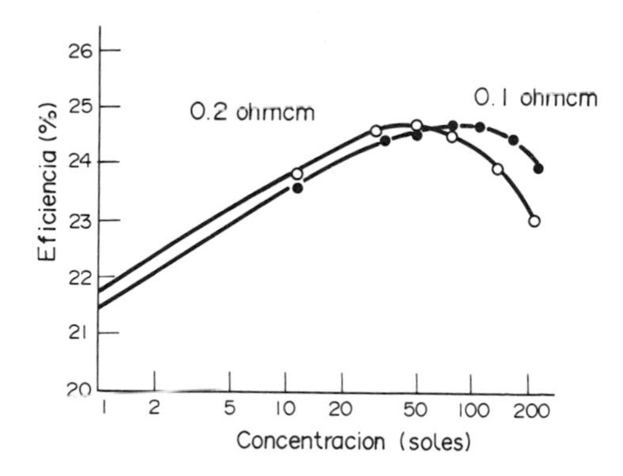

Fig. 12.6(b). Efficiency vs. concentration for the UNSW cell.

One drawback of this structure is the enlarged emitter area implied by the groves that increases the recombination but this drawback is well balanced by the mentioned advantages of this structure. Another drawback is associated to ability of metal fingers to reflect the incoming rays onto the groves. This feature that is very clear for normal incidence of light can fail for incidence in certain directions. As concentrators produce incidence of rays in a more or less wide angle it results that some rays can be reflected out of the metal fingers and the light transmission is somewhat reduced.

The results are given in the table included in the figure and we see that efficiencies above 25% have been achieved at concentrations of 30-80 suns.

Stanford University point contact cells

The highest efficiency so far achieved with a single solar cell is 28.3% and has been achieved with a cell of revolutionary structure developed by Richard Swanson and co-workers at the Stanford University, whose basic features are presented here.

A schematic drawing of the cell is presented in Fig. 7. Two are the basic features which are different in this cell. One is the fact that all the emitters, of p+ and n+ type, are formed on the cell back side so that no shadowing is produced in the cell front. The other is that the emitters are reduced in area to that of small dots so that the recombination in them is greatly reduced.

Celula de Si de concentracion

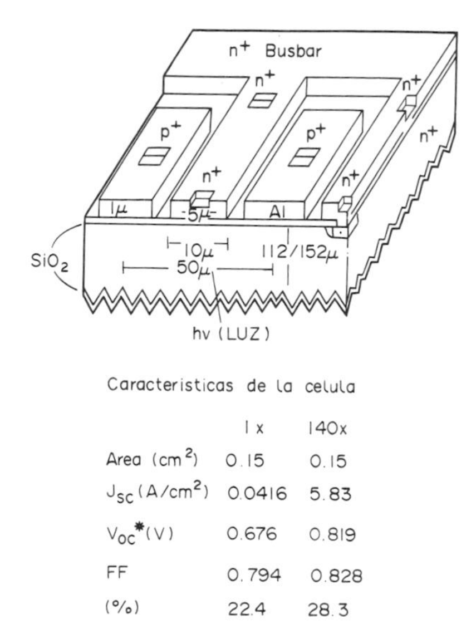

Caracteristicas de la celula

	1x	140x
Area (cm^2)	0.15	0.15
J_{sc} (A/cm^2)	0.0416	5.83
V_{oc}* (V)	0.676	0.819
FF	0.794	0.828
(%)	22.4	28.3

Fig. 12.7. Stanford University (SU) Back Point Contact (BPC) cell.

With regard to the location of the emitters in the back of the cell the antecedent is, to our knowledge, that of the works of Schwartz and co-workers (25), who developed the cells with interdigitated back p+ and n+ emitters located at the back face. The drawback of all these structures is that the carriers are to be collected at the face opposite to the front face where they are mostly collected but this has proven not to be a big drawback, not only by the results presently achieved but also by the fact that bifacial cells, able to be illuminated on either face, have been successfully developed by us, and commercialised, using cell thicknesses similar to those of conventional cells (320 μm is typical for commercial bifacial cells) and back quantum efficiency of about 90% of its front value, while using commercial Cz wafers (26) of standard quality (98% is found in Fz wafers (27)). Therefore, the back collection of light is not a major problem unless we operate at very high concentration as we shall see later.

The Stanford University cell uses a high resistivity (its value is not important; ≅150 Ωcm) FZ high lifetime base. In this cell this is a better choice than the low resistivity base used also in conventional cells, because although it gives lower voltage, permits a better transference of the carriers to the back side (this is also necessary in bifacial cells as the industrial experience confirms). On the other hand we shall see that even in voltage the results are outstanding due to the drastic reduction of emitter recombination achieved with this structure.

Also in this case the base is made conveniently thin, up to some 100 μm, to reduce base recombination, and a regular array of pyramids is patterned in the front face to produce double incidence of light (as it occurs in normally textured surfaces) and AR coated, so that almost total light transmission is produced. Then back contacting layers produce a good light reflection at the back so that almost isotropical (optimal) light confinement is produced in this structure. The whole front and back surfaces are covered with a thin layer of passivating silicon dioxide that also acts as AR coating for the front face and as an optical matching for the back mirror.

As no trade-off with shadowing is needed here the interdigitated metallisation at the rear surface is so dense that the series resistance is negligible.

Table 2. Performance of several high efficiency measured at one sun

Institution	SU	UNSW*	UNSW	UPM	UPM
Area (cm^2)	0.15	.09	2.00	2.00	2.00
Base resistivity (Ωcm)	150	0.2	0.2	21	0.33
J_{SC} (mA/cm^2)	41.6	40.2		38.9	36.6
V_{OC} (mV)	676	651		623	645
Fill Factor (%)	79.4	82.1		78.4	80.9
Efficiency (%)	22.4	21.6	20.6	19.0	19.1

All measured by Sandia at AM1,5 global spectrum except * measured by SERI

SU Stanford University; UNSW University of New South Wales; UPM Universidad Politecnica de Madrid

The results are given in Table 2 and in Fig. 8. We can observe that at one sun the short circuit current is higher than in Green cells mainly because of the use of a very high resistivity base with higher lifetime; also the opencircuit voltage is higher due to the reduction of recombination by reducing to a small volume the heavily doped regions (emitters). The efficiency is thus higher than in the Green cells and reaches 28.3 at 140 suns.

We have seen that the most effective way to reduce the emitter recombination is to reduce its volume and this can be done by reducing its size to that of small points densely packed. Such a technique can be applied to the back side only but also to the front side or to both sides and connect these emitters to metal

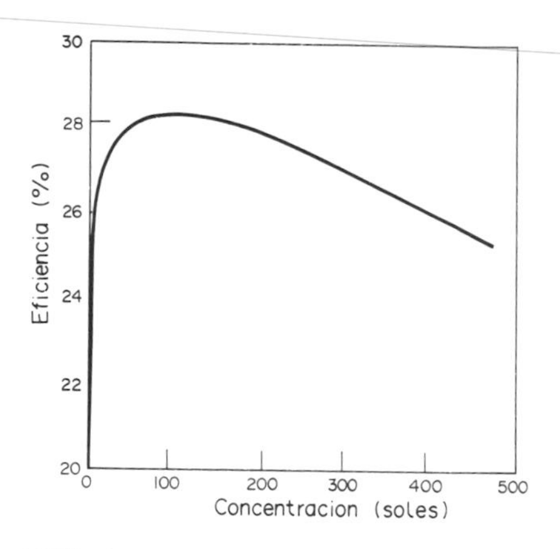

Fig. 12.8. Efficiency vs. concentration for the SU-BPC cell.

fingers or layers isolated from the cell base by a passivating oxide; these points cover only a fraction of the cell surface. We present in Fig. 9 several schematic of a structure with point shaped emitters in both faces.

When the surface is placed in contact with the high density of minority carriers existing in a lowly doped base, a strong recombination can take place there. The problem is more serious than the recombination in surfaces of emitters that, because of their high doping, present very small minority carrier densities. It is clear that reducing the emitter coverage implies to uncover the base surface and the additional recombination caused cannot exceed the reduction of emitter recombination.

Two methods (11) can be followed to reduce the surface recombination: one is to reduce the density of surface states (28) by oxidizing the surface or by other techniques (densities down to 5×10^{9} cm^{2} have been observed to compare to the 10^{14} cm^{-2} in a bare surface); the second method is to reduce the density of minority carriers in the cell so reducing the recombination rate.

If a high density of majority carriers is added to the surface then the minority carrier density is reduced accordingly. A procedure for it is to make a very shallow emitter. Heavily doped zones have often been used as surface passivating regions. The use of these shallow emitters require an already reasonably well passivated surface because otherwise (as it happens when a metal is at the surface) the emitter must be thick.

However the best way, I think to produce a surface passivation based on minority carrier reduction is by adding electric charge to the cell surface so as to produce a strong induced junction (29). A method to achieve this has been developed by Hezel and co-workers (30).

Phenomena of bandgap shrinkage are likely to be present in the

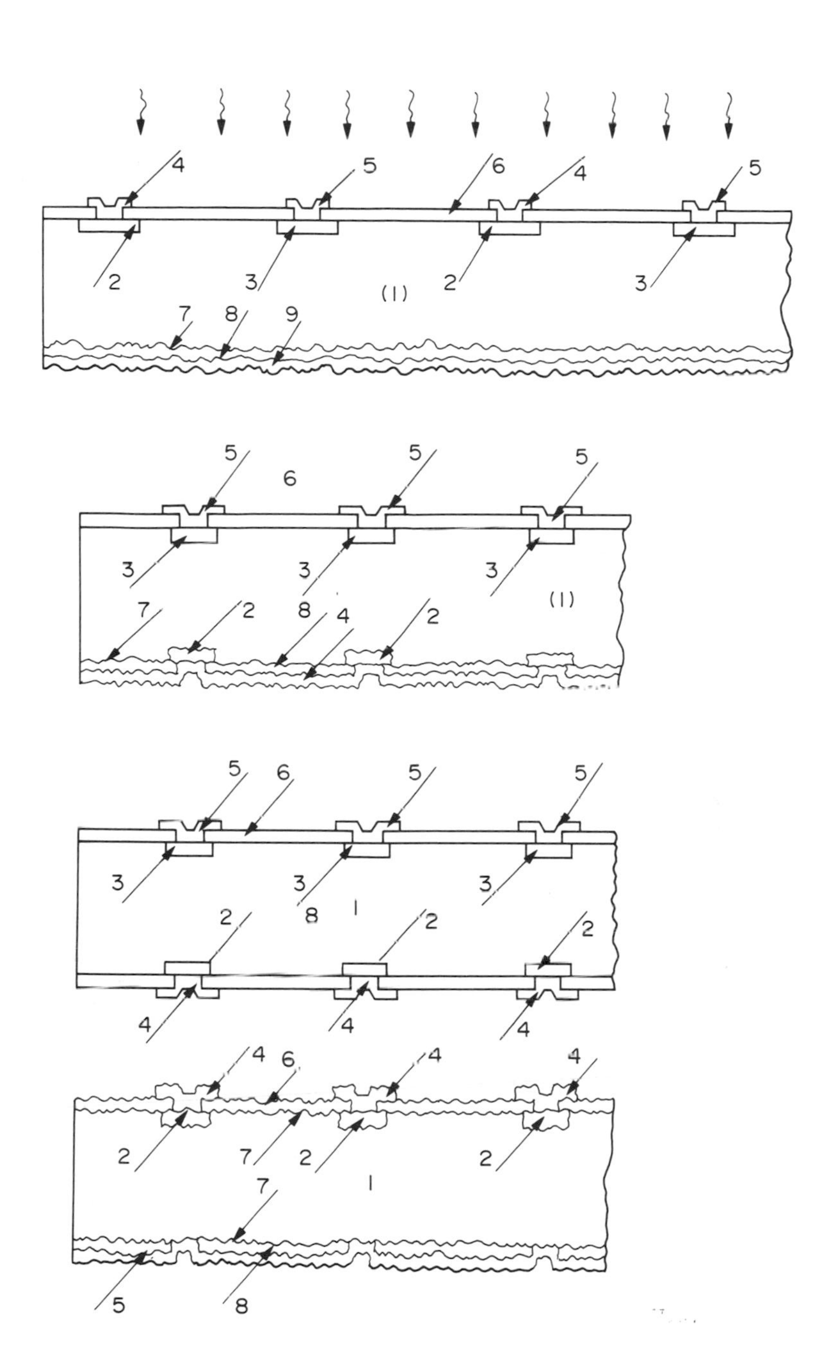

Fig. 12.9. Schematic of several point contact cells.

minority carrier density of this induced junction, in this case due to the sole influence of exchange terms (i.e. weaker than when caused by heavy doping) so that any charge above a certain level will probably saturate the minority carrier density. Theoretical calculations of exchange terms in Si and GaAs are given in the literature (31).

With either method: reducing the surface density of states or reducing the minority carrier density, or with both, the base surface recombination can be decreased to a level below the Auger recombination.

The use of point shaped emitters cannot be done without some harm. As matter of fact a density well appears around every point shaped emitter that is necessary to cause the carriers to flow towards the point shaped emitter. The height of this well is given by (29)

$$p - p_{U,H} \cong \frac{Jr_{p+,n+}}{4fqD_{p,n}} \qquad (16)$$

where p_U or p_H are the minority carrier densities (the base is n in this case, although this is irrelevant in a first approximation) at the pn and homopolar junctions respectively and p is the density at the end of the density well. Actually a link between p_U and p_H is produced by requiring n to be the same at the top of the well at the n-p and homopolar junction vicinity. The densities p_U and p_H are directly related to the voltage at the pn and the homopolar junction respectively, and thus to the voltage applied to the cell while the recombination in the cell depends on p outside the well. Thus the effect of the density wells is to increase the recombination so decreasing the current at a given voltage.

In Eq. (16) r_{p+} or r_{n+} are the radii for the n+ or p+ dots and f is the coverage factor of each set of dots over the total cell area. The density well heights are clearly increased when this coverage is reduced but if their radii are made very small, while keeping f constant then the density well height becomes negligible. So a dense array of small dot is the best solution to avoid the deleterous effects associated to point contact emitters. To make them negligible the emitter radius has to be of the order of tenths of a micron, and present technologies do not reach this situation so experiencing a somewhat increased recombination in base.

Another peculiarity of point shaped cells is in their resist behaviour. In such cells no emitter resistance is to be considered but it is substituted by the lateral resistance in base. For good point contact cells a worst case estimate of this resistance is a spread resistance between the point shaped emitters and the base volume. The voltage drop caused by this resistance is clearly related to the depth of the wells (29) in accordance to the Eq. (16)

$$U_{U,H} \cong -(kT/q) \ln (p_\infty/p_{U,H}) \qquad (17)$$

where p_∞ is the minority carrier density outside the well. Therefore the same techniques for avoiding the density wells, i.e. making small and closely spaced emitters, decreases this resistance to a zero limit.

The vertical resistance can be avoided if a cell is made with all the contacts in the same face. This is very interesting because it removes the only component of the series resistance that, being unrelated to the shape of the metallisation grid, seemed to be fundamental. We thus conclude that none of the series resistance components is fundamental and therefore an ideally complex and sophisticated structure can be resistance free. In practice this is not true, and the high concentration behaviour of most cells is limited by series resistance, but we mean that this is not a limitation that an improved technology could not remove.

Limits of efficiency in silicon cells

We analyze now the limit of efficiency of silicon solar cells operating at high concentration. In this analysis (11) we assume that the only remaining recombination is the Auger recombination in base i.e. all the emitter and surface recombination have been avoided. Also we assume that all avoidable series resistance has been avoided so that only remains the vertical resistance in those structures that require it.

For the purpose of avoiding the emitter recombination, point contact structures are advised but the resource to a recombination reducing technology is not discarded. Both concepts can be applied to bifacially contacted cells as well as for cells contacted only on one side. In this case the dots are assumed to be so small and closely spaced that no wells of concentration and consequently no series resistance is produced by them. A schematic of point emitter bifacially contacted cells is presented in Fig. 10.

The cells that have been analyzed are the following (1) the Back Point Contact cell (2) the same cells but illuminated on the front of it, and called by us Front Point Contact (FPC) cell, and (3) conventional cells with the n+ emitter on the front and the p+ emitter on the back or viceversa. With the assumptions used the cell behaviour is totally one-dimensional in spite of the apparently two-dimensional (or three-dimensional) nature of the point contact cells and thus this analysis is also valid for conventional bifacially contacted cells with emitters recombination avoided. A schematic of the FPC is also presented in Fig. 10.

The suggestion of using a Front Point Contact illuminated on the face where the contacts are located requires some justification. In the Back Point Contact cell the light impinges on the face opposite to that where the emitters are made. In consequence a strong carrier density is produced in the front face which is needed to cause the flow of carriers towards the back face where electrons and holes will be collected by the n+ and p+ emitters respectively. This high carrier density will produce a strong recombination so reducing the current. At the same time the voltage is related to the carrier density at the rear face, which is smaller so that a lower voltage is extracted (the voltage is $V \cong (kT/q) \ln (n_H p_U / n_i^2)$ at the back surface). To avoid these drawbacks the cell is reversed and illuminated on the face where the emitters are made.

The placement of emitters in the front surface forces us to locate metallisation grids that produce a shadowing on this face, as discussed before. However this is not a necessary consequence because some kind of device, that we call microconcentrator, can also be placed there deflecting the light rays from the metal grids to the uncovered semiconductor area, an example of these devices, but not the only one is the surface structure in the Green cell. The use of light deflecting devices is done at the only expense of the achievable concentration as discussed later. Another option is to place the cell in a cavity without absorption losses that prevents the light to escape except through a small entry aperture. This device makes the cell more insensitive to the fraction of cell metal covered, but again this is done at the expense of reducing the concentration (16,17).

High injection behaviour (implying the use of high resistivity bases) and low injection behaviour are considered in this analysis for all the structures. The main conclusions are deduced from Figs. 10, 11 and 12. These are:

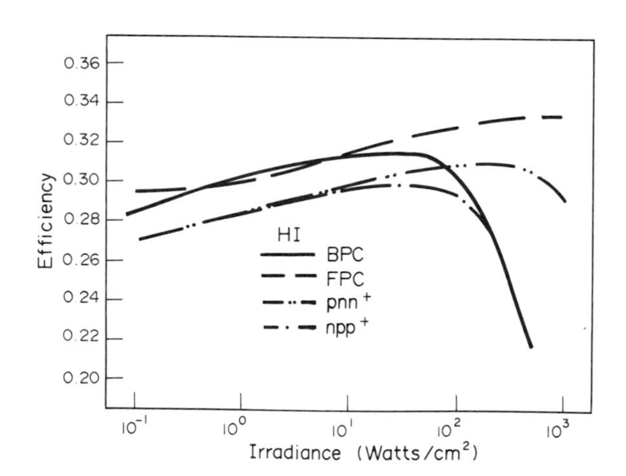

Fig. 12.10. Efficiency limit vs. concentration for the BPC cell, the FPC cell and p+nn+ and n+pp+ BC cells, all in high injection.

(1) high injection BPC cells are the best for moderate concentration but their top efficiency, of 30-31%, occurs (due to the non linear Auger recombination) at 200-500 suns.

(2) Low injection FPC cells present its top efficiency, of some 34%, at a higher concentration, above 5000 suns.

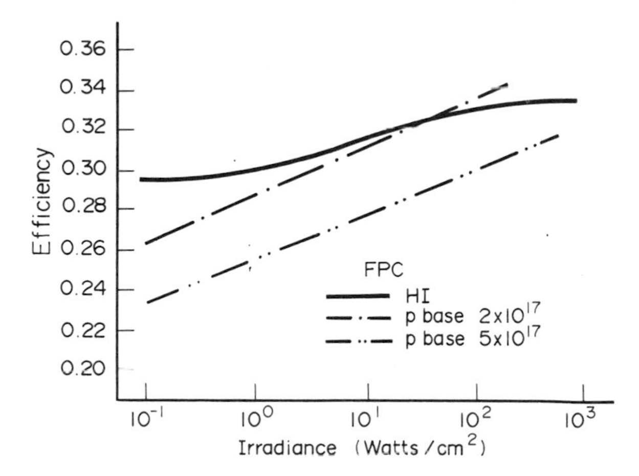

Fig. 12.11. Efficiency limit vs. concentration for the high injection FPC cell sand two low injection p-base FPC cells of 2 and $5x10^{17}$ cm^{-3}.

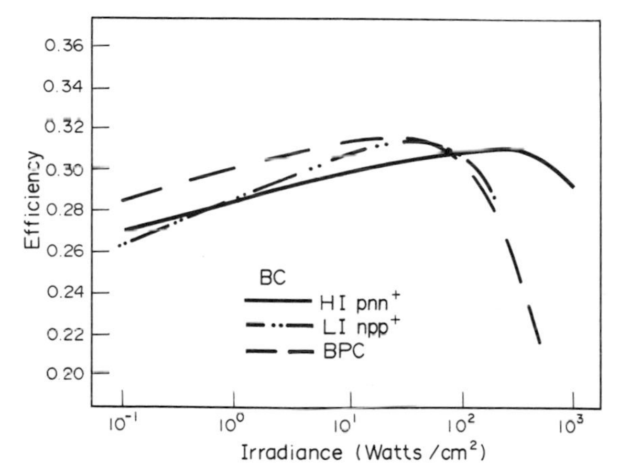

Fig. 12.12. Efficiency limit vs. concentration for high injection and low injection $2x10^{17}$ cm^{-3} p-base BC cells, compared with the BPC cell.

(3) High injection bifacially contacted p+nn+ cells present an efficiency almost as good as that of BPC cells at 2000 suns. High injection n+pp+ cells behave more poorly.

(4) Low injection bifacially contacted n+pp+ cell present its top concentration at some 200-500 suns, very similar to that of BPC cells. Low injection p+nn+ cells present its top efficiency at higher concentration, not as high as that of high injection cells but the efficiency is poorer than for the high injection cells.

(5) All the cells, excepting BPC, cells require microconcentrators (not yet existing) or light confining cavities to pair the efficiency of BPC cells.

Two phenomena that have been considered in high concentration is the possibility of having bandgap shrinkage effects due to highly injected n and p densities silmultaneously (31). This mechanism will increase the recombination and reduce the extracted voltage. However this mechanism is not likely to occur at the concentrations examined (up to 5000 suns) where the top efficiencies take place.

The second phenomenon is the breakage of charge neutrality in the base zones neighbouring the point contacts. This effect would produce a strong voltage drop. However this effect is found to occur only when the I-V curve is already saturated and consequently in does not affect the cell efficiency.

The results of the preceding analysis must not induce us to consider that the achievement of cells operating at concentrations above 1000 suns is already feasible. It rather shows in which cases it is unreachable. For the cases where it can be reached, a strong effort, in particular in the development of novel metallisation techniques, must be applied. However we must be confident on the creativity of the mankind to solve the remaining problems. We can announce that we have already seen several novel metallisation patterns (32) that shows that there is very much to expect from this creativity in the forthcoming years.

2.5 Low concentration cells

In the preceding lines we have devoted most of our attention to the problem of the high efficiency and high concentration cells. Another approach that we already mentioned in the introduction is the one of flat panel cells, somewhat modified to meet the requirements of low concentration. In this respect the only consideration is that the metallisation pattern must be optimized either using the explained procedure, or the cell must be fitted with microconcentrators that, for low concentration, are actually available (33). Also the use of light confining cavities might be used for this purpose. These points will be discussed further later.

2.6 Bifacial cells

Finally an interesting type of cell, needed for static concentrators, is the bifacial cell. This cell has been discussed sufficiently elsewhere (5). It has been developed at our Institute and industrialized by the spun off company Isofoton. Essentially it is (today) a double diffusion n+pp+ cell with metal grids and AR treatment in both faces. This cell, schematically shown in Fig. 13, is made with the same low cost techniques that flat panel cells, and therefore its cost is almost the same. They have also shown to operate satisfactorily with large grain multicrystalline silicon (34) but there are some doubts regarding the reliability of the manufacturing process if this material is used.

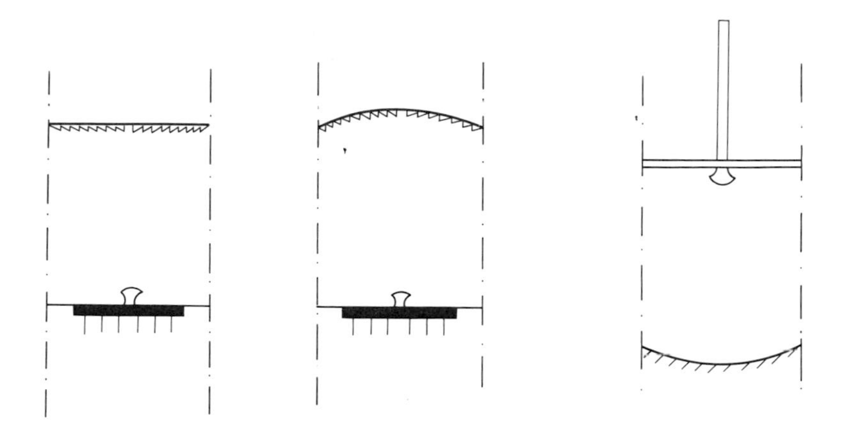

Fig. 12.13. Bifacial cell structure showing the double grid.

This cell uses a p-base with any value above 2-3 cm. In this way the lifetime is not affected by Auger recombination and very good back quantum efficiency is achieved: 70-90% of the front value is typical in production with cepplls 320 μm thick (26 . In research cells almost 100% has been achieved (27) with thicknesses of 240 μm. The cells with lower resistivities tend to present better voltages but poorer symmetry, and the opposite is true for the higher voltages but even for the higher resistivities the one sun open circuit voltages in good cells are around 600 mV.

A technique of simultaneous p and n diffusion has recently been developed at our Institute (35) that if successful in production will make the processing of this cell as cheap as that of conventional flat panel cells.

2.7 GaAs solar cell

The GaAs solar cell, although manufactured quite differently, has a similar behaviour than that of Si solar cells. Major differences are that the stronger absorption of GaAs (a direct gap semiconductor) requires a thinner active cell. As a consequence this cell is very sensitive to surface recombination and special arrangements have been used to avoid or reduce it. The bandgap of this cell is larger than that of silicon (1.43 ev. vs. 1.12 in Si) so that GaAs cells give less current and more voltage. The efficiency tends to be somewhat higher although the highest efficiency achieved today with GaAs ($\cong$ 26.1% (36)) is lower than the highest with Si ($\cong$ 28%).

Two types of cells are common today. One uses a window layer (37) of GaAlAs with higher bandgap and presenting a good lattice matching with the GaAs underneath, to avoid the surface recombination: since almost zero minority carrier density takes place in the larger bandgap window layer the surface recombination becomes negligible. Apart from this effect the window layer plays a virtually passive role for generation as well as for recombination. This cell is depicted in Fig. 14.

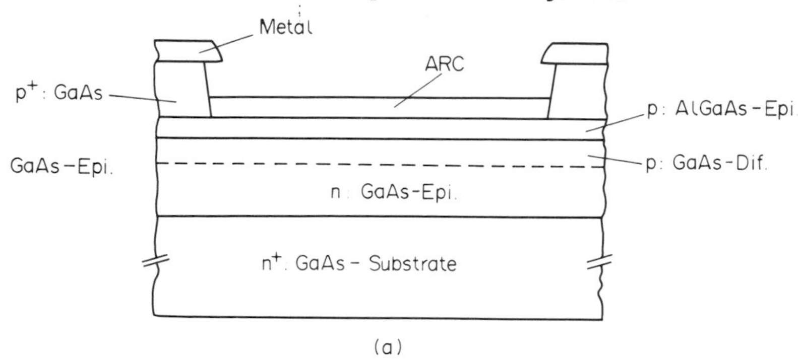

(a)

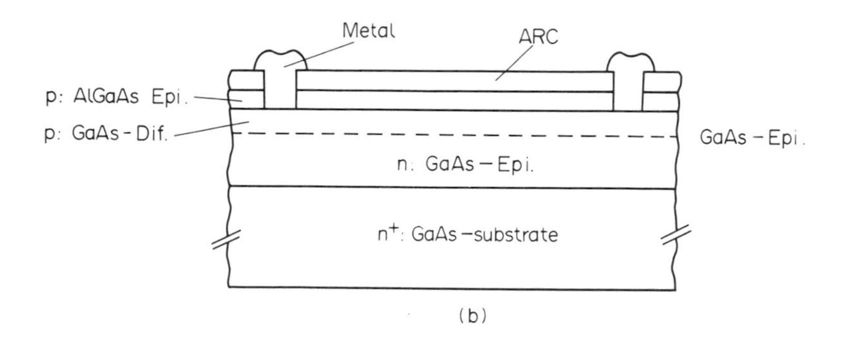

(b)

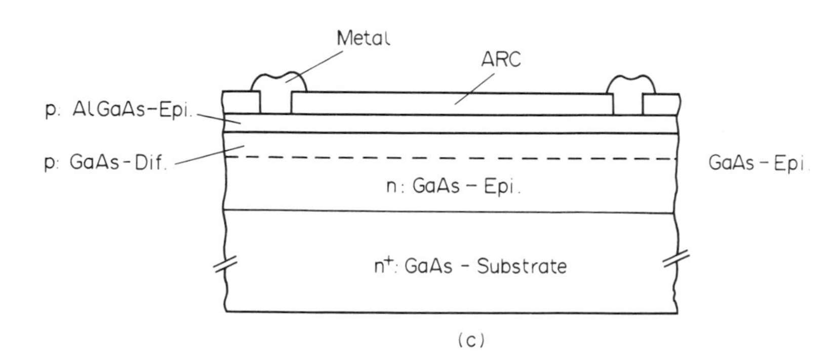

(c)

Fig. 12.14. GaAs cell structure with window GaAlAs layer.

The junction is formed in a n substrate or, alternatively on a n layer grown epitaxially a n+ substrate. In this approach the n and p regions might be interchanged.

The best technology for the fabrication of these cells is liquid epitaxy that leads to layers of excellent quality with a big flexibility but this technology is usually very slow and many experts consider MOCVD (metalorganic chemical vapor deposition) as more promising for industrial manufacturing.

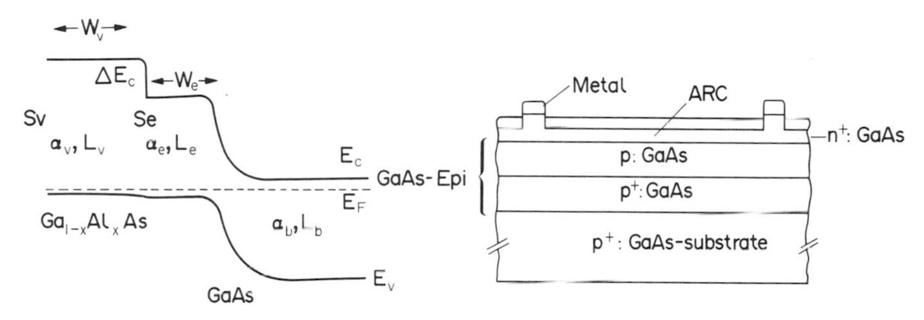

Fig. 12.15. Shallow emitter GaAs Cell structure.

Another approach, less used and represented in Fig. 15, is the so-called shallow homojunction cell (38). In this case a very shallow heavily doped n+ layer is grown at the surface, without any window heteroface layer on top. The reduction of surface recombination is here produced by increasing the doping of the n+ layer but the current photogenerated inside it is virtually lost. Therefore the n+ layer must be very shallow to permit most of the light to pass through it. In this case it is compulsory to use an n-type doping for the layer on the top because the higher electron mobility leading to lower series resistance. Also it is convenient that the underlying region, where most of the generation is produced be p-type so that its minority carriers are electrons with higher mobility to reach more easily the junction (they have a higher diffusion length).

The later cell structure is conceptually very similar to the silicon low injection cell. The former one can also be treated in a similar way but in this case the doping of the emitter is similar to that of the base and the light absorption is produced in both regions. No heavily doping effects appear in this cell and its behaviour is quite perfect from a theoretical point of view. In particular under concentrated sunlight the radiative recombination becomes important and the cell starts to approach to its ideal behaviour.

The higher mobility of electrons in GaAs and the small thickness of the active layers changes drastically the weight of the different components of the series resistance of these cells. While in Si cells the major component tends to be the base resistance here this component is unimportant because the bulk of the wafer is totally passive and therefore can be very heavily doped. Cells of n and p type are preferable because in GaAs the lateral emitter series resistance becomes the most important and it is of interest to take profit of the higher electron mobility. Nevertheless in window cells the relatively low doping that the high lifetime requires, and also the moderate thickness needed for effective collection makes the emitter component of the series resistance a major drawback for high concentration operation. In shallow homojunction cells with the n layer on the top the emitter has to be so shallow, to avoid light absorption, that the series resistance is also high. Notwithstanding these difficulties GaAs cells have done always with very fine grids and of small size and in consequence many cells have had its maximum of operation at concentrations of some 500-750 suns.

G.L. Araujo, of our Institute has proposed (39) a cell structure, that he calls Back Emitter Contact cell, represented in Fig. 16 that prevents the preceding drawbacks. In this cell the upper p-layer goes through a via in the n-layer underneath to contact the p+ substrate and so it is contacted on the bottom; since the vias do not produce shadowing they can be very closely spaced and thus produce a low emitter series resistance. The n-layer underneath can be deposited on a n+ layer that now is not an absorbing layer. In this way its sheet resistance can be very small. Metal fingers are deposited on this layer to form a metallisation grid after the proper window has been opened on the p-layer above. If the cell is small enough this metal grid is not really necessary (the n-layer can be contacted laterally) and this cell has no metal grid on top. The calculations with a good technology, similar to that of the cell that achieved 26.1%, yield efficiencies of 30.5% for operation at 2000 suns, as represented in Fig. 17.

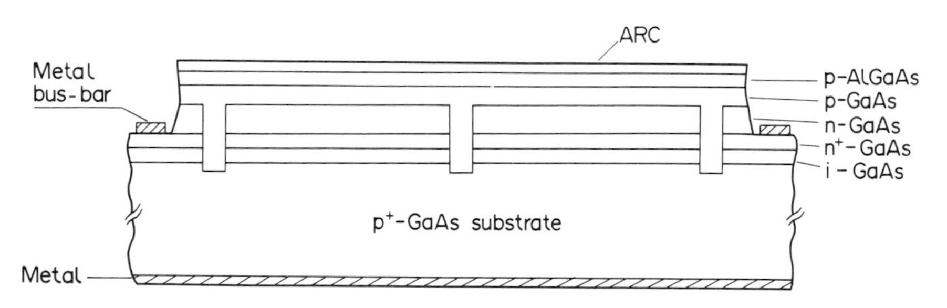

Fig. 12.16. Proposed Back Emitter Contact (BEC) GaAs cell structure for high concentration operation.

The recent improvements on Si cells has stirred big interest on the applicability of the novel concepts generated in the Si cells field to that of GaAs. I think that texturization of surfaces, and light confinement, although difficult to apply (being probably two separated operations in GaAs), might be beneficial for GaAs cells. Actually making them thinner will reduce the recombination and not the generation if the confinement is perfect. On the other hand the concepts of point contact emitters are related to avoiding recombination on heavily doped zones and these zones are not actually used in GaAs. In addition back contacting both regions as in Si-cells, to leave the front surface free of metal, requires diffusion lengths higher than the cell thickness that in GaAs are very difficult to reach.

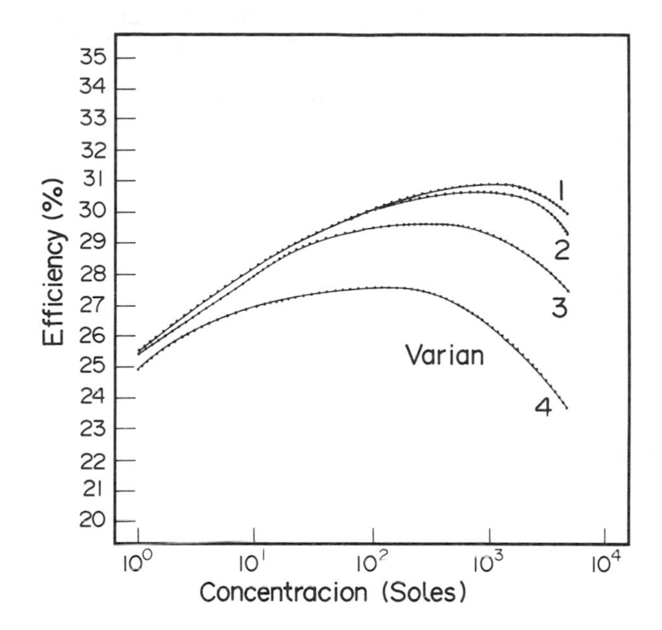

Fig. 12.17. Calculated efficiency vs concentration of the BEC GaAs cell structure.

In conclusion, GaAs cells are today behind the silicon cells, but if sufficient effort is done on them, I foresee that they might be more promising than Si cells in the range of high concentration and high efficiency, with the added advantage that they open the way for multigap developments to be undertaken in the future.

3. TRACKING CONCENTRATORS

3.1. Physical limits

The sun light cannot be concentrated indefinitely. Actually in the cell surface, if placed in the air, a radiance (or luminous power density per unit of solid angle) above that of the sun photosphere (as seen after the atmospheric absorption) cannot be achieved. If the cell is surrounded by a medium of index of refraction n the achievable power density per unit of solid angle (now not called radiance anymore) is increased by n^2.

In fact any concentration consists in illuminating the receiver with a solid angle greater than the one that subtends the sun at bare eye. This is very evident if a lens is used.

Thus it is concluded that the upper limit of concentration is reached when the receiver, a bifacial cell, is isotropically illuminated by the concentrator that in this case cannot be a simple lens but some cell-encircling reflecting structure; for a monofacial cell hemispherically isotropic illumination is needed and concentrations half of those in a bifacial cell can be reached. In the case that this (hemispherically) isotropic illumination is achieved on a monofacial cell the irradiance (or power flux) concentration can reach the limit of $C = n^2/\sin^2\phi$ being $\phi \cong 0.265°$ the semiangle of the sun so that the upper limit of concentration in earth is 1 ≅ 46.747 for n = 1. An upper limit is obtained for n equating the silicon index of refraction (n ≅ 3.7). This limit is in a monofacial cell is 639970 and twice this value in a bifacial cell (1.27×10^6) the last being an absolute limit on Earth (that is higher in planet Pluto, although the limit of luminous power density on the cell remains the same).

The concentrator receiving this limit has to (a) collect all the rays from the sun disc and to reject all the rays from outside it; in addition it has to (b) distribute these rays in an isotropical (or hemispherically isotropical) way without leaving darker any impinging direction, nor any darker spot in the cell.

I believe that both conditions cannot be fulfilled simultaneously with real optical devices unless if we use position dependent indices of refraction (40,41). The reason is that we do not have enough degrees of freedom with a finite number or even with a numerable infinity of lenses and mirrors although it has recently been shown that one can approach indefinitely (in theory) to this goal (42). Therefore, even from a theoretical point of view no practical device can produce perfect isotropic illumination on the cell and at the same time to collect strictly the rays in the solar disc (43).

In addition technical reasons make clear that collecting the cell rays from the solar disc is just impossible. In fact an allowance has to be given in order to accout for errors in tracking, aiming and manufacturing. Present concentrator technology probably insures that all the rays in a circle of radius 1° are effectively collected (44) (and perhaps of a circle somewhat smaller). Concentrators that aim to collect all these rays cannot achieve a concentration above $1/\sin^2(1°) = 3283$ for n=1.

3.2 The design of two stage concentrators for very high concentration

Concentrators encircling the cell are necessary to achieve isotropic radiation on it. This leads, in practice, to two-stage concentrators in which the first stage may be a Fresnel lens or a parabolic mirror, all of them with symmetry of revolution, and the second stage of small size, can be made of a dielectric so that the cell becomes surrounded by a medium of index of refraction n ≅ 1.5. Thus the limit concentration is increased to 7387 for the 1° allowance case, but yet quite fundamental reasons reduce this value in practical devices as we shall briefly explain now. Secondaries can be made following design rules explained in the literature (45,46) on which we shall insist later.

Our suggested strategy when designing tracking concentrators is to try to increase as much as possible the concentrations at the secondary entry aperture. This strategy does not ensure that we are approaching to the highest concentration in the cell because concentrators with a higher secondary entry aperture might provide a higher overall concentration, but some additional reasons support our strategy.

If we accept it we find the highest first stage concentration is achieved when this stage illuminates the secondary entry aperture as if the primary were a Lambertian radiator (radiating homogeneously with a consinus angular distribution). Actually this condition is equivalent to that of isotropic illumination that we required for receiver encircling concentrators, in the case that the receiver is not actually encircled. This condition can be achieved with a very big mirror that produces the proper illumination on the receiver but spills many of the rays that receives from the sun (47).

However if we want to use a concentrator joining to the preceding property that of not spilling any ray, in order to have simultaneously highest irradiance concentration and efficiency the problem becomes impossible in practice. Such an ideal concentrator has to have a very specific ellipsoidal shape approaching to a hemisphere of well determined radius and having certain very specific ray deflection law (48). This concentrator might be made with a curved Fresnel lens of infinite index of refraction.

Indeed actual spheric shaped Fresnel lenses approach to this condition. Any other Fresnel lens or mirror intended to avoid ray spillage produces a concentration that is limited by its specific law of bending rays. This is very easy to see in the case of a mirror in which the reflection does not change the angular spread of the beam (only the direction) and thus it has to be located inside the sphere produced by rotation of the arc subtending the ±1° in our example, as represented schematically in Fig. 18. The deflection law of Fresnel lenses is still more restrictive. Procedures to design the concentrators so to achieve this upper limit of concentration without ray losses are given in the literature (43, 47-49). Their rims must lay on the loci in Fig. 18.

In addition chromatic aberrations change the index of refraction of a Fresnel lens for the different wavelengths and in consequence its design must take into account this effect that again reduces the achievable concentration (50).

As a matter of fact the achievable concentration of the primary depends on its f/number (defined here, in non imaging concentrators as the distance of the concentrator rim to the receiver divided by the primary diameter). For very small aperture diameters all the concentrators approach to their ideal behaviour but lead to very small primary concentration, (that reverts in not having any primary at all). The relationship of concentration achievable with a well designed primary concentrator and its angular aperture D (the inverse of the f/number) is given in Fig. 19 as compared to the maximal one achieved with the lambertian radiator of the same shape. The superiority of the mirrors is due to the absence of chromatic aberration, otherwise optimally shaped Fresnel lenses should have been better.

As a final comment the concentration in reduced by the need of

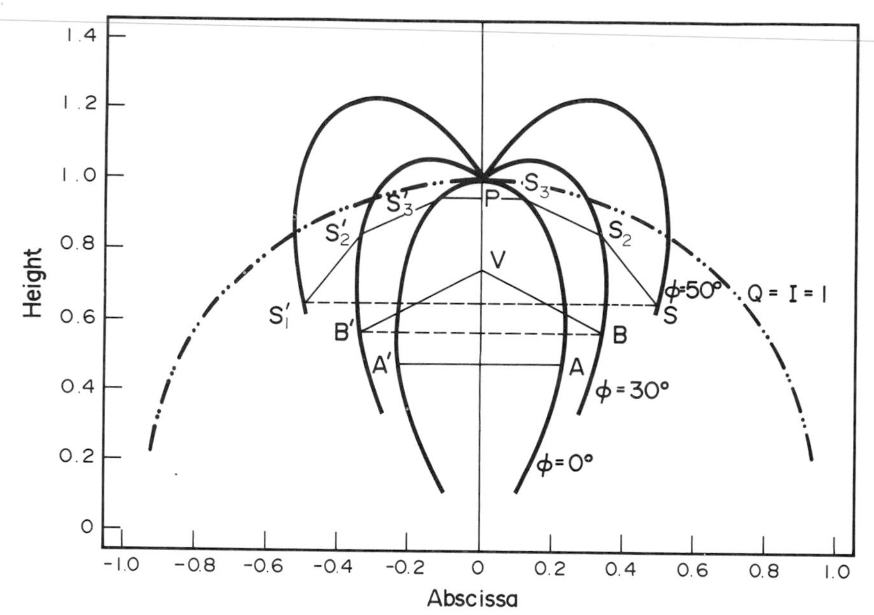

Fig. 12.18. Locus where a primary concentrator rim must lay to give maximal concentration in the following cases: (a) Theoretical upper limit based on conservation of radiance; (b) limit for mirrors; (c) limit for optimally shaped Fresnel lenses; (d) limit for flat Fresnel lenses. All the lenses are considered without chromatic aberration, (which leads to very optimistic results).

covering the whole space with lenses or mirrors. For it the theoretical circular shape has to be cut into inscribed squares (or hexagons) and the achievable concentration is still reduced.

In the strategy we mention here secondaries are simply designed to collect the rays of a Lambertian primary with a specific f/number. The nature and the characteristics of the primary are irrelevant in the design and this is a very important flexibility, but yet a more important characteristic is that they produce a strong homogeneization of the radiation in the cell. We think that this is a general prooperty of this type of design.

Thus Compound Parabolic Concentrators (CPC's) or better Compound Elliptic Concentrators (CEC's), both well known types of the Winston-Welford Nonimaging concentrators (51), must be used as secondaries to produce additional concentration, by the means of illuminating isotropically the cell. It is advisable to use them dielectric filled because in this case the concentration can be increased in n^2 although an economical consideration enters in this decision. Note that even if the first stage is hemispherical, producing isotropic illumination on the secondary entry aperture a new additional concentration can be achieved

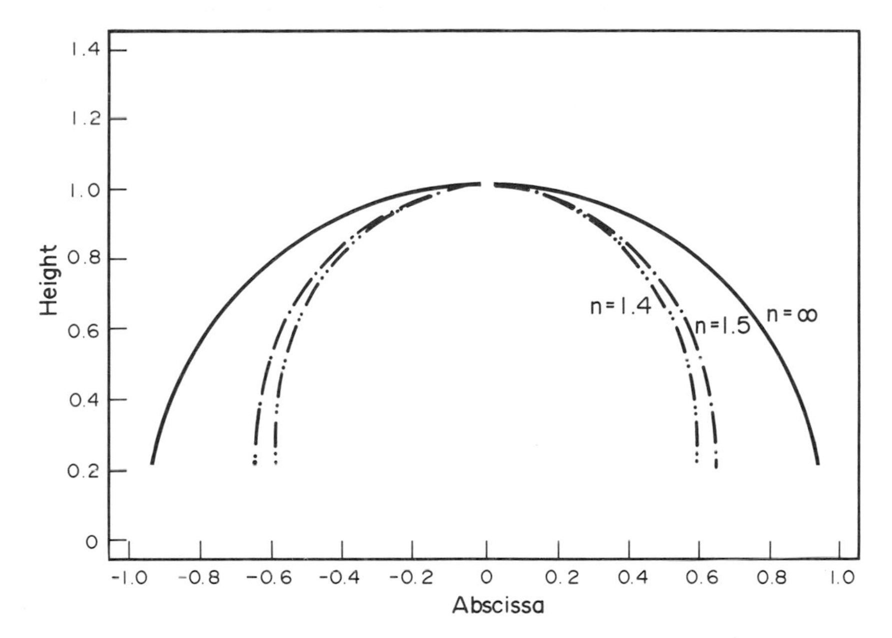

Fig. 12.19. Normalized bidimensional concentration vs. angular aperture defined as the ratio of the rim diameter to its distance to the receiver (the inverse of the f/number) for a mirror, an optimally shaped Fresnel lens and a flat Fresnel lens, all of 1° of allowance angle, takin taking into account the effect of chromatic aberrations. The normalization is with respect to 1/sin 1° = 57.3. These curves refer to the bidimensional outline of the axisymmetrical concentrator; the tridimensional concentration is obtained by squaring these bidimensional values. Curves starting from zero are the primary concentrations. Curves starting from one are the two stage concentration assuming an ideal secondary and measure the deviation of the primary from Lambertian behaviour.

with a dielectric filled CPC that converts this isotropic illumination into a new isotropic illumination in a cell n^2 times smaller but surrounded by dielectric, as stated in the preceding subsection.

To properly speak the CPC or CEC does not produce isotropical illumination except for meridian rays (i.e. in a bidimensional analysis): we have already commented on the impossibility of achieving this result in three dimensions. In addition non-meridian rays can be rejected after some reflections inside the secondary in the CPC if it is devised to collect strictly the meridian rays within the angle subtended by the primary. This means that a certain angular allowance of 2-3° must be adopted in the secondary whose angular acceptance is high, of the order of arc tan (D/2), to insure the virtual collection of all the monmeridian rays. This leads to some additional reduction of the achievable concentration in the secondary that becomes (approximately) $n^2/\sin^2[2°+\arctan(D/2)]$. Taking this factor into account we find that there is an optimal f/number for the primary-secondary system as results from the drawing in Fig. 14.

Finally we have to mention that levelling rays impinging the cell are strongly reflected at its interface. Because of it the output angle of the concentrator is to be restricted to some 60-70°. CPC's or CEC's can be designed for this purpose. This design is often called an angle transformer.

Table 3. Some selected high concentration two-stages concentrators of revolution, (secondary filled with a n=1, 49 dielectric) taking into acccount chromatic aberration (n_u=1, 50, n_I=1, 48), shape factors, output angle limitation (70°) and allowance for secondary skew rays collection (2°).

	Flat Fresnel		Shaped Fresnel		Mirror	
Extended sun semiangle	1°	0.265°	1°	0.265°	1°	0.265°
Aperture	f/1.98	f/2.12	f/2.03	f/1.97	f/1.96	f/1.96
Secondary concentration	26.2	28.2	26.3	25.0	24.9	24.9
Total concentration:						
Round cell/round conc.	1989	13018	3783	30931	4681	66661
Round cell/hex. conc.	1645	10766	3129	25579	3872	55128
Round cell/squared conc.	1266	8288	2409	19691	2980	42438

We give in Table 3(52) the achievable concentration for the f/2 concentrators formed with flat Fresnel lens, spheric Fresnel lens and parabolic mirror and the maximal concentrations achievable with the optimal f/number are given in Table 4. We note that practical flat Fresnel lenses with 1° of angular allowance can only achieve a concentration of 1266, and this with the secondary filled with an index of refraction of n=1.5, that would be reduced to 562 if a dielectric empty secondary is used, although this is dependant on the angular allowance that we can achieve without increasing the concentrator cost in an inconvenient way. We also mention that the highest concentration that can be achieved with a single stage flat Fresnel lens is 485; these considerations reveal the importance of high quality secondaries to achieve concentrations above 500.

A schematic of different concentrators is given in Figure 13.

Table 4. Maximally concentrating two-stages concentrators of revolution, (secondary filled with a n=1.49 dielectric) taking into account chromatic aberration (n_u=1.50, n_I=1.48), shape factors, output angle limitation (70°) and allowance for secondary skew rays collection (2°).

	Flat Fresnel		Shaped Fresnel		Mirror	
Extended sun semiangle	1°	0.265°	1°	0.265°	1°	0.265°
Aperture	f/4.14	f/6.14	f/2.03	f/3.13	f/1.51	f/1.51
Secondary concentration	82.0	146.0	26.3	53.3	16.3	16.3
Total concentration:						
Round cell/round conc.	2742	23489	3783	33375	4752	67666
Round cell/hex.conc.	2268	19425	3129	27601	3930	55959
Round cell/squared conc.	1746	14953	2409	21247	3025	43078

The suggested secondary is not a CPC but a related concentrator with curved entry aperture and total internal reflection that reduces the secondary size and improves the reflection losses (45,46).

We note that in the preceding calculations we have not considered losses other than those implied by the concentrator shape. The actual concentration is to be corrected by the amount of these losses.

3.3. The design of concentrators for low concentration.

Low concentration devices to be fitted with low concentration cells is of interest, we think, for short term development of attractive devices of interest in certain applications.

In these concentrators one sun cell, slightly modified to meet the series resistance requirements, must be used gaining profit of the low cost at which these cells can be fabricated. Low concentration is only compatible with these cells and low cost concentrating panels of small size (around 100 Watts peak) must be thus used for such a purpose.

In most cases they have to be single stage concentrators and from the drawings of Fig. 20 and 21 it is clear that an upper bound of concentration corresponding to every type of concentrator for a given angular acceptance, or, what is equivalent, the f/number allowing highest angular acceptance must be used in this approach. We give in Table 5 some combinations that we consider to be adequate.

Table 5. Some selected low cost tracking concentrators, (non dielectric secondary) taking into account chromatic aberration (n_u=1.50, n_I=1.48), shape factors and skew rays.

	Flat Fresnel Revolution Two stages Round cell Squared lens	Flat Fresnel Revolution One stage Round cell Hex. lens	Flat Fresnel Revolution One stage Squared cell Squared lens	Mirror Linear One stage Squared cell Rect. mirror
Extended sun semiangle	5°	3°	3°	2°
Aperture	f/1.22	f/1.25	f/1.06	f/0.5
Concentration (1st.stage)	4.15	13.99	12.92	14.3
Two stage concentration	29.3			
Secondary concentration	7.1			

3.4 Microconcentrators

We have mentioned in preceeding sections that finger shadowing cannot be considered as an unavoidable drawback. Today there are devices that deflect the light from the metal fingers as the structure exemplified in Fig. 20. Also the engravings done by Green in the cells avoid, to a certain extent the loss of the light incident on the metal fingers. Therefore for low concentration there are solutions to avoid finger shadowing and for higher concentrations its development is a problem of technology. We call microconcentrator to this type of devices. For high concentration where very fine structures are used in the metal grid, these devices can be conceptually designed, and examples are also given in Fig. 20, but their practical implementation remains problematic, as well as their concept itself since the range of validity of geometrical optics can be overreached.

A microconcentrator is intended to cast the light on the uncovered area fraction F. This actually means that the real concentration defined (for the loss-less structure) as entry aperture to the receiver area is increased in the factor 1/F and if F is not close to one this concentration can exceed the achievable one, as presented in Table 4. As a matter of fact, because of the linear geometry used for the metal fingers, a more realistic factor to use is $1/F^2$ (and yet this requires bidimensional ideality, possessed by the CPC's in the example of Fig. 23b). The result of it is that for a limit of achievable concentration with a given technology of say 1300 X, if we want to operate at 1000 X the fraction F of metal fingers coverage that can be allowed with the micro-concentrator is of 0.88. A novel efficiency-concentration optimization takes place in this case instead of the one taking place when cell covering and series resistance

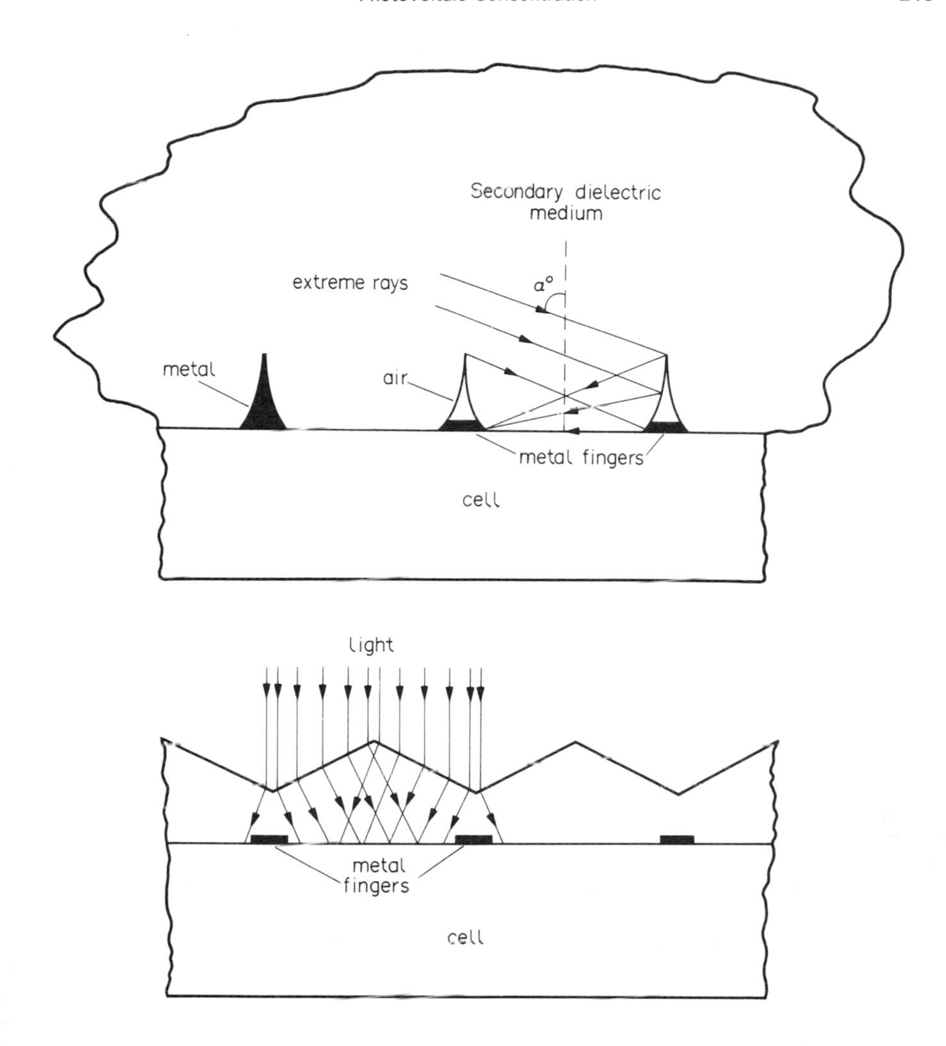

Fig. 12.20. (a) Schematic of the microconcentrator invented by McNeill (33) and (b) concept of a more perfect microconcentrator engraved in a dielectric secondary using total internal reflection or a metal-reflecting CPC-like concentrator.

are the counteracting effects. In this case a certain concentration allows a given coverage, as explained above, and no optimization of F has to be done (the smaller the F the better for decreasing the series resistance) but when the concentration can be varied then at smaller concentrations the open-circuit voltage is smaller but F can be small and the series resistance negligable. On the other hand when F approaches to unity then the allowed concentration is higher and the open-circuit voltage is increased but the series resistance is decreased. Thus a concentration for optimal efficiency is obtained. This optimization requires a detailed knowledge of the microconcentrator behaviour and this is not yet available since we do not know exactly how to make these microconcentrators for the higher concentration case.

3.5. External light confinement

Light confinement can be obtained inside the cell, as explained in the preceding section, but also outside the cell by placing it into an internally reflecting cavity (17), as the one shown in Fig. 21. The main practical interest of it is that the light reflected by the cell, either in their metallization grid or in the semiconductor surface that is not totally absorbing, is not lost, but reflected into the cavity that in part reconduct this light again into the cell.

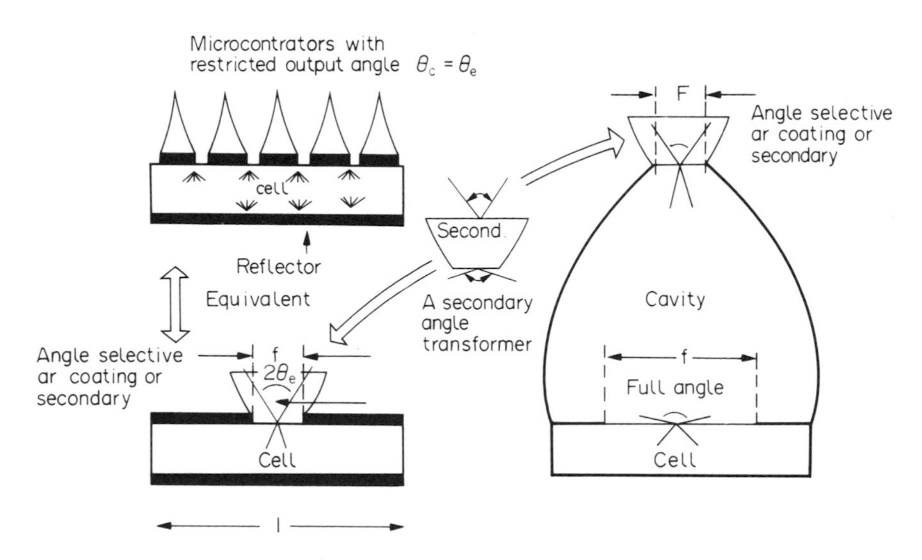

Fig. 12.21. A cavity for external confinement of sunlight.

This cavity must be designed in such a way that the rays entering its entry aperture are first cast into the cell and those reflected by it are fed into the cavity. Once inside it they suffer so many reflections that we can consider that an isotropic radiation is achieved inside the cavity. Under this condition a balance of photon flux is produced so that the only losses of photons are produced by either escape through the entry aperture, absorption in the cavity reflecting internal walls or absorption by the cell. As a result the transmission T of light into the cell located inside the cavity is

$$T = t \frac{1 + F + \alpha S}{t + F + \alpha S} \tag{18}$$

where t is the actual transmission into the cell when located outside the cavity F is the cavity aperture area related to that of the cell S the inner cavity surface also related to the cell area and α the absorption of light by the cavity walls (which is supposed to be small). Note that T is always bigger than t and becomes unity when $F + \alpha S << t < 1$, i.e. when the cavity aperture is much smaller than the cell area and the cavity losses are small. In any case the role of the cavity is positive in the increase of the cell transmission.

What is lost in using such a cavity is concentration because now we must direct all the incoming light into the cavity aperture of smaller area than the cell and we know that this task has some theoretical and practical limits. Furthermore if the light has to fall into the cell once it enters the cavity aperture it has to enter into the cavity within a certain cone of light and thus under a restricted angle which limits even more the achievable concentration. Indeed we do not have many requirements for the cavity shape so that it can be placed very close to the cavity aperture but still we must assure that full randomization of light is produced into the cavity if the preceding formula is to be applied. What we gain is a strong light absorption even if cells of heavy front metallization are used so leading to a small t. This configuration avoids the use of any microconcentrator but at the expense of greatly reducing the achievable concentration on the cell. Its use is thus advised if the level of concentration that the optical system allows is bigger than that permitted by the cell.

More concentration can be achieved if we allow the light to fall partly or totally outside the cell in the first reflection into the cavity. In the last case the equation to be used is

$$T = t \frac{1}{t + F + \alpha S} \tag{19}$$

that gives a smaller absorption. Only if t is low ($t+F+\alpha S < 1$) this configuration leads to a $T > t$. However it can be useful in some cases.

We note that the reflecting coverage of the cavity walls can be a mirror or a white paint producing diffuse reflection and little absorption, even more, the use of such a paint leads to smaller concerns regarding the achievement of the full light randomization. The cavity can be filled by an optically dense medium to permit the combination with a dielectric secondary but t and α are modified by the use of such a medium. In this case a dielectric secondary can be integrated easily.

Thus we conclude that this device is of high interest any time we can have a concentrator with more concentration than the one required by the cell. For it the use of non-imaging optics becomes, in most cases, a need.

In the limit of idealized cells the use of such cavities reduces the efficiency limit (16) but such idealized cells are possibly not achievable.

3.6. Heat removal

Against what it is intuitively thought, heat removal does not seem to be a major limitation of concentrating solar cells. As a matter of fact it must be considered that the operating cell temperature is produced by the balance of the input and output energy. The input energy entering in the PV system (not in the cell) is rather small, received at a rate of 1 kW/m^2 in standard conditions. This power is evacuated by radiation and convection in a heat dissipator (heat sink). In flat panels the energy entrance is more or less the same and the dissipator can be worse than in concentrating panels, which in theory can be done cooler. Thus the problem, not easy from a technological point of view, is not as serious at it appears at a first glance.

Actually what is done in a concentrator is to cast all the energy entering the system entry aperture into a set of cells of reduced dimensions. Each cell must have a heat sink that defocuses this energy again and transports it towards the dissipating surface of the heat sink. This transport can be done by conduction (or with a heat pipe) and (in the first case) if the cell is very small this conduction is controlled by the thermal spread resistance, proportional to $1/r_C$, where r_C is the radius (size) of the cell. However the luminous power on the cell is $P_L \cong \pi r_C^2 C$ (of which a fraction is extracted as electrical power) and if C is conserved the temperature drop is proportional to the product of the preceding magnitudes, i.e. proportional to r_C. Using small cells the temperature drops will be reduced consequently and thus the use of small cells allow these to operate at conditions similar or better to those of a flat panel. As a matter of fact the concentration of a given family of concentrators can be optimized and so that the efficiency gains due to concentration balances out the reduction due to temperature decrease leading to an optimal operation concentration that in some examples can be above 2000 suns (53). We present in Fig. 22 an example of such optimization and of the cell temperature (for resting wind, which is a pessimistic condition). Note that in this example the cell is not a very high efficiency one so that higher efficiencies than those in this figure can be sought.

3.7. Practical experience

The greatest experience on concentrators has been achieved in the United States of America with the relatively strong involvement of the government of this country. As a consequence a certain number of demonstration projects, among which the two bigger have been the 350 kWp central in Saudi Arabia (in cooperation with the government of this last country) and the 225 kWp in the Phoenix Sky Harbor Airport, both with modules based on point focus single stage Fresnel lens with concentrations around 50 X. They were built in 1981-82 and no similar size system has been built since then. Outside the United States and Saudi Arabia

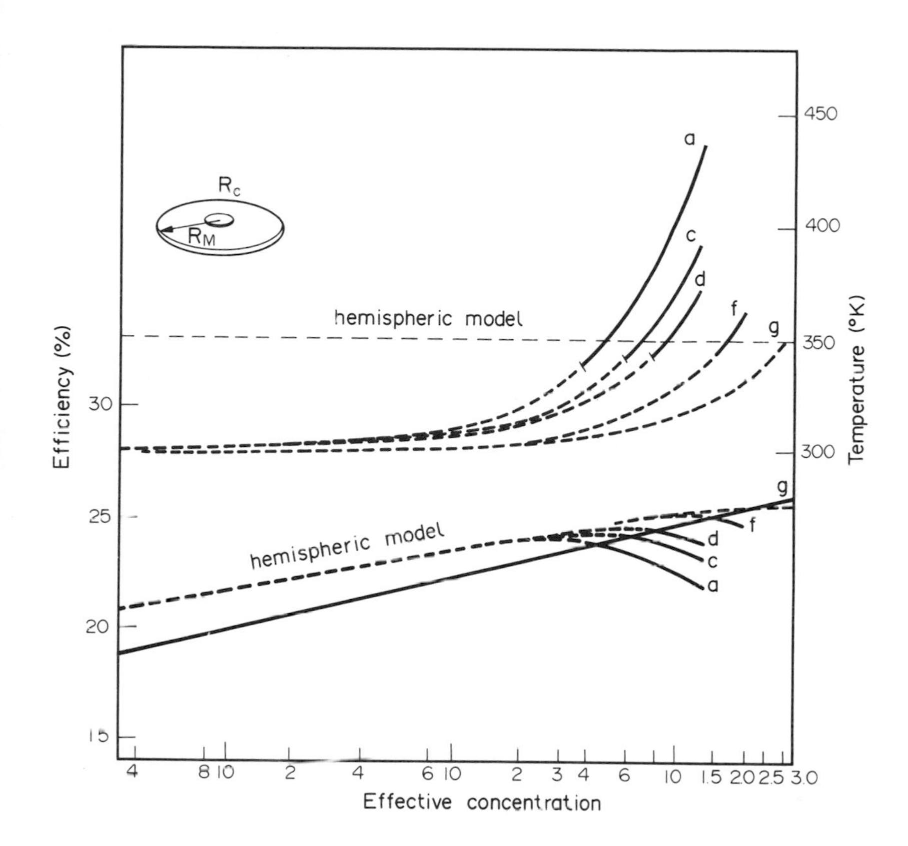

Fig. 12.22. Temperature and efficiency vs. concentration for a heat sink formed of a plate of the same size than the concentrator. Curves (a) and (d) correspond to a Fe heat sink and (c) and (f) to an Al one. Curves (a) and (c) correspond to a cell diameter of 2 cm and the (d) and (f) of 0.5 cm. Curve (g) refers to an isothermal plate of infinite radius, valid for this case in the solid region of the curves. (CUU).

no similar size demonstration system has been built, and the interest for this technology has always been lower. It is argued that among the industrialized countries USA is by far the one receiving more direct radiation, crucial for the use of these systems. Annual electrical efficiency in the range of 6-8% has been achieved and this is not lower than the efficiencies with flat panel installations of similar size. The installations have experienced minor problems of little concern so that it can be affirmed that for central applications the concentration complexities do not represent a real hindrance. A large variety of smaller installations have been built with different approaches, both inside and outside the USA, and some conclusive opinions have been developed based on this experience. However I think that this experience is too restricted to consider these opinions as strongly supported. Nevertheless I want to express those more widely accepted.

Fresnel lenses have been considered as the best suited concentrator for photovoltaics. Point focus are the most popular and they tend to be developed in groups of several lenses to cover with it a whole module.

Linear curved shaped Fresnel lenses have been developed by the firm Entech and they have presented the biggest efficiency in a full module of big size. It proves the adequacy of the curved scheme we have presented before, but there are reasons to consider point focus optics more resistant than line focus optics to imperfections. Probably the curved shape compensates this drawback of linear optics.

Early experiences with linear reflective optics were very disappointing. However in recent time a company is pushing strongly this approach with improved thin film (laminated) mirrors with results not yet fully available, but apparently promising.

Experiments with bulky lenses for very small cell were examined at the earlier times but the results were probably poor because the technology was abandoned. Recent attempts to follow this pattern are considered promising. Bulky lenses are in principle more efficient than Fresnel lenses (because of the light scattering with the step edges in the latter).

No sizeable spontaneous market has developed with concentrators as it has happened with flat panels. The main reason, in my opinion is the too big size of the basic panel, compared with the real present applications, but tracking complexity has certainly affected to the acceptance of this scheme, as well as the inability to collect diffuse light that has restricted its geographical applicability.

Recent second generation of point focus and of curved lineal Fresnel lens panels have been studied at Sandia National Laboratory in installations of 22-25 kWp and have demonstrated 15% peak efficiency for the point focus system, while the demonstrated peak efficiency of the line focus concentrator has been 11% electric plus 50% thermal (54).

A histogram of the cell and concentrator efficiencies (12) is shown in Fig. 23. Note the breakthrough produced in cell efficiency in 1985, corresponding to the Stanford cell previously discussed. A delay of some three years takes place before an improvement of cell efficiency is translated to the concentrator efficiency. Efficiencies of 17% have been already achieved with cells of 20% efficiency. We are expecting in this days concentrator efficiencies of 20% at 300-500 suns with cells of 25%. We know that this goal is being attempted at Sandia Laboratories in USA.

4. STATIC CONCENTRATORS

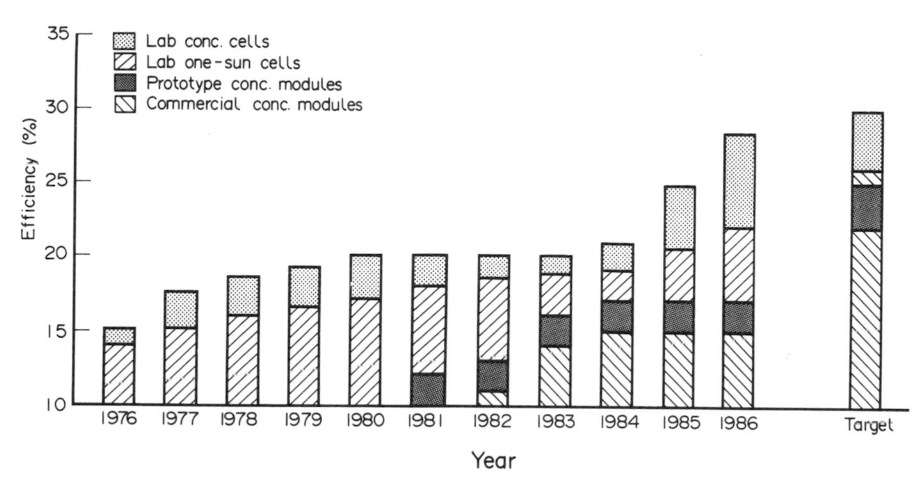

Fig. 12.23. Histogram of the efficiency of laboratory concentrating and one sun cells, prototype and commercial concentrating modules.

4. STATIC CONCENTRATORS

An important objection that has been commonly made to concentrator systems, are the need of a tracking structure and the waste of the diffuse radiation, very important in most climates. Both problems result simultaneously solved in static or stationary concnetrators. For it a very wide angular acceptance is required so that rays from different sun positions are cast on the receiver. In this way also a wide portion of the sky is collected resulting in collection of diffuse radiation.

The consideration of an extended source helps in the study of static concentrators. This source can be the yearly averaged radiance (power received per unity of receiver area and of solid angle in a given direction) of the sky at given locations. This is what is ideally obtained if we take a photographic camera with an eyefish objective and leave it with the shutter open for a whole year. The preferred orientation for this camera is that of an equatorial telescope, i.e. pointing to the intersection of the celestial equator and the local meridian. The result of such experiment (as calculated (56) from recorded data) for the sky of Madrid is presented in Fig. 24. Note that the darkest zone with zero radiance, corresponds to the reflection of the ground, visible when looking to the south. The region of intermediate radiance, towards the north, is the region of the sky never travelled by the sun. The brighter region is the part of the sky travelled by the sun sometime in the year. In this representation the sun travels a horizontal line every day, and this line changes of the coordinate v (direction cosine with the Earth axis, i.e. arc cos δ, (δ the sun declination) every day.

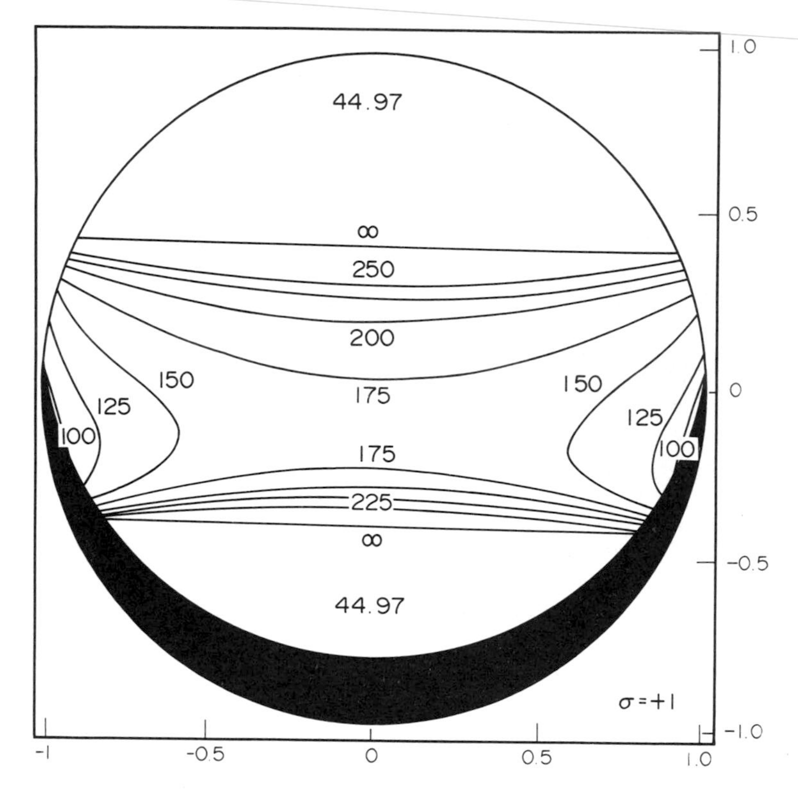

Fig. 12.24a.

4.1 The limits of static concentration

A static concentrator can be designed to collect all or most of the direct radiation (i.e. the brighter region in the drawing) or a wider portion of the sky. In both cases it will collect a substantial fraction of the diffuse radiation or even all of it. Unfortunately the irradiance concentration is limited to $C=2\pi n^2/A_S$, where A_S is the area of the collected region in the map of Fig. 25. This is a direct consequence of the theorem of the conservation of radiance of, mentioned in preceding section and for a concentrator collect ng all the rays in a cone of semiangle ϕ $A_S=\pi \sin^2\phi$. The factor 2 is applicable to bifacial cells and accounts for the fact that even if all the sky is collected, this represents only a hemispherical isotropy and full spherical isotropy at the receiver can, in principle, be attempted if bifacial cells are used.

The high value that is aimed for A_S in static concentrators make the achievable value of the concentration small and stresses the need of utilizing strong theoretical backgrounds to scratch the theoretical limits.

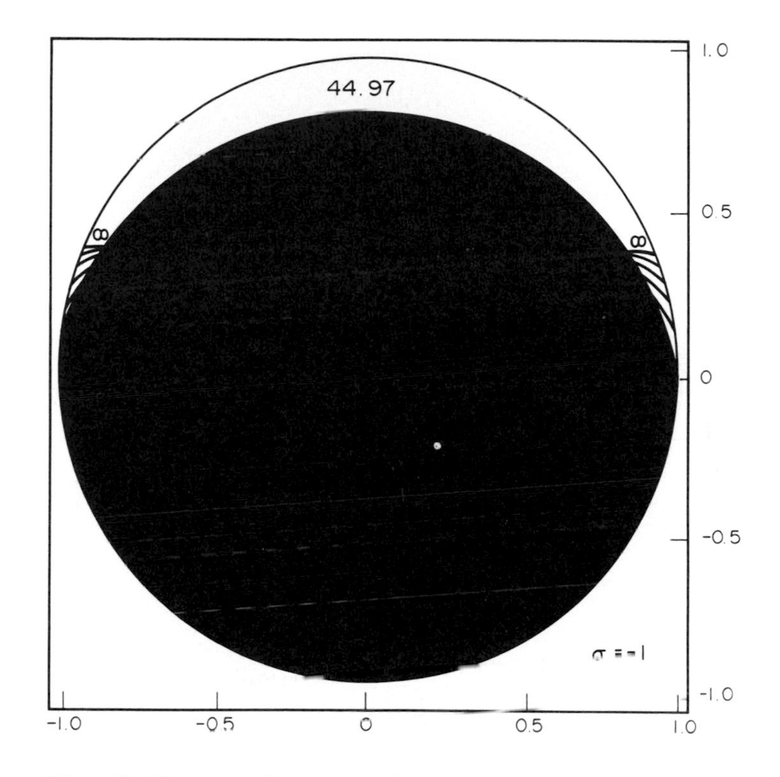

Fig. 12.24b. Yearly averaged radiance of the sky in Madrid. The numbers in the figure are radiances ($wm^{-1}sr^{-1}$). The dark zone is the one looking to the ground. The coordinates are direction cosines with respect to the NS (u) and EW (v) lines Map (a) corresponds to the sky hemisphere and (b) to the ground one.

For instance, if the whole sky hemisphere is to be collected, just as it (almost) does a flat panel, then the use of bifacial cells immersed in a medium of high n (index of refraction) allows for a concentration limit of $2n^2$, that for n=1.5 becomes 4.5. Sometimes we hear that diffuse radiation cannot be concentrated. As we see this is not strictly true in photovoltaics (in thermal devices it can with our definitions of concentration but, in a first approximation, it becomes useless).

We can understand the effect of the index of refraction (in a specific class of configurations) if we consider that the refraction converts the incoming hemispherical beam of rays into a refrated one, with the critical angle (of total internal reflection: arc sin 1/n) as apex semiangle. Thus the angular spread of the incoming ray is reduced and some concentration can

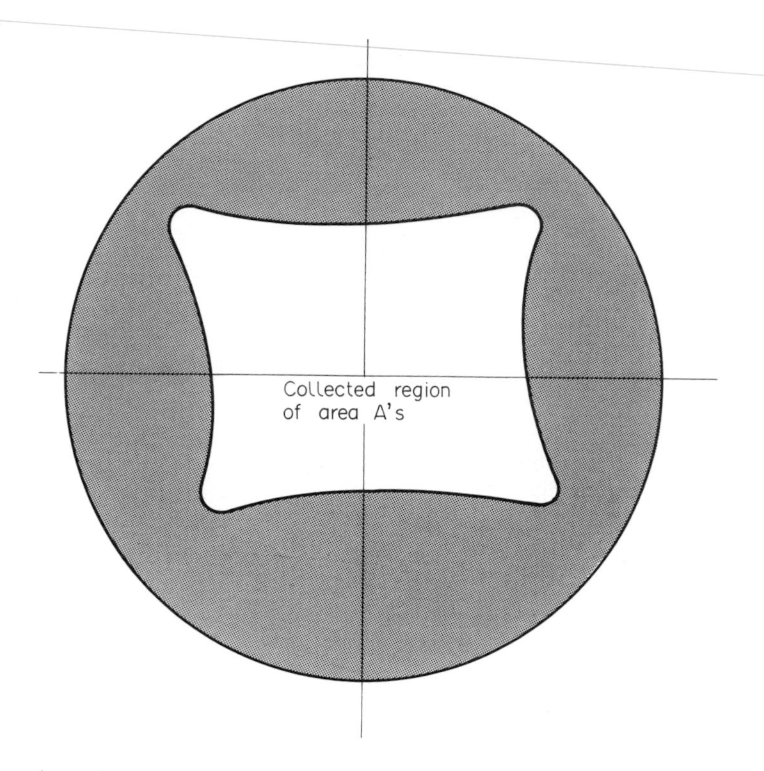

Fig. 12.25. Source collected area.

be achieved. Theoretically the formula we use becomes invalid if n is higher than the index of refraction of the semiconductor.

Actually a flat panel does not collect all the rays. Those very oblique are certainly reflected. If we assume that only rays with 65° of incidence are collected, the concentration of the "whole diffuse light" can be improved in an additional factor of 1.1, up to 5. Higher concentrations can be achieved with some sacrifice to diffuse light.

The aspects concerning the limits of static concentration have been analyzed in detail by Minano and myself (58). We have made clear that the higher the fraction of rays collected the lower the achievable concentration or gain of irradiance but this concentration can be increased if we sacrifice the amount of light we intend to concentrate, i.e. the efficiency of the concentratin panel. However note that we are talking of limits and therefore the concentrators considered are loseless. This reduction of efficiency is thus due solely to impossibility to directing the rays towards the target. In addition it must be noted that we are talking of an efficiency related to the yearly averaged source which is not the instantaneous source commonly used in experiments. This is the reason we prefer to call intercept factor to this efficiency with respect to the averaged source.

The limit of irradiance gain vs intercept factor is given in Fig. 26 for the source in Fig. 24.

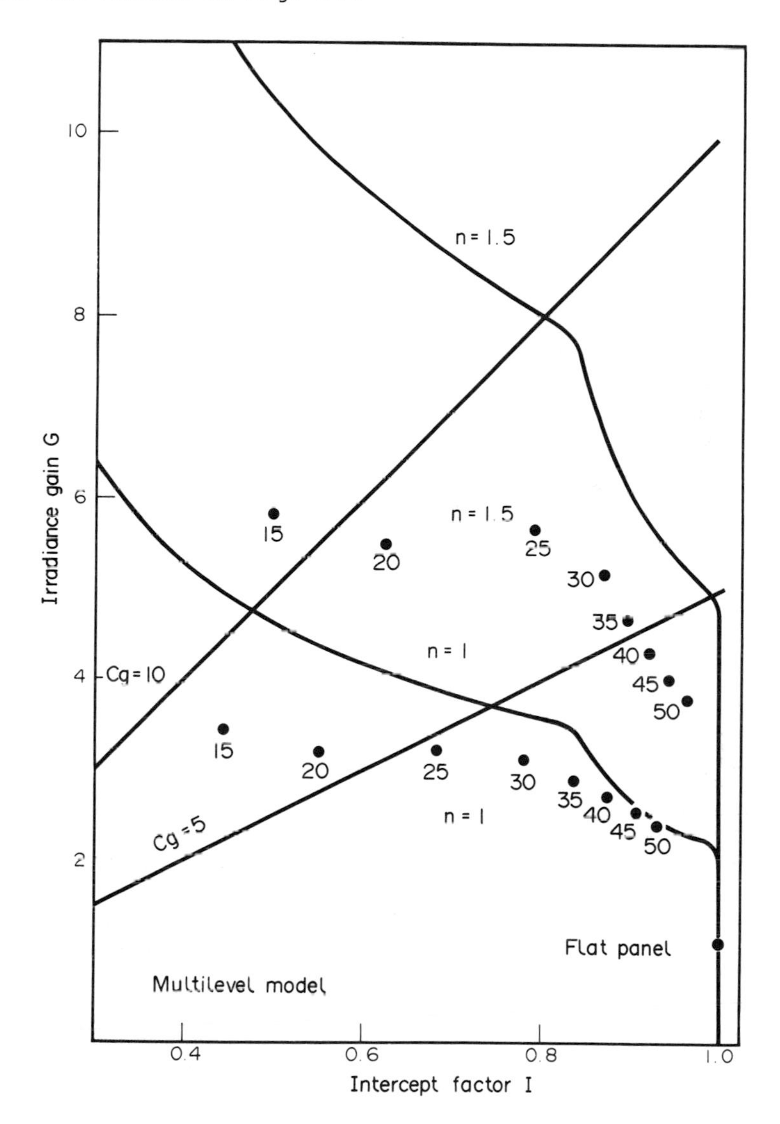

Fig. 12.26. Limit of irradiance gain vs intercept factor for dielectric empty and filled (n=1.5) concentrators. Dots correspond to dielectric filled and empty CPC's with the semiangle in the drawing. The dotted line is the limit for dielectric filled linear concentrators.

To reach the limits in the preceding curve one must fulfil two conditions:

Condition (a) is to illuminate the receiver, a bifacial cell, isotropically if the concentrator itself is illuminated isotropically. The limits are increased in n^2 if the cell is immersed in an optically dense medium.

Condition (b) is to collect all the rays in a region bounded by a line of isoradiance but only these rays. In this way the source collected by the concentrator is reduced in size and bigger concentrations (irradiance gains) can be achieved, while keeping the intercept factor high because we are only collecting the brightest parts of the sky. The line of isoradiance used determines the intercept factor.

5. CONCLUSIONS

From the travel around the problems and solutions concerning photovoltaic concentration several points appear, we think, clear.

In an attempt to achieve high efficiencies, solar cells have experienced in the last three years a tremendous convulsion. They are not to be considered any more as specialized diodes but as devices specifically devised to convert into electricity. A number of novel concepts like those associated to light confinement have been incorporated in the cell.

To date Si cells have given the best efficiencies but I expect that GaAs related cells will soon reproduce these efficiencies and overreach them forecasting a rapid explosion of results with multiple bandgap cells in a near future. Efficiencies in the range of 40% will too soon be available.

To pay these advanced cells need very high concentration , as far as efficiency is concerned for which sophisticated optics are necessary. The situation is today so that we can freely interchange complexity in the cell for complexity in the concentrator, by e.g. by using cavities that require increased concentration at the aperture but allow for the use of more dense metallization patterns in the cells. We advocate in this manner to cease considering that concentrators are a simple theoretical matter, whose difficulty is in the engineering and start to consider in the optical designs how to reach the limiting compromises of optical allowances and achievable concentration.

If this is done I think that the photovoltaic energy production will start to be considered by utilities in central plants by mid 90's. This will occur first at the USA due to their industrial and climatic situation.

The same optical background must be also considered for the production of scattered electricity to make possible the combination of the relatively low cost cell-manufacturing techniques in flat panels and the concentrating techniques.

In many countries diffuse light made a strong component of the available luminous resource but it has to be abandoned to belief that concentration cannot be applied to it. On the contrary the use of sophisticated optic panels can yield a sizeable reduction of the panel cost and this with very small investment so that these techniques are not so strongly hindered by the small market volume that prevents the commercialization of other technologies based in harder technology.

Along this chapter we have not presented any cost analysis. It has been intentional. A cost analysis is related to a product and we have not been describing any. In this concentration is different from other PV technologies where the product is more or less known and the only totally unknown variables are the efficiency and the equipment ammortization costs, both of them easily considered as single parameters that permit an easier parametrization of the energy cost. We have done and published (78) such a type of calculation when the very numerous options of concentration were not yet apparent but here we have preferred to rely on the simple common sense qualitative ideas on cost that we have used along the chapter. It should be a matter for the industrialist to make their own calculations of the cost of their selected solution.

We wish that this chapter might open the eyes of some of them as well as of some scientists to incorporate some of these ideas to their work.

Chapter 13

RECENT DEVELOPMENTS IN SINGLE AND POLYCRYSTALLINE SILICON CELLS

W. SCHMIDT

AEG AAG Electronic, Germany

1. Introduction

The first and today still commercially important application of crystalline silicon solar cells has been for powering satellites in space. For this application, the requirements of reliability, high performance and radiation hardness led to the development of specially adapted and rather expensive single crystal silicon solar cells. After the first 'oil crises', when the search for renewable energy sources was intensified, silicon solar cells have been widely studied and developed as a renewable energy source for terrestrial use.

Thus, research has been carried out from the point of view of materials, fabrication processes, systems and operation conditions (1 - 5). First efforts mainly aimed at strong cost reduction. This resulted in new low-cost and high-throughput cell fabrication technologies as screen-printing and the use of polycrystalline silicon as base material (6, 7). Some of these technologies are already implemented in production lines. The status of these new fabrication technologies will be presented in the following chapter.

However, the cost analysis for the first pilot production lines revealed, that the potential for cost reduction on a module/system base was smaller as expected. This was due to the slow progress in the development of a new low-cost 'solar grade' silicon sheet material and the high area-related module and system costs. Consequently, various efforts to improve the efficiency have been undertaken (8, 9). The basic design principles and the recent experimental realizations of highly efficient single crystal solar cells are the subject of chapter 3. In chapter 4, recent achievements in polycrystalline silicon solar cells and the ways to improve their efficiencies will be pointed out.

Since the module is the final product that will be integrated in a photovoltaic system, and its fabrication process has much impact on solar cell design, important developments in module fabrication, performance and application-oriented aspects will be highlighted in chapter 5. In chapter 6, an outlook will be given on efficiency goals that may be achieved, and on possible future development directions.

2. Status of Low-Cost Fabrication Technologies

The commonly used n+pp+ (back surface field, BSF) solar cell structure is schematically shown in Fig. 1. The usual substrate material is a 0.5 - 10 Ωcm resistivity p-type silicon wafer of 200 - 600 µm thickness. The processing sequence generally consists of the steps: surface cleaning (including saw damage removal) / n+p and pp+ junction formation / back and front side metallization / antireflection (AR) coating formation. The conventional processing sequence as used for space solar cells was a batch process sequence characterized by multi-step wet cleaning procedure, gaseous source open tube diffusion of phosphorus (n+) and boron (p+) or alloying of evaporated aluminium (p+), evaporated Ti(Pd)Ag multilayer contacts and evaporated titaniumoxide or tantalumoxide AR coating layers. Approaches for low-cost process sequences aimed at the development of simple, high-throughput and continuous processes with high potential for automization (6, 9 - 14).

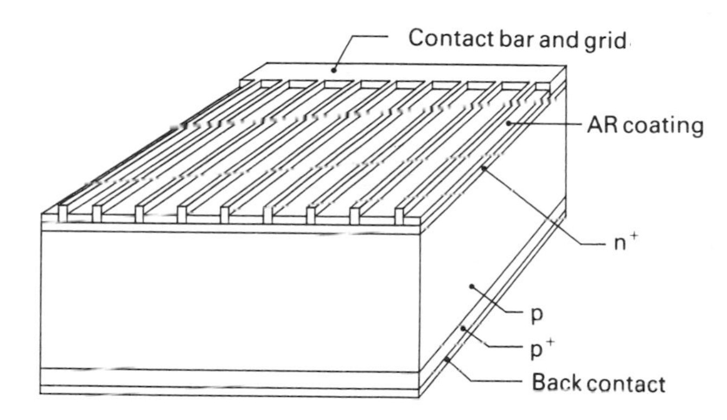

Fig. 13.1.

- Surface Cleaning

Until now, no economic alternative to wet processing has been found. However, the number of cleaning steps is normally reduced to only one or two. Etching in diluted alkaline solutions is a standard means to obtain pyramidally textured (non-reflective, NR) surfaces on <100>-oriented single crystal wafers for reduced reflectance.

- Junction Formation

Open tube diffusion at 800 - 900 °C with a $POCl_3$, PH_3 or PBr_3 source to achieve n+ regions with 25 - 80 Ω/square sheet resistivity is still in widespread use. Doping films, applied by screen-printing, spray-on or spin-on with subsequent dry and drive-in step in a belt furnace have also proved their reliability (6, 12). Doped silicon oxide layers formed by chemical vapour deposition are another alternative source. Rapid thermal annealing equipments which can be used instead of a furnace are also under investigation (14). For the p+ region formation, similar methods are used to dope with boron. To avoid counterdoping, diffusion masks of undoped silicon oxide layers can be applied by the same means as described for doping films.

To obtain the n+pp+ structure, another simple way is to diffuse phosphorus all around the wafer, use alloyed Al contact for compensation, p+ formation and possibly gettering on rear side and remove parasitic edge diffusion either by laser scribing or dry plasma etching of stacked wafers (7). In Fig. 2, the lay-out of the plasma etching equipment is shown in principle.

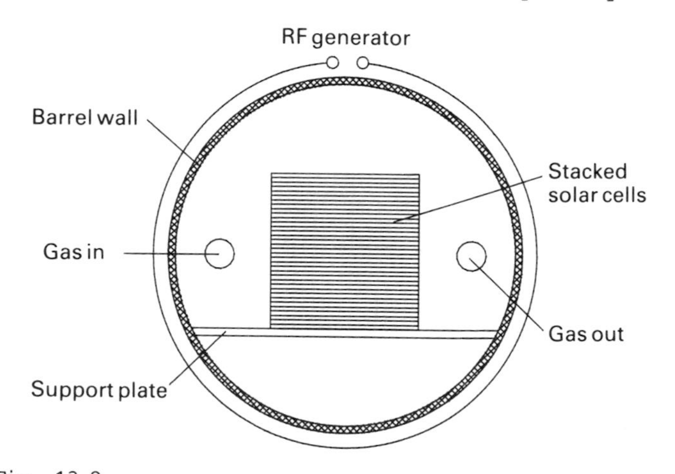

Fig. 13.2.

Non-mass analyzed ion implantation and subsequent annealing with pulsed high intensity light or electron beams can also be used for automated junction formation (13, 15 - 17). Potentially being a lifetime-conserving 'cold' process, it could be of special advantage for future solar-grade silicon materials.

- Contact Formation

For large-scale production, screen-printed back and front contacts are used by many manufacturers (6, 7, 11, 18). The pastes to be printed are mainly composed of metal and glass powders dispersed in an organic vehicle that is adjusted to the right viscosity. Burn-out of organics and sintering are usually performed in a conveyor belt furnace with peak temperatures in the 600 - 850 °C range for a few seconds or minutes. Normally, this is requiring larger junction depths in order to avoid shunting paths. Flame-sprayed Al is also applicable for back metallization (7, 19). Another metallization technique is electroless plating of Ni or/and Cu which requires an oftenly screen-printed plating resist pattern on front side and a subsequent anneal step to increase contact adhesion (20, 21).

Finally the contacts can be solder-dipped to reduce contact series resistances and facilitate interconnection process.

- Antireflection Coating

Low-cost deposition techniques use a metalorganic titanium or tantalum compound mixed with suitable organic additives. Such a solution is deposited on the solar cell by screen-printing, spin-on or spray-on, followed by a temperature treatment at about 500 °C to drive-off organics and form the metaloxide antireflective layer (6, 13). Such solutions can be used in a CVD conveyor belt furnace or for spray pyrolyses on hot substrates (18, 22). Another method is chemical vapour deposition of silicon nitride (7, 23).

For a complete process sequence, different processing steps have to fit with each other and with module fabrication techniques to obtain best results. Therefore different junction formation and metallization techniques could be used for front and rear side. It could also be advantageous to rearrange the process steps, e.g. to perform the AR coating first and then to fire through it the screen-printed contact in order to avoid short-circuiting of a shallow pn junction. Or a screen-printed AR coating could be used as plating resist pattern. A really simple process sequence is obtained by using screen-printing (thick film) technology for all process steps except cleaning. This has the advantage that the fabrication equipment consists of identical modules as those depicted in Fig. 3, allowing fully automatic, continuous wafer processing. The solar cells fabricated with this technique almost reach the quality level of those made by conventional processing, as demonstrated in Fig. 4 for the AR coating, while fabrication coats are distinctly reduced (24).

However, with the stringent goal to drastically improve the cell

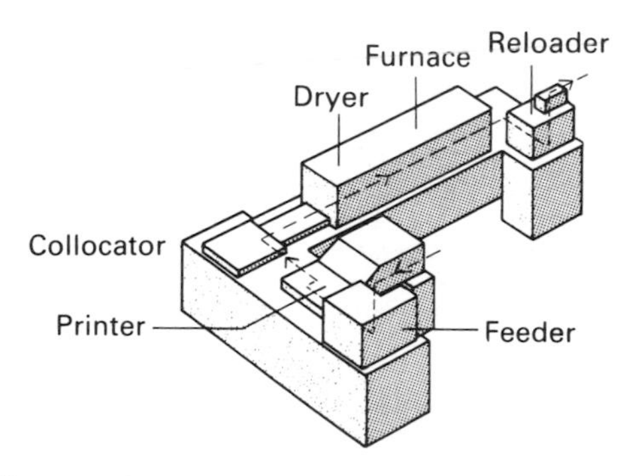

Fig. 13.3.

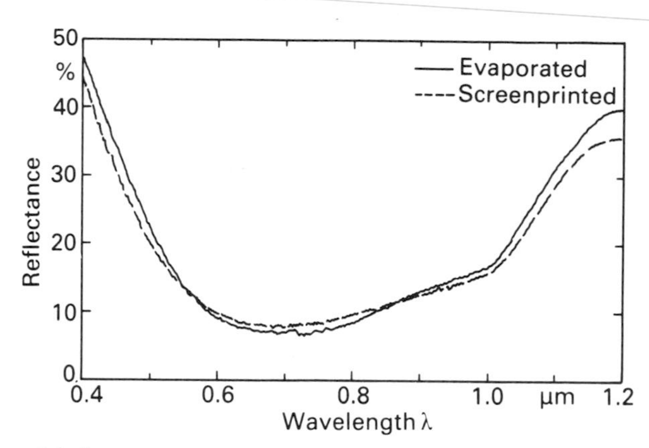

Fig. 13.4.

efficiency, those processes have to be selected out of the broad spectrum outlined before that can match the high efficiency cell design features. These features will be discussed in the next chapter for single crystal silicon solar cells.

3. High Efficiency Single Crystal Solar Cells

3.1 Basic Design Principles

A fast uplift of the upmost conversion efficiency of single crystal silicon solar cells was reported by many groups since 1983 (25, 26), culminating presently at values close to 22 % under AM 1.5 (100 mW/cm²) conditions and 28 % at 150 suns concentration (27). Before presenting the different technological approaches to the high efficiency cell, it is useful to demonstrate the general rules how the efficiency can be enhanced.

The starting point should be the most widespread commercially realized version of the cell structure shown in Fig. 1. It is characterised by the use of CZ-grown <100>-oriented 1 - 10 Ωcm p-type substrates with textured surfaces. Rather heavy diffused front regions about 0.5 μm thick and with a surface dopant concentration in the 10^{20} - 10^{21} cm^{-3} range are imposed by the screen-printed contact formation, including a p+-layer formed by firing of screen-printed aluminium paste. To further minimize reflection losses, a single antireflective coating layer (TiO_x) with a refractive index of 2.3 is deposited on front surface. These low-cost , textured BSF solar cells have efficiencies of typically 14 - 15 % AM 1.5.

To improve such a cell, it is necessary to look at the different loss mechanisms. They can roughly be divided into optical and electrical losses. Optical losses are due to reflection at the front surface, shadowing by front contact grid, light absorption in the reat contact and coupling out at the front surface. Electrical losses are due to recombination processes in the bulk and at the surfaces, but can also result from high ohmic resistances and shunt leakage currents.

The reflection of light incident on the front surface can further be reduced by applying a double layer antireflective coating (DLARC). A marginal reduction can be achieved by applying V-grooved surfaces instead of the pyramidally textured ones (28), as pointed out in more detail later on. Absorption of a considerable amount of long wavelength light in the rear contact can be avoided by depositing a reflective metal layer on the rear surface which acts as back surface reflector (BSR) (29). Reflection values of up to 98 % are possible, if a quarter-wavelength dielectric layer is inserted between the silicon surface and the metal layer (30). In order to trap the light in the bulk of the cell, optical confinement based on the randomization of the light direction by diffuse back reflectors (31) or other randomizing schemes as well as geometrical light trapping schemes have been proposed. The result of theoretical calculations done by Campbell and Green (28, 32) is given in Fig. 5. It presents the upper limit for the attainable short-circuit current at a given cell thickness. It demonstrates that very effective light trapping is provided by randomizing (Lambertian) surfaces and the random pyramid layout, while the perpendicularly grooved structure is shown in Fig. 6 is found to give even superior performance.

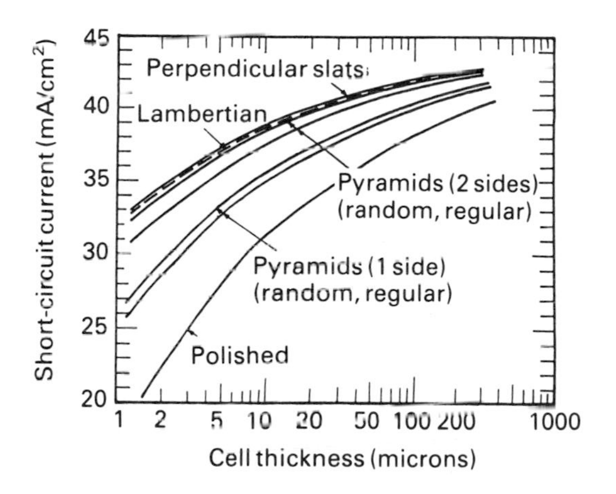

Fig. 13.5.

Electrical power losses due to low shunt and high series resistances generally depend on cell fabrication technology. They can be reduced by applying proper edge treatments and contact annealing to avoid shunting or by using optimized contact patterns and materials especially for the front contact formation (33).

To enhance the short-circuit current, recombination of minority carriers has to be reduced. Recombination can occur in the bulk

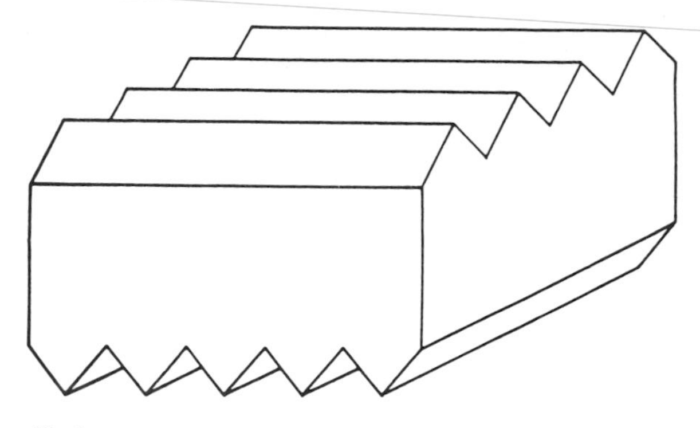

Fig. 13.6.

and emitter region due to intrinsic processes as Auger and radiative recombination or bandgap narrowing effects. Extrinsic recombination occurs at states with energies in the forbidden gap that are generally attributed to impurities and crystallographic defects in the bulk and the surfaces. However, the open-circuit voltage is also effected by such recombination mechanisms since it is related to the short-circuit current density roughly by the familiar relationship

$$V_{OC} = Kt/q \ln(J_{SC}/J_O) \tag{1}$$

where J_O is the reverse saturation current density. Thus, J_O has to be minimized. Assuming constant doping levels in emitter and base, J_O can widely be expressed as a function of intrinsic carrier concentration n_i, doping density N of donors (D) and acceptors (A), width W of base (B) and emitter (E), minority carrier diffusion length L, diffusion constant D and surface recombination velocity S of electrons (n) and holes (p) (Sn can include BSF effect) (34):

$$J_O = J_{OB} + J_{OE} = \frac{qn_i^2 D_n}{N_A L_n} \times \frac{S_n L_n/D_n + \tanh(W_B/L_n)}{1 + (S_n L_n/D_n)\tanh(W_B/L_n)} + \frac{qn_i^2 D_p}{N_D L_p} \times \frac{S_p L_p/D_p + \tanh(W_E/L_p)}{1 + (S_p L_p/D_p)\tanh(W_E/L_p)} \tag{2}$$

For commercial low-cost cells on medium resistivity (2 - 5 cm) substrates with relative high W_B and low L_n values, calculations using eqn. (2) reveal that the open-circuit voltage is dominated by the base component of the saturation current.

However, if higher quality material and lifetime enhancing processes are used, than the emitter contribution J_{OE} due to recombination in the highly doped thick diffused layer becomes dominating the open-circuit voltage. For space solar cells, this has been improved by reducing junction depth and doping level, leading to the so-called 'violet' cell (35) with a 'transparent' emitter (36). It should be mentioned, that the correspondingly higher sheet resistance requires fine grid design, generally to be defined by photolithographic techniques, to obtain optimum small values of resistance and grid shading.

Now, when recombination in the emitter is reduced by improved emitter profiles, the minority carriers are lost at the front surface due to high surface recombination rates. The recombination at the free silicon surface can effectively be reduced by appropriate passivation. To-date, best results are attained by thermally growing a SiO_2 layer thereon (37) as it will be discussed in detail in the next section. For free surface passivated cells, recombination losses beneath the ohmic contact grid become important (38). This is due to the fact, that metals in contact with silicon induce a barrier to majority carriers, giving rise to a contact resistance. To keep this resistance at acceptable low values, doping level immediately under the contact region has to be high to allow majority carrier flow via quantum-mechanical tunneling. But since there is no barrier to minority carriers, the recombination rate at such a metal contact is high as pointed out in Fig. 7 and its contribution to the saturation current density of the emitter can generally be calculated from:

$$J_{OC} = A_C \, J_{OE} (S_P \quad 5 \times 10 \quad cm/s) \qquad (3)$$

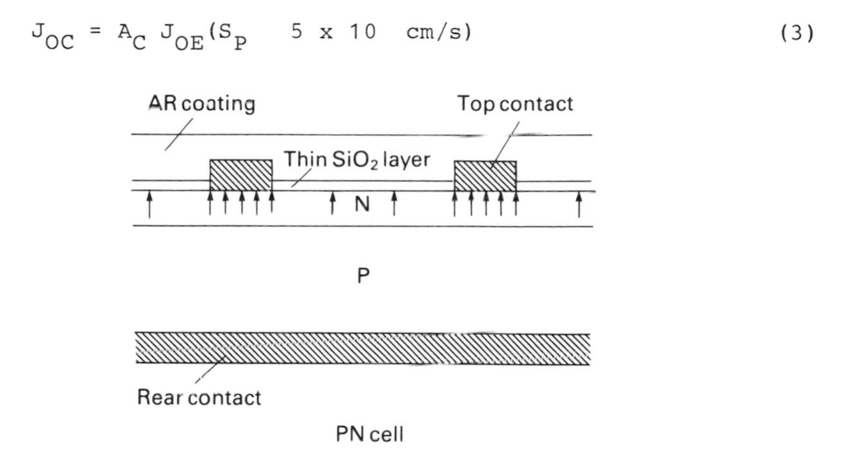

Fig. 13.7.

where A_C is the contact coverage factor. Equation (3) together with eqn. (2) show that minimizing of J_{OC} is achieved by high doping levels and thick doped layers which is inconsistent with the requirements for the non-contacted areas.

An obvious approach to overcome this problem is the incorporation of a higher doped region into the contact area as realized in the laser grooved, buried contact cell structure (39), given in Fig. 8. Another approach is the insertion of a thin insulating oxide layer between the contact and the silicon as pointed out in Fig. 9. This approach is the co-called metal-insulator-np junction

(MINP) cell structure (40). However, this oxide layer should be thin (about 2nm) to keep the contact resistance low, while thicker layers of 10 nm or more are favourable to the non-contacted surface area (41, 42) as can be deduced from Fig. 10. Therefore complicated processing is required to obtain optimum results with this structure.

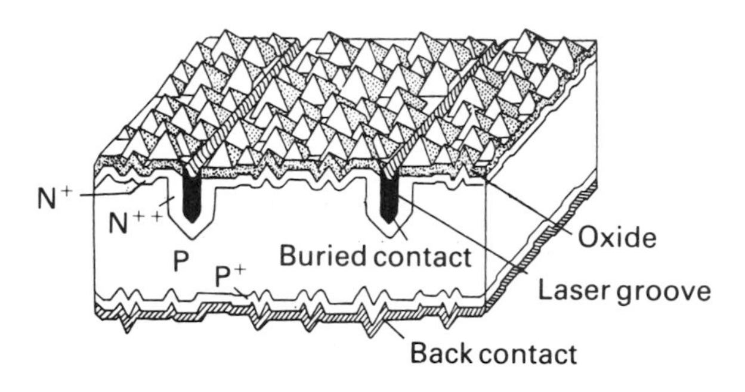

Fig. 13.8.

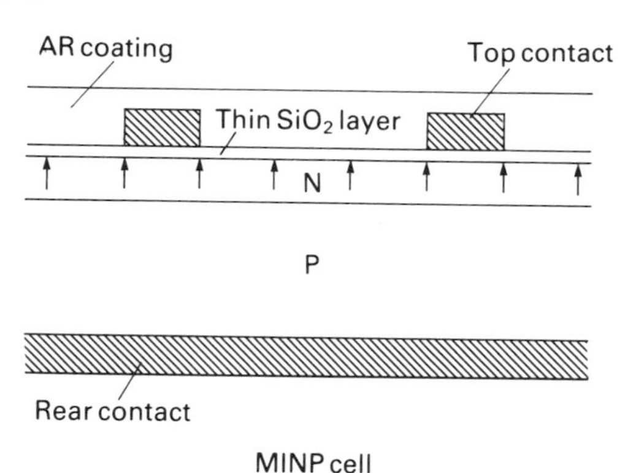

Fig. 13.9.

The passivated emitter solar cell (PESC) design (43, 44), shown in Fig. 11, bypasses this problem by restricting the area of metal contact to the diffused layer to very small openings in the passivating oxide layer. With this structure, efficiency values exceeding 20 % AM 1.5 have been realized (39, 45).

Another recent development has been to replace the metallic layer by a layer of degenerately doped polysilicon (46, 47) or semi-insulating polysilicon (SIPOS) (48). These layers induce a barrier to minority carriers while allowing flow of majority carriers. Open-circuit voltages of 720 mV have been demonstrated with small test structures (48). Especially for concentrator applications, structures with n+ and p+ diffused regions on the back side have been evaluated (49 - 51), following the interdigitated back contact solar cell structure (IBC) (52) of Fig. 12, which completely avoids contact shadowing.

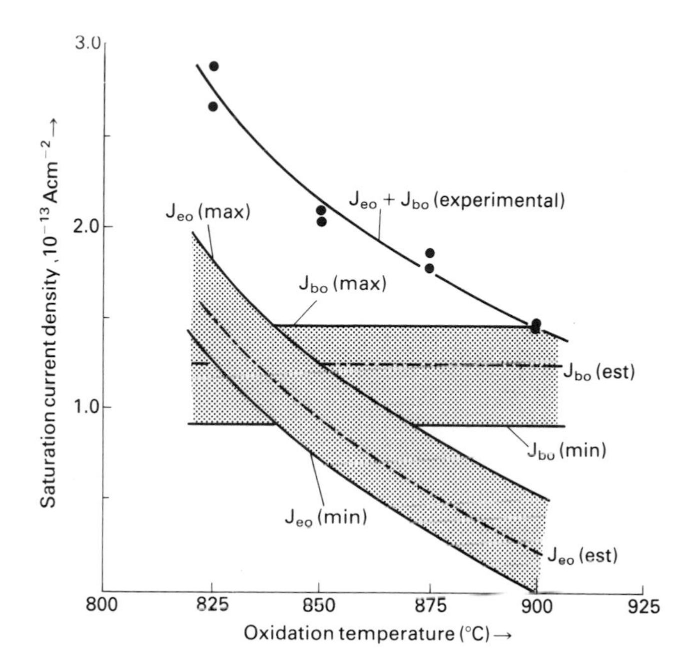

Fig. 13.10.

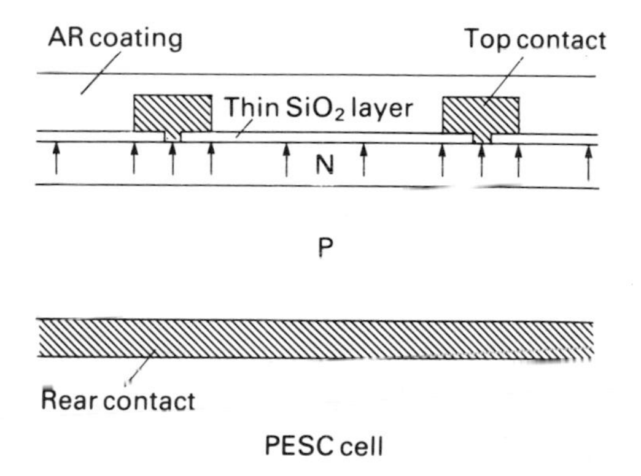

Fig. 13.11.

To find the optimum design parameters for high efficiency crystalline silicon solar cells, it is necessary to quantify the aforementioned design considerations. For this purpose, solar cell simulation programs have been developed by various groups (53 - 56) to investigate how the design parameters influence solar cell performance.

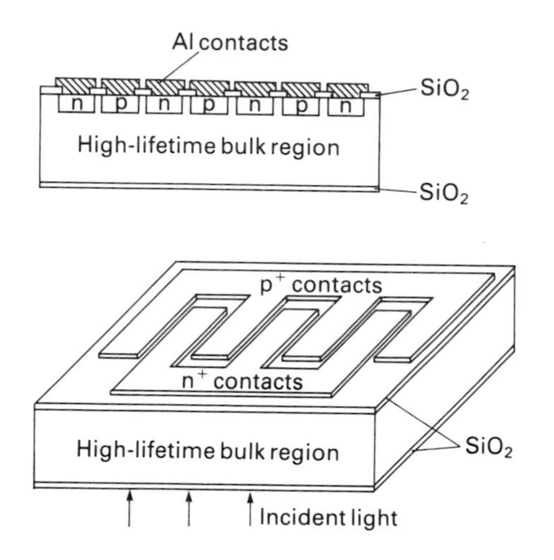

Fig. 13.12.

Representative results were reported by Saitoh et al (57). They have utilized the one-dimensional semiconductor device analysis program PC-1D developed at Iowa State University (55) to demonstrate the influence of front surface recombination velocity S_F, surface doping density N_S and junction depth x_j (Erfc impurity profile) on the performance of an n^+pp^{++}, 2 Ωcm, 400 μm thick cell. The p^+ layer parameters were held constant with 1×10^{19} cm^{-3} doping density, 14 Ω/sq. sheet resistivity and 10^7 cm/s surface recombination velocity. Front and back surface reflection values were 0% and 90%, front contact area was 3.2% of total cell area.

Figures 13 to 16 show the dependence of open-circuit voltage, short-circuit current and efficiency on doping profile in the emitter region for the two different cases of high (S_F = 10 cm/s) and low ($S_F = 10^3$ cm/s) surface recombination velocity.

For high recombination velocity, shallow highly doped emitters are expected to give the best performance due to relative low recombination in the emitter and shielding of the surface. However, for low surface recombination velocity corresponding to well passivated surfaces, rather high efficiency values can be expected over a wide range of carefully selected combinations of doping profile parameters. For dopant surface concentrations of about 2 - 5 x 10^{19} cm^{-3}, good results are achievable for the usual junction depth of about 0.5 μm. Higher surface dopant concentrations require shallower junctions, lower concentrations deeper junctions to attain optimum performance. Best results are expected in the latter case. But it should be pointed out,

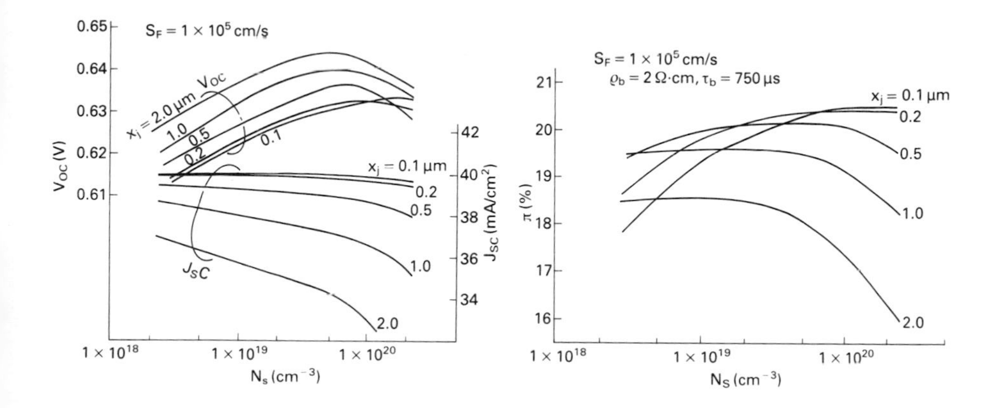

Figs. 13.13 and 13.14.

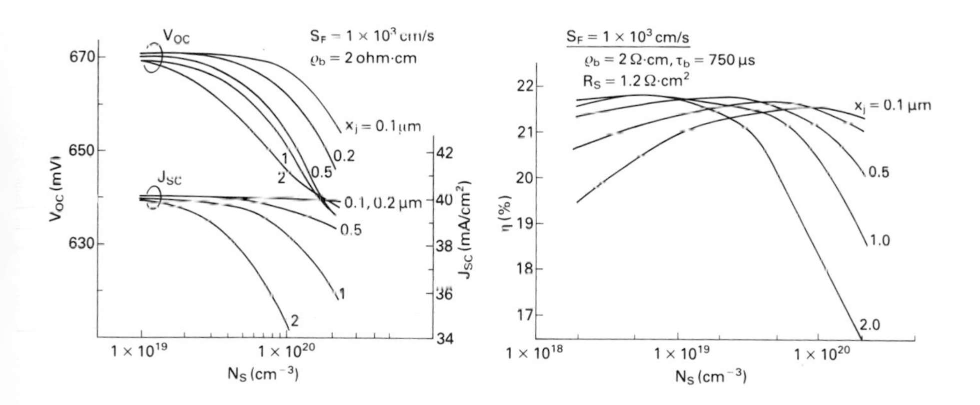

Figs. 13.15 and 13.16.

that efficiencies exceeding 20% seem to be feasible even for the higher surface recombination value of 1 x 10^5 cm/s. Therefore, different design approaches can lead to rather high efficiency values.

3.2 Experimental Status of Concentrator Cells

Under an economical viewpoint, the most promising application of highly efficient single crystalline silicon solar cells seems to be in concentrating devices. Observing the basic principles for high efficiencies and adjusting them to high illumination conditions (50, 58, 59), various approaches to high efficiency concentrator solar cells have been realized. Measured peak efficiencies are in the 25 - 28% range. The results are summarized in Table 1.

- Interdigitated Back Contact (IBC) Solar Cell

The Universitè Catholique de Louvain working with the Universitè d'Aix-Marseille III have realized an improved version of the IBC structure (60), shown in Fig. 12. Starting with high carrier lifetime (1 ms), <100>-oriented FZ 10 Ωcm p- or n-type silicon wafers, p^+ and n^+ collecting junctions are formed by ion implantation of boron and phosphorous predeposition, respectively. An n^+ region is also formed on the front side, which is then texture-etched, leaving remnants of the n^+ layer and thereby forming a polka-dot floating tandem junction for p-type and front surface field for n-type substrates as shown in Fig. 17. An 1000 Å thick SiO_2 layer acts as antireflection coating, contacts are formed with 2 μm thick Al layers. The final n^+ and p^+ regions are 4.4 μm and 2.3 μm deep, sheet resistance is 5.7 and 57 Ω/square respectively. Best results were obtained for p-type substrates, with peak efficiency of 25.6% at 100 suns and over 24% for concentrations exceeding 300 suns. Rather interestingly, poorer performance has been found with island diffusions which are applied in the following approach.

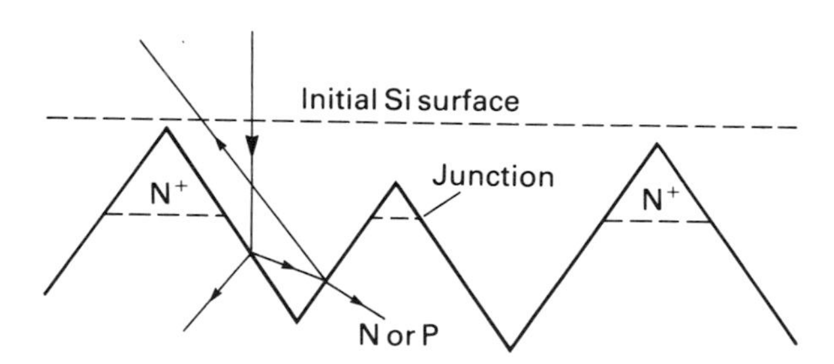

Fig. 13.17.

- Point Contact (PC) Solar Cell

This structure, developed at Stanford University, is shown in Fig. 18. The major difference between the PC and the IBC cell is the restriction of diffused regions to small pockets rather than to fingers, which are arranged in a checkerboard fashion over the rear of the cell. Best results have been obtained for a 92 μm thick cell on high-resistivity n-type <100>-oriented substrate with 1120 Å SiO_2, 3 μm thick Al metallization, 45 μm spacing of

Table 1: High efficiency concentrator cells (AM 1.5, 1 sun equal to 100 mW/cm^2, 25 °C, (28 °C:PESC))

Cell structure	Substrate	Concentration (suns)	Responsivity (mA/W)	V_{OC} (mV)	FF (%)	Eff. (%)	Area (mm^2)	Source
IBC	<100>, FZ p, 10 Ωcm 280 μm	104 207 311	445	756 775 789	76.0 71.6 69.4	25.6 24.7 24.4	2 x 2	U. Louvain/ U. d'Aix- Marseile /160/
	dto. 160 μm	1	417	636	80.0	21.0	10 x 10	
PC	<100>, FZ n, 390 Ωcm 92 μm	1 52 79 105 178 206 408 697	415 414 414 412 411 407 402 389	681 807 814 820 831 834 845 852	78.6 84 84 83 83 82 79 74	22.3 28.0 28.4 28.4 28.3 28.0 26.6 24.3	3 x 5	Stanford U. / 61/
		1 100	413 408	663 789	79.4 81.4	21.7 26.2	8 x 8	/ 63/
μg PESC	<100>, FZ p, 0.2 Ωcm 280 μm	1 11 31 50 76 133 216	402	653 717 740 753 751 773 780	82.9 82.6 82.5 81.4 80.2 77.3 73.7	21.8 23.8 24.6 24.7 24.6 24.0 23.1	3 x 3	UNSW / 64/
		≈1000				>20		/ 66/

doped regions and a photolithographically defined texturized front surface. An efficiency greater than 28% was measured at Stanford University for 50 - 200 suns, peaking at 28.4% at 100 suns. Measurements at Sandia Nat. Lab. indicate efficiencies of 22.3% at 1 sun, 28.2% at 95 suns and 21.9% at 914 suns for a cell from the same wafer. Detailed cell analysis has shown that different recombination components dominate with varying incident power density. As shown in Fig. 19, surface recombination dominates up to 5 suns, the emitter contribution up to 200 suns and Auger recombination beyond 200 suns. However, the cells discussed are only small laboratory cells that can not be integrated in a concentrating device. More recently, larger 8 x 8 mm² cells with multilevel metallization have been developed that can be soldered on alumina mounts (63). Efficiencies of 26.2% at 100 suns and 21.7% at one sun were measured at Sandia Nat. Lab.

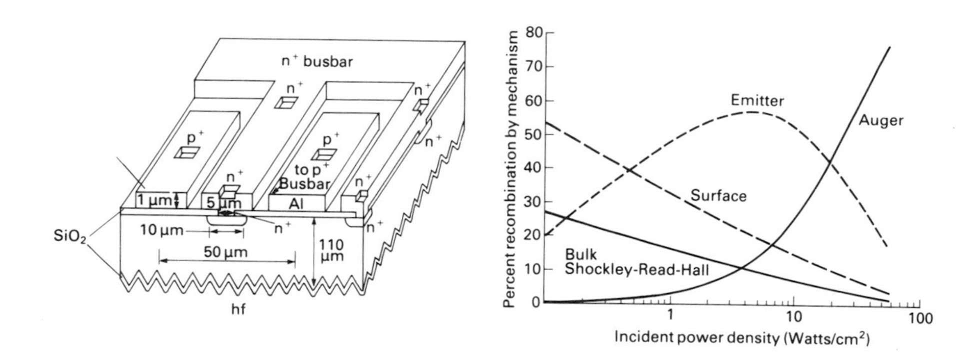

Figs. 13.18 and 13.19.

- Microgrooved Passivated Emitter Solar Cell (µg PESC)

The µg PESC structure (64, 65), shown in Fig. 20, is the most conventional approach. Characteristic features are the photolithographically defined V-grooves with a typical depth of 5 µm on a pitch of 10 µm, the metallization with gridline spacing, width and thickness of 150, 6 and 1.5 µm, respectively, whereas the fingers run obliquely to the direction of the grooves. As a consequence, demonstrated in Fig. 20, light is partially reflected from the metallization onto the active cell surface. Combined with a DLAR coating, shading and reflection losses are reduced to about 3%. Compared to a textured surface, lateral series resistance in the n^+ layer is also reduced in this design.

The BSF region was formed by evaporation of 0.5 - 1 µm Al and alloying at 950 - 1000 °C. The final top layer sheet resistivity is about 50 Ω/square. For 280 µm thick cells on <100>-oriented FZ 0.2 Ωcm p-type Si substrates, efficiencies of 22% at one sun and 25% in the 80-100 suns concentration range (T = 25 °C) have

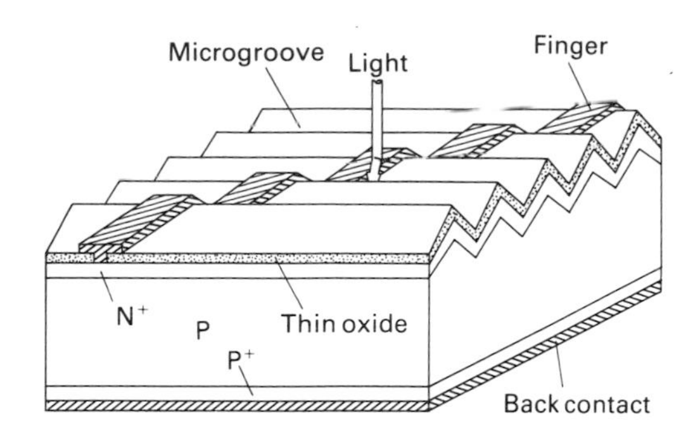

Fig. 13.20.

been measured (66). With more heavily plated top contact metallization, efficiencies of above 20% to beyond 1000 suns concentration were achieved.

3.3 Experimental Status of Non-Concentrating Cells

The previously described concentrating cells have shown good performance at one sun with efficiencies of about 22%. However, their size is much smaller than the traditionally accepted minimum area of 4 cm² for comparing non-concentrating silicon cell technology. Nevertheless, these 22% must now be considered as a technologically achievable goal. The highest efficiencies for 4 cm² area non-concentrating silicon solar cells (given in Table 2) have been realized at the UNSW with the µg PESC structure.

- University of New South Wales (UNSW) Approaches

Pioneering work on high efficiency cells was performed by the group of M. Green at the UNSW. In the beginning, their approach focussed on obtaining high voltages. They used low-resistivity (0.2 Ωcm), high quality FZ Si as substrate material, giving L_n values of about 200 µm in finished cells. Solar cells with MINP (40) and PESC (43, 44) structures, shown in Fig. 9 and 11, have been realized. Maximum efficiencies of 18.7 and 19.8 % and open-circuit voltages over 640 and 660 mV, respectively, were achieved (AM 1.5, 100 mW/cm², 28 °C). The first cells with over 20% efficiency have been produced using the µg PESC approach (39, 45). With an improved BSF, cells of greater than 20% efficiency have been demonstrated on p-type substrates right across the range from 0.1 to 2000 Ωcm (27). A maximum efficiency of 21.4% (AM 1.5, 25 °C) has been demonstrated recently on 0.2 Ωcm FZ Si substrate (67). This is the highest efficiency value reported up to now for non-concentrating cells (see Table 2).

A more practical approach is the laser grooved cell structure (39) outlined in Fig. 8. Unlike all other high efficiency approaches, the fabrication needs no photolithography or other

Table 2: High efficiency non-concentrating cells (AM 1.5, 100 mW/cm^2)

Cell structure	Substr. type/ resist. (Ωcm)	V_{OC} (mV)	I_{SC} (mA/cm^2)	FF (%)	Eff. (%)	Test Temp. (°C)	Cell Area (cm^2)	Source
MINP	FZ p 0.2	641	35.5	82.2	18.7	28	4	UNSW /43/
PESC	FZ p 0.2	662	36.5	81.9	19.8	28	4	/39/
μg PESC	FZ p 0.2	669	38.6	82.9	21.4	25	4	/67/
	100	630	41.7	79.2	20.8		0.8	/27/
laser grooved	FZ p 0.25	653	36.7	80.8	19.4	25	12	/68/
	0.5	648	37.5	80.6	19.6			
	10	634	39.3	79.4	19.8			
	100	627	39.5	75.5	18.7			
	CZ p 1.0	621	38.7	78.6	18.9	25	47	/27/
μg PESC	FZ p 0.2	647	39.3	80.8	20.5	28	4	Hitachi /42, 69/
	2.0	639	39.5	80.4	20.3			
	6.0	621	39.6	78.4	19.3			
textured PESC	FZ p 0.2	660	36.8	82.7	20.1	25	4	JPL /70/
intrinsic pass.	FZ n 0.3	634	36.1	81.3	18.9	25	4	ORNL /71, 72/
extrinsic pass.		657	36.0	81.6	19.3			
hybrid		663	36.2	81.7	19.6			
textured PESC	FZ p 0.3	635	36.3	81.6	18.8	28	4	Spire /73, 74/
passiv. BSF	FZ p 1.5	61	38.0	78.1	18.1	25	54	/76/
passiv. BSF	p 0.33	645	36.6	80.9	19.1	25	4	Univ. Polit. de Madrid /78/
	21	623	38.9	78.4	19.0			

expensive processing steps. In the texture-etched, lightly phosphorus diffused and oxidized surface, 15 - 20 μm wide and 40 - 100 μm deep slots are scribed with a laser, cleaned by etching and then heavily diffused. Aluminium is deposited on the rear surface and sintered at high temperatures. Finally, Ni and Cu are electroless plated to slotted areas and the rear of the cell using the front oxide as plating mask. The heavy diffusion in the grooves shields the high recombination of the contact area and reduces contact resistance, while the metal filled grooves give fine line width metallization with only about 2% front contact shading. On 200 μm thick FZ silicon substrates, efficiencies over 19% have been demonstrated over a wide resistivity range (0.2 - 10 Ωm) for 12 cm^2 cells (68). Even more important, large area (47 cm^2) cells on 1 Ωm commercial CZ silicon wafers with up to 19% efficiency have been fabricated, demonstrating that this approach is well suited for large area cells, as well.

- Hitachi Approach

Silicon solar cells with the μg PESC structure have also been fabricated by Hitachi Central Research Laboratories (42, 69). Exceptionally high efficiencies of over 20% have been obtained with 300 thick FZ p-type substrates of 0.4 and 2.0 Ωm resistivity. Main features of this approach are the formation of BSF by rapid annealing of screen-printed Al paste in a lamp furnace at 750 °C, the 24 nm thick passivating oxide, evaporated Ti (2 μm thick) Ag grid with 6 μm electroplated Au covering 2.5 % of cell area and a 50 nm TiO_2/110 nm MgF_2 DLAR coating (42). In Fig. 21, the change of diffusion profile during oxidation is shown, while Fig. 22 exhibits the improved blue response for a thicker passivating oxide due to reduced surface recombination velocity.

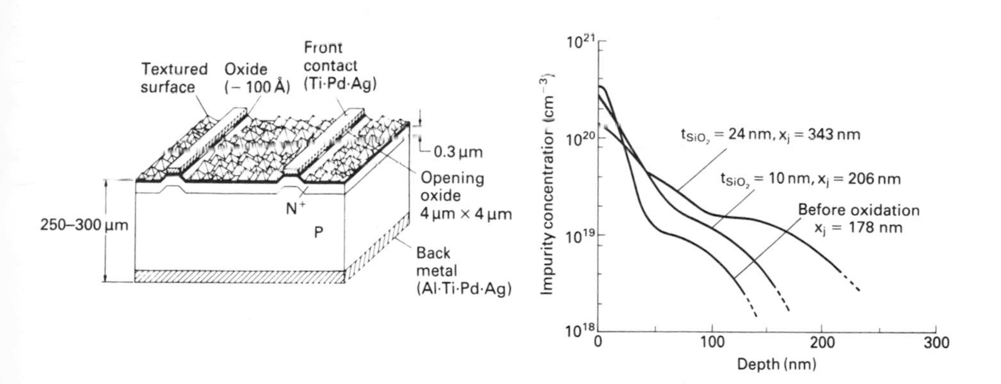

Figs. 13.21 and 13.22.

- Jet Propulsion Laboratory (JPL) Approach

This approach combines the PESC structure with contacts resting on mesas above a textured surface as outlined in Fig. 23. With TiO_x/Al_2O_3 DLAR coating, 20% efficient solar cells have been realized on 0.2 Ωcm p-type FZ substrates.

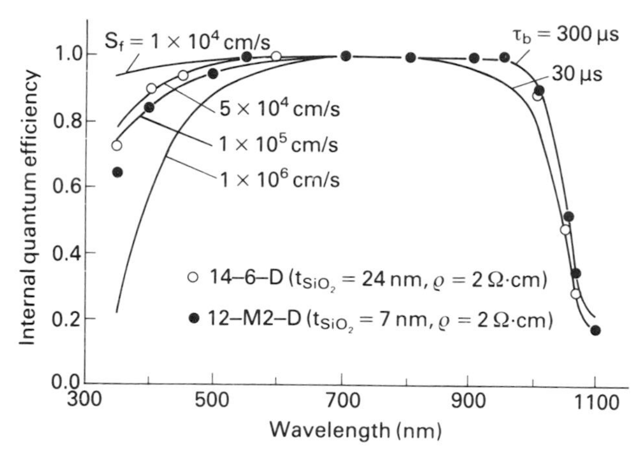

Fig. 13.23.

- Oak Ridge National Laboratories (ORNL) Approach

Outstanding feature of this rather unconventional approach is the formation of the doped layers by glow-discharge (≈ 1 keV) non-mass analyzed ion implantation. Subsequent pulsed excimer laser annealing results in a highly doped narrow emitter with a strong built-in electrical field that shields the high recombination at the surface, acting as intrinsic passivation. With TiPdAg contacts covering 3 - 4 % of cell area and ZnS/MgF_2 DLAR coating, about 19% efficiency for 0.3 Ωcm n-type FZ base material were achieved. The main drawback of intrinsic passivation is that heavy doping effects in the emitter limit the open-circuit voltage. Consequently, better results have been obtained for thermal annealing combined with oxidation, referred to as extrinsically passivated cells. However, best results were achieved for hybrid cells, combining excimer laser annealing and oxide passivation. Maximum efficiency of 19.6 % was obtained, though front surface was polished.

- Spire Approach

The work on high efficiency silicon cells conducted by the Spire Corporation (73 - 77) predates most of the other ones. Using the cell structure outlined in Fig. 24, efficiencies close to 19 % have been obtained on low-resistivity both p-type and n-type front textured FZ substrates (74, 75). Characteristic feature of these cells is the formation of n^+ and p^+ regions by low-energy (5 keV) mass analyzed ion implantation and thermal annealing. The front contact grid covers 4% of cell area, but ohmic contact area through holes opened in the oxide is only 0.1%. The improvement of open-circuit voltage by surface passivation and reduced ohmic contact area is clearly demonstrated in Fig. 25 for different emitter doping values.

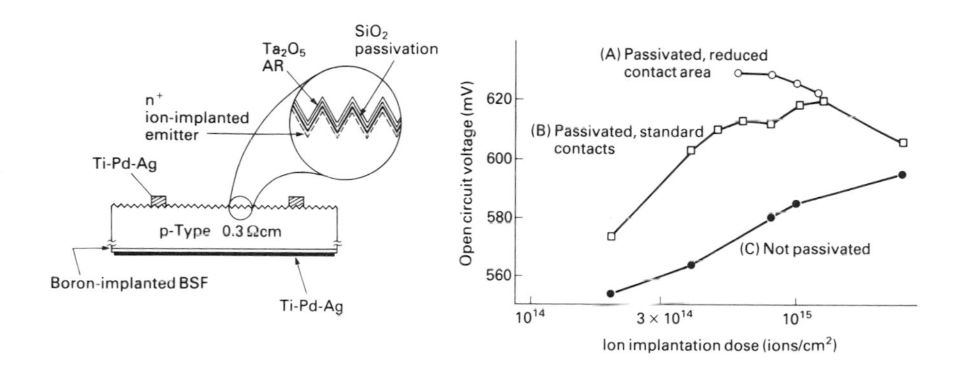

Figs. 13.24 and 13.25.

As already demonstrated by theoretical calculations in chapter 2 and shown in Fig. 15 and 16, deep and lightly doped passivated emitters are suited to achieve high efficiency solar cells. For a constant dose, emitters of different thickness have been performed by varying anneal temperature and time. As can be seen in Fig. 26, nearly constant blue response was observed for surface passivated cells while it degrades without a passivating oxide with increasing junction depth (77). Same results hold for the efficiency. An important feature of the cells with the deepest junction is an excellent high temperature contact stability, at least up to 600 °C for 15 minutes in N_2.

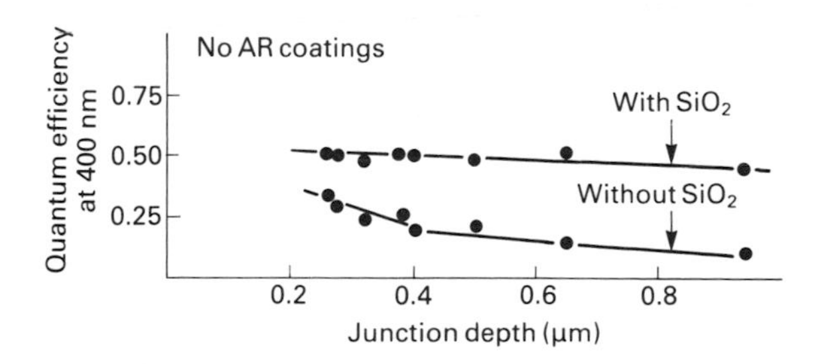

Fig. 13.26.

The developed processing sequence has been applied to the fabrication of large-area cells and modules with only little loss in performance (76). Using 1.5 Ωcm p-type FZ substrates, batches of cells of 53 cm^2 area with average efficiency of 18.1% have been obtained (76), with efficiencies up to 17.3% for encapsulated cells. Similar results were obtained for 2 Ωcm CZ material (17.6% and 16.8%, respectively). A module with 84 FZ cells has shown an efficiency of 15.2% (76).

- Universidad Politecnica de Madrid Approach

Based on theoretical calculations for optimized emitter dopant profiles (78), this approach concentrates on thick passivated emitters with large metal-contact interaction area. 19% efficient cells have been realized on low- and high-resistivity p-type substrates with a simple process sequence including phosphorus diffusion (70 - 150 Ω/square W_E = 0.8 - 1.6 μm, about $10^{19} cm^{-3}$ surface concentrations, annealed aluminium BSF, single layer Ta_2O_5 ARC and a metal grid pattern covering 6% of the surface and having the same contact area (79). It has been concluded that efficiencies of 18.5% should be achievable for a 10% metallized surface and 1.5 - 3 μm deep junctions on 1 Ωcm substrates. This would permit to form the contacts by applying low-cost techniques such as screen-printed metal pastes. Beside the laser grooved cell approach of the UNSW, this thick emitter approach clearly demonstrates, how the performance of today's industrial cells can significantly be improved.

4 Cast Polycrystalline and Ribbon Silicon Cells

4.1 General Features

Polycrystalline silicon substrates for commercial solar cell fabrication are mainly produced by solidifying semiconductor-grade silicon melt within a crucible to ingots, which are then sliced to square 10 cm x 10 cm wafers. The most well-known materials are SILSO (Wacker-Chemitronic) (80, 81), SEMIX (Solarex) (82, 83), HEM (Crystal Systems) (84) and Polyx (Photowatt) (85), though there are some other proprietary materials of solar cell manufacturers or materials still under development (Osaka Titanium (86)). A more unconventional approach still under development is the direct casting of the silicon sheets either by the spin cast process developed at Hoxan Corp (87) or by the RAFT process of Heliotronic (88).

Since it was demonstrated more than 10 years ago, that large-area multicrystalline silicon solar cells with efficiencies exceeding 10% Am 1.5 can be fabricated (89), many experimental and theoretical research was undertaken to characterize the multicrystalline silicon and to investigate the impact of material properties on solar cell performance (89 - 102).

The effective minority carrier diffusion length (L_{eff}) is the primary parameter which limits the short-circuit current and open-circuit voltage (97 - 102). The value of L_{eff} is determined by the value of the intragrain diffusion length and of the effective grain size which can largely vary within the ingot. The effective grain size, different from optical grain size, is limited by the electrically active incoherent grain and subgrain (low-angle) boundaries. Coherent boundaries (as twins) are normally electrically inactive. However, the value of grain boundary recombination velocity can strongly be enhanced by temperature treatments during processing (103, 104), as demonstrated in Fig. 27, due to impurity gettering (104, 105). The intragrain diffusion length is mainly limited by point defects and impurity content. Especially the oxygen and carbon content is very critical (106, 107) and should be kept as low as possible.

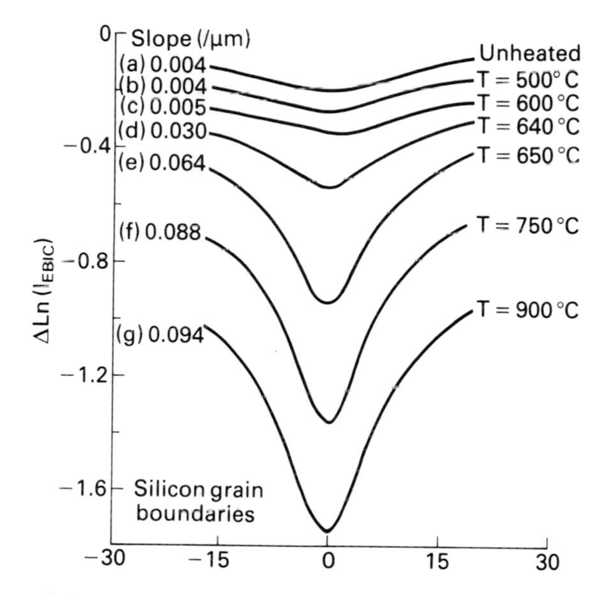

Fig. 13.27.

With growing knowledge of the correlations between polycrystalline silicon material properties and solar cell performance, silicon melt preparation and solidification processes were continually improved and have now reached a level, where efficiencies fairly over 11% are routinely achieved for large-area cells with n^+p structure, processed in mass production lines (106 - 109). For the present standard large-grain polycrystalline ingot materials, the short-circuit current and open-circuit voltage are controlled by the value of the intragrain diffusion length and not by the grain boundary recombination.

Another approach to low-cost silicon substrates is to solidify molten silicon directly in large-area sheet form using one of several ribbon growth processes, reviewed in (110). In the first commercially used approach, the edge-defined film-fed growth (EFG) technique (111) of Mobil Solar Energy Corp., hollow nonagonal tubes with 5 cm wide faces are grown and separated into 5 x 10 cm^2 blanks by cutting at high speeds using CO_2 lasers. The development of 10 cm face octagons is underway (112). Another advanced method is the 'dendritic web' approach of Westinghouse (113). Albeit both materials could be considered as crystalline with (110) and (111) surface orientation respectively, the large amount of crystallographic and chemical imperfections causes similar electronic properties as for the mentioned cast polycrystalline material of large grain size (114).

It should be pointed out that all commercially used polycrystalline and ribbon silicon as well as single crystal silicon materials still base on expensive semiconductor-grade silicon. The investigation of new low-cost processes for the preparation of 'solar-grade' silicon by new purification methods has shown encouraging results. The procedures recently developed at Bayer (115), Siemens (116) and Heliotronic (117, 118) lead to silicon material, that allows the fabrication of over 10% efficient large-area cells. However, those materials have not yet reached a production status.

4.2 Gettering and Hydrogen Passivation

For the standard multicrystalline and ribbon silicon materials based on semiconductor-grade silicon and especially for those under development based on solar-grade silicon, it is important to find solar cell processing steps that can reduce the detrimental influence of the material imperfections.

- Gettering

One possibility is the introduction of impurity gettering which is well-known in semiconductor device manufacturing. Effective gettering processes have proven to be high temperature annealing of wafers without or with mechanical damage (119, 120), phosphorus diffusion (120 - 122), and annealing with aluminium layers of back side (123 - 127).

The gettering effect of mechanical back side damage and subsequent annealing on solar cell parameters was studied by Saitoh et al (120) for single crystal and polycrystalline wafers prepared from CZ-grown ingots using refined metallurgical-grade silicon. Improvements in efficiency resulted from increases in V_{OC}, J_{SC} and FF and were found to require relatively lengthy annealing periods in N_2 ambient at 1100 °C as shown in Fig. 28.

Gaseous phosphorus diffusion was applied to polycrystalline silicon wafers grown from semiconductor-grade material by means of a casting technique (83). In Fig. 29 is shown that the photovoltaic parameters (except FF which remains constant) increase after a 30 min annealing treatment at 900 °C and even more at 1100 °C while decreases after annealing at 700 °C. Gettering experiments with single crystal wafers, intentionally doped with Ti (120, 121), Mo and Fe (121) revealed that Ti can be gettered in $POCl_3$ and O_2 ambient or by introducing back side damage. However, a significant reduction of bulk concentration requires unreasonably long gettering times (many hours) at high temperature. $POCl_3$ gettering is very efficient to remove Fe, while Mo-doped cell performance is insensitive to this procedure.

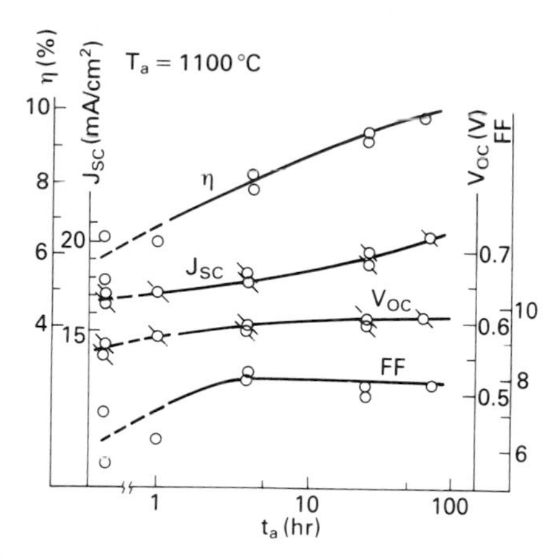

Fig. 13.28.

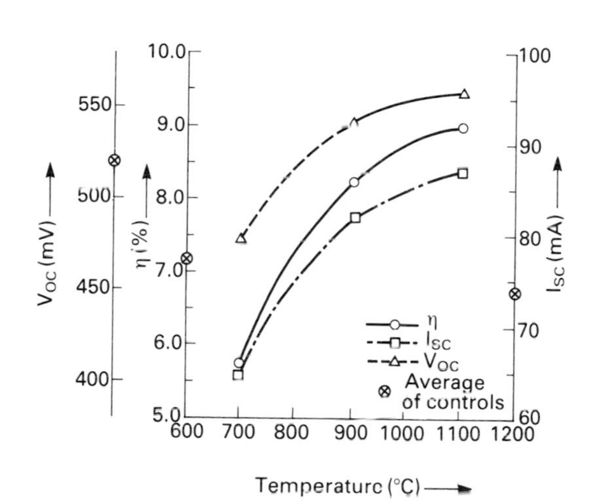

Fig. 13.29.

Improvements of solar cell parameters have also been observed for p-type polycrystalline Si wafers coated with an Al layer and annealed at medium (about 450 °C) or elevated (700 - 800 °C) temperatures (123 - 127). Enhancement of gettering effect was observed for samples pregettered by phosphorus diffusion (127). In some samples L_n degraded after an annealing at 700 °C/2 h and restored after 450 °C/2 h (126).

In summary, it could be concluded that the effectiveness of all these gettering techniques depends on the specific material properties. Figure 30 shows that high temperature annealing in oxidizing ambient can improve or degrade the efficiency of solar cells fabricated from p-type polycrystalline silicon wafers that were taken from different positions within an experimental ingot cast by Heliotronic.

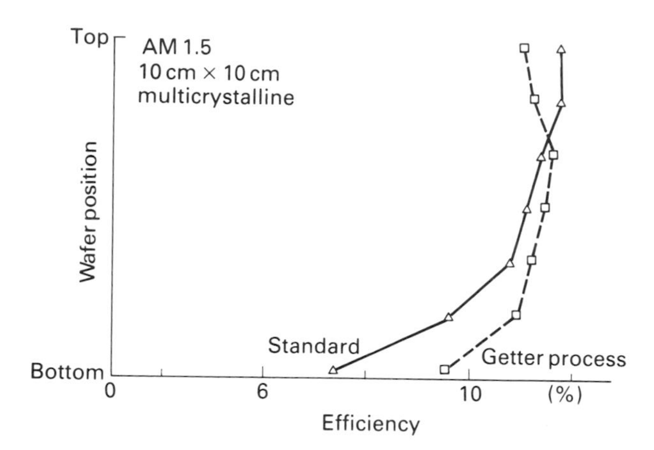

Fig. 13.30.

- Hydrogen Passivation

Passivation of material defects with atomic hydrogen is by far the most usual and successful procedure to improve any kind of polycrystalline or ribbon silicon solar cells (105, 111, 114, 128 - 140) and has already been introduced in production lines (111, 139, 140). The passivation can be performed at 200 - 400 °C by low-energy ion implantation with a broad-beam Kaufman-type (141) ion source as schematically shown in Fig. 31, by exposure to a hydrogen plasma discharge as outlined in Fig. 32 or a combined method (138). Annealing in a molecular hydrogen gas flow also works, but needs exposure times of several hours (137). A diagram of a production oriented system for continuous mode hydrogen implantation is shown in Fig. 33, allowing a throughput of 5 m^2 cell area per hour in an optimized configuration. A pilot-line equipment with similar throughput but working in batch mode has also been realized for passivation by hydrogen plasma discharge (140).

The hydrogen passivation improves the long wavelength spectral response (see Fig. 34) of the solar cells due to passivation of grain boundaries and particularly intragrain recombination centres as revealed by light-beam-induced-current (LBIC) measurements (Fig. 35). Beside the short-circuit current, the open-circuit voltage is improved, too, while the fill factor is usually not effected. The resulting enhancement of solar cell efficiency is the more as the efficiency is less before the passivation procedure. As a consequence, efficiency distribution becomes narrower on a higher level, as shown in Fig. 36. Hydrogen passivation also works very well for aluminium-doped p-type or

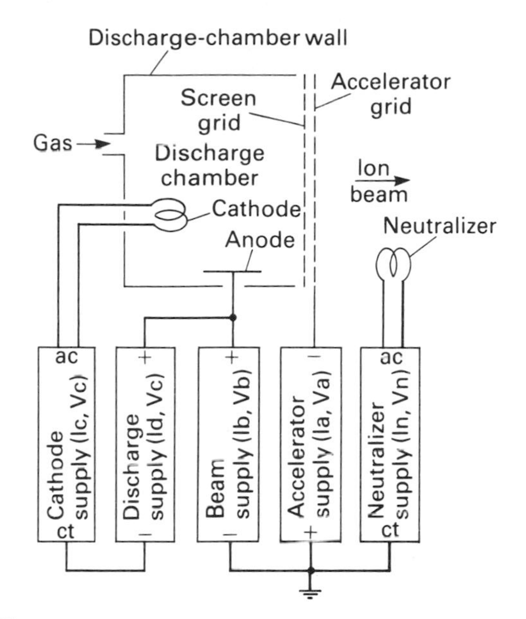

Fig. 13.31.

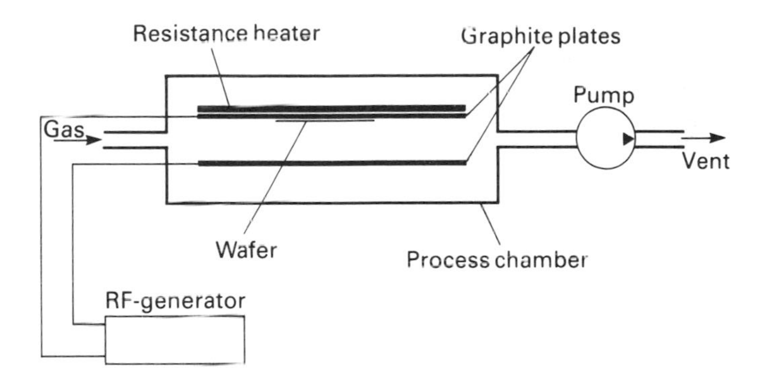

Fig. 13.32.

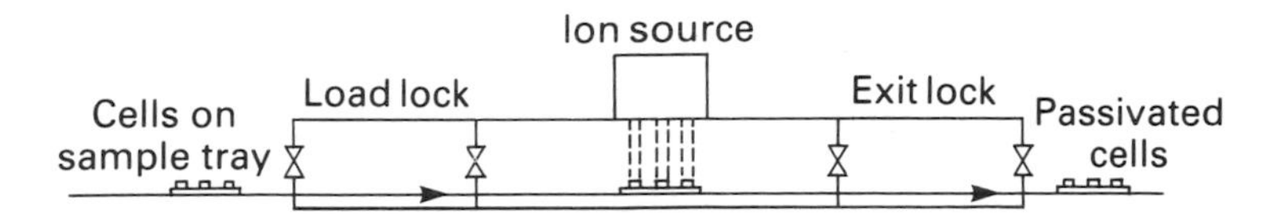

Fig. 13.33.

phosphorus-doped n-type base material and for experimental low-grade material with high impurity content (131). Though best results are obtained for hydrogenation treatments on the solar cell front side, improvements have also been observed for back side hydrogenation (137, 140). As shown in Fig. 37, the increase of minority carrier diffusion length by back side annealing in H_2 gas flow is higher for the thinner polycrystalline silicon solar cells. The hydrogen passivation has proven to be stable at operation temperatures for many years.

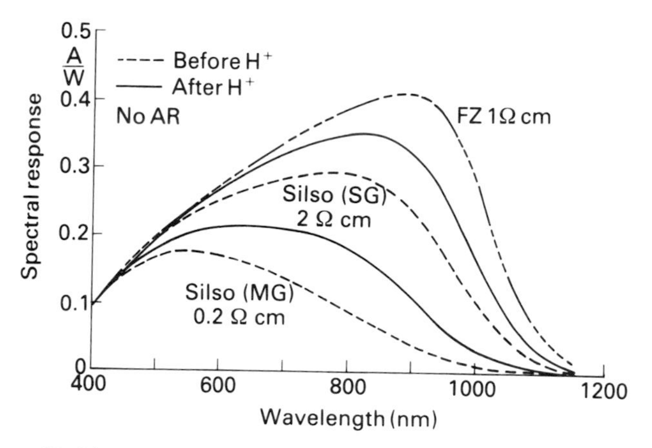

Fig. 13.34.

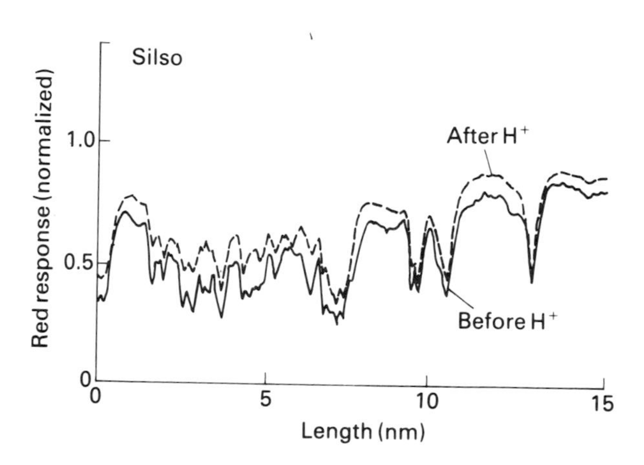

Fig. 13.35.

The chemistry of hydrogen interaction at grain boundaries in B-doped polycrystalline Si has directly been studied at SERI by Volume-Indexed SIMS (secondary ion mass spectrometry) and AES

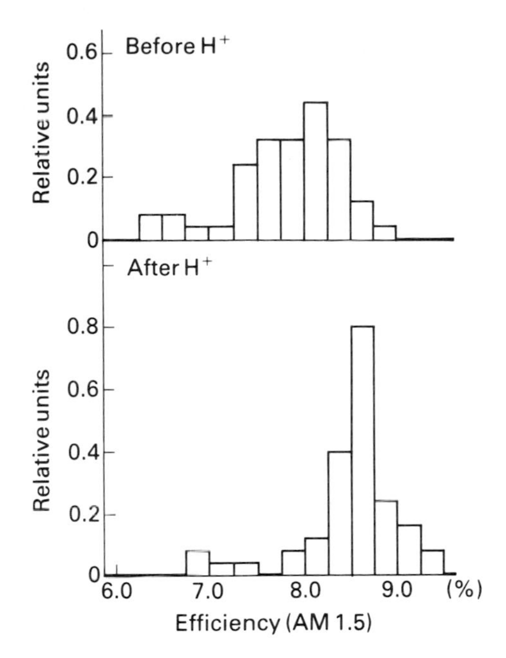

Fig. 13.36.

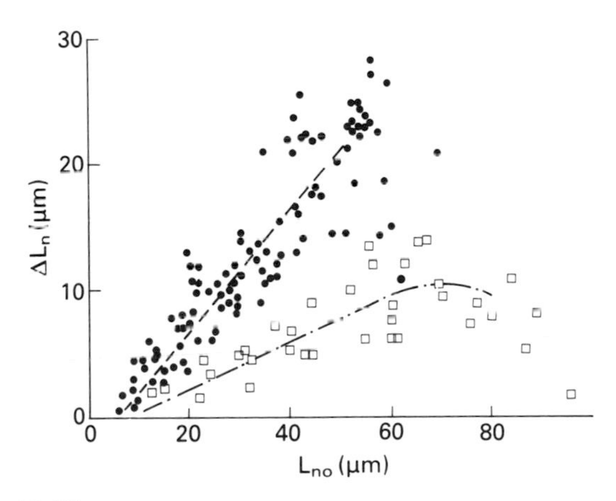

Fig. 13.37.

(Auger electron spectroscopy) measurements (104, 142 - 144) as illustrated in Fig. 38. Figure 27 shows an increasing electrical activation of a grain boundary with increasing annealing temperatures above 600 °C. SIMS mapping reveals increasing oxygen segregation in the boundary layer with increasing temperature. During passivation process, hydrogen penetrates the grain boundary and forms electrically inactive SiOH whereas in unannealed oxygen-deficient grain boundaries, Si-H bonding is found. These results are shown in Fig. 39 and were confirmed by IR spectroscopy measurements (144). From the studies of hydrogen penetration of bulk silicon and grain boundaries, the dependence of the diffusion coefficient upon T can be extracted. The intragrain diffusion follows the relationship $D = 4.2 \times 10^{-5} \exp(-0.56\ eV/kT)\ cm^2/s$, the grain boundary diffusion fits the function $D' = 8.2 \times 10^{-5} \exp(-0.28\ ev/kT)\ cm^2/s$ (142).

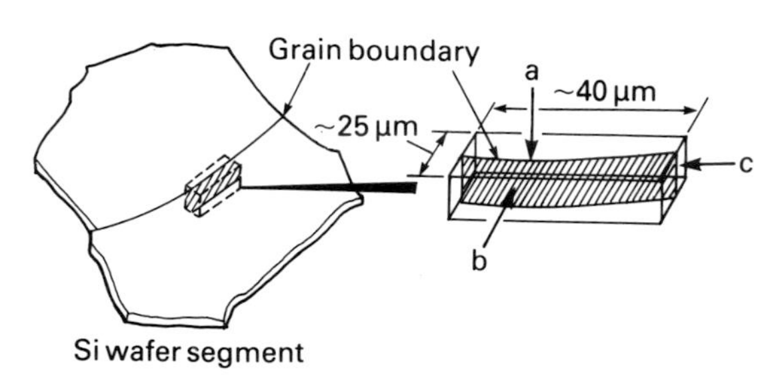

Fig. 13.38.

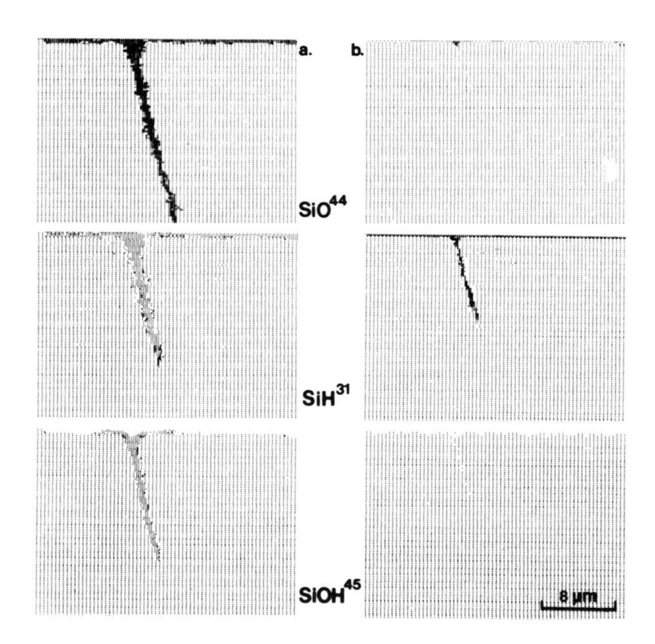

Fig. 13.39.

The electrical neutralization of shallow impurities (B, Al, Ga, In) in silicon grain boundaries was also studied at SERI (145).

Auger difference spectroscopy confirms Si-H bonding to dominate B-Ga- and In-doped grain boundary cases, while Al-rich boundaries show Al-H and Al-hydroxyl bonding. This is demonstrated in Fig. 40 for B- and Al-doped grain boundaries.

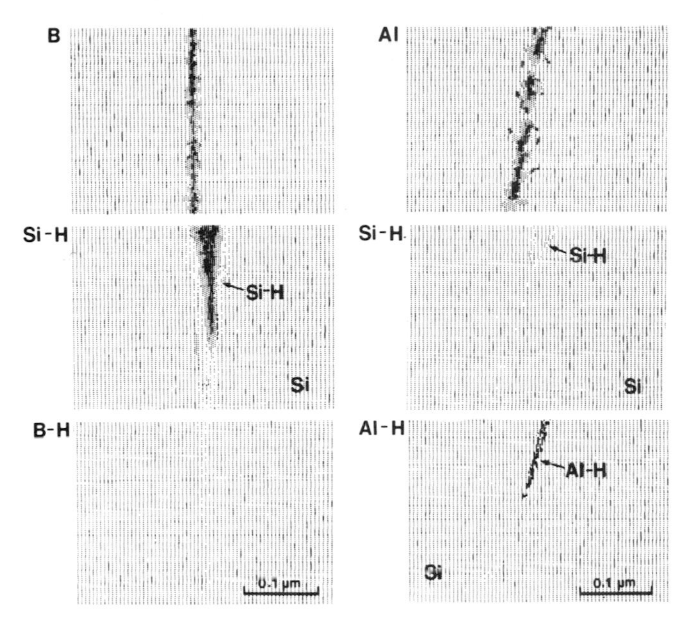

Fig. 13.40.

To summarize, hydrogen passivation can be considered as a widely understood, commercially favourable mean to improve the efficiency of polycrystalline and ribbon silicon solar cells.

4.3 High Efficiency Approaches

At least with hydrogen passivation and/or gettering, it may be possible to obtain effective minority carrier diffusion length values in the base which allow the beneficial application of high efficiency features such as BSF. A precondition for the effectiveness of BSF is a diffusion length in the range of the cell thickness. If this is fulfilled, efficiency is enhanced over 1% absolute compared to n^+p cells, leading to average values over 13% (AM 1.5, 100 mW/cm², 25 °C) without any other high efficiency features as presented in Fig. 41 for SILSO and in Fig. 42 for HEM as substrate materials. This enables now the fabrication of 12-13.5% efficient polycrystalline silicon solar cells under production conditions (23, 86, 140, 146, 147). Similarly, the efficiency of ribbon silicon cells was improved. Baseline dendritic web silicon cells reach average efficiency of about 14%, allowing the fabrication of modules with 12.5% efficiency (148) (AM 1.5, 100 mW/cm², 25 °C). EFG solar cells efficiency from pilot production runs averages about 13% with large batch averages of 14% (112).

Maximum efficiencies scarcely above 15% were reported for both ribbon materials (112, 148). For laboratory high efficiency approaches, the status is as follows.

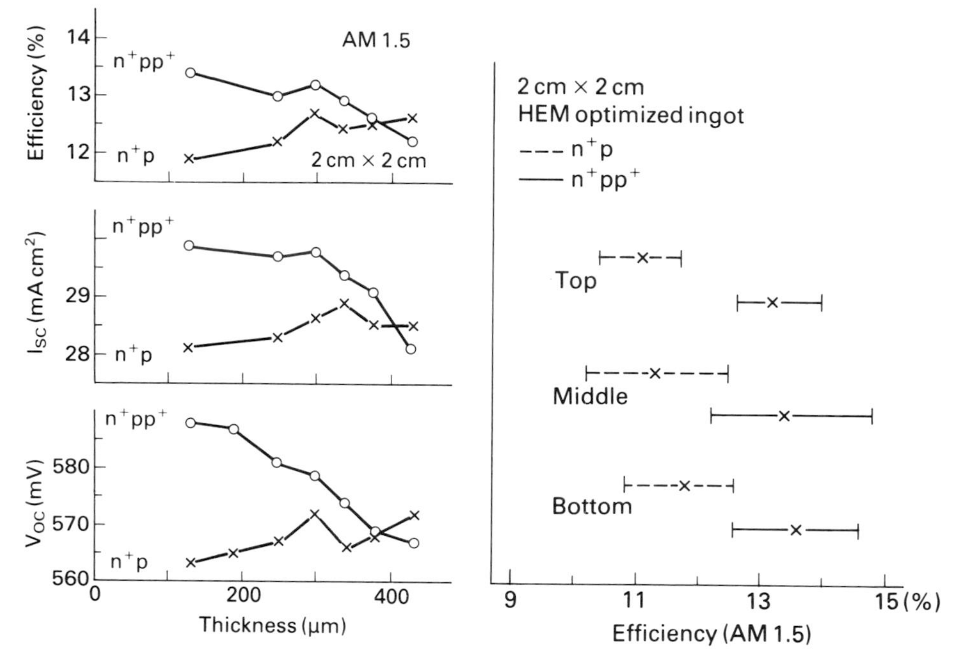

Figs. 13.41 and 13.42.

10 cm x 10 cm multicrystalline silicon (SILSO) solar cells have been fabricated by TELEFUNKEN electronic GmbH on 1 Ωcm, 350 μm thick substrates. Gaseous phosphorus diffusion, metal mask defined evaporated contacts and TiO_x SLAR coating, but no BSF were applied. Batch conversion efficiencies of encapsulated cells of 14% on the average and 14.5% at the maximum were measured relative to a SERI calibrated cell (global AM 1.5, 100 mW/cm², 25 °C) (149).

Maximum conversion efficiencies of 14.5%, 15% and 15.1% were achieved by Kyocera Corp. using p-type 10 cm x 10 cm cast multicrystalline substrates from Wacker, Osaka Titanium Corp. and Kyocera Corp., respectively (146). The process sequence includes etching in NaOH solution, gaseous diffusion using $POCl_3$, BSF formation with Al paste, deposition of plasma CVD silicon nitride film (P-SNF) at 400 °C on front and back side, patterning of the film, screen-printed back and front metallization using fine grid pattern and finally covering of metallization with solder. The P-SNF acts as passivation layer probably by ejecting hydrogen in the bulk during electrode firing. The improvement of SILSO solar cell characteristics achieved by applying the P-SNF is demonstrated in Fig. 43. For the 15.1% efficient cell, V_{OC} of 595 mV,

I_{SC} of 3.42 A and FF of 0.742 were measured (AM 1.5, 100 mW/cm², 25 °C).

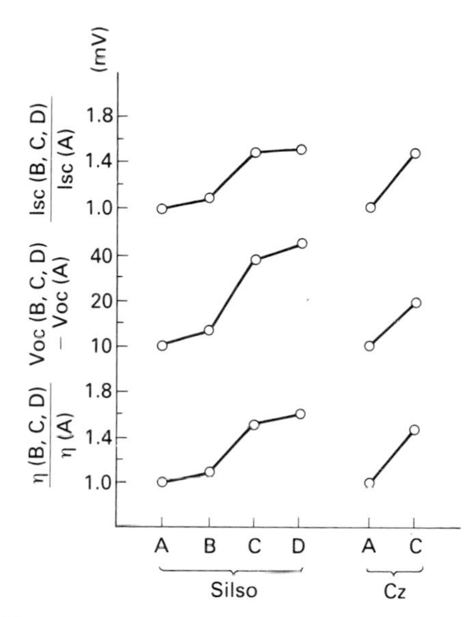

Fig. 13.43.

Selected 1.7 Ωcm, 225 µm thick, proprietary semicrystalline silicon wafers with large effective grain size and high intragrain diffusion length were processed to solar cells by Solarex with the application of high efficiency features as shallow junction, DLAR coating on textured surface and photolithographically defined front metallization for minimized shading (97). Such large area (100 sqcm) polycrystalline silicon cells exhibit average efficiency of 13.5% and a maximum value of 14.1%. Hydrogen passivation could provide further enhancement.

Similar approaches with 4 cm² cells on their proprietary multicrystalline silicon by Solarex and recently with 2 sqcm cells using Polyx by Laboratoires de Marcoussis (107) gave 16% efficient polycrystalline solar cells. It should be mentioned, that the Polyx cell was fabricated with screenprinted contact metallization.

Applying their PESC process sequence, 4 cm² SILSO cells with 15-16% efficiency were fabricated by the UNSW on low-resistivity (0.5 - 1.0 Ωcm) substrates (149). To achieve these values, it was necessary to include a phosphorus pretreatment step in the PESC process sequence. This pretreatment consists of a heavy phosphorus diffusion with sheet resistance of 20 - 50 Ω/square followed by etching the surface of the diffused region until the sheet resistance increases to the 100 - 150 Ω/square range usual for PESC structure. The pretreatment increased the effective diffusion length, probably due to gettering, and correspondingly

V_{OC}, I_{SC} and efficiency were enhanced as shown in Fig. 44. Recently, a 16.6% efficient cell (V_{OC} = 605 mV, J_{SC} = 35.2 mA/cm², FF = 0.782) was reported, fabricated in the same way on 1 cm SILSO substrates (67).

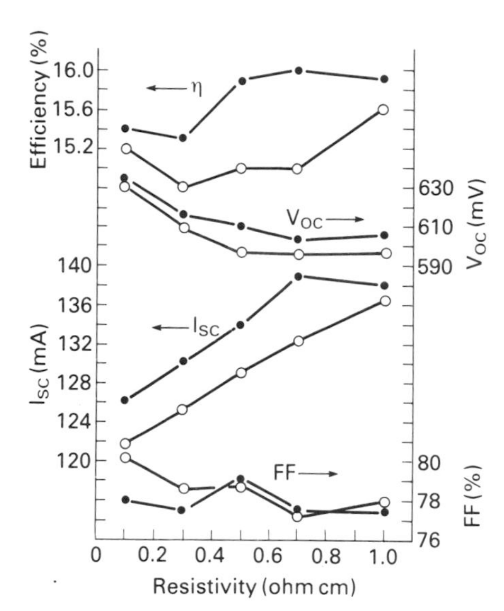

Fig. 13.44.

Solar cells with BSF structure and both front and back surfaces oxide passivated were fabricated on 40 cm p-type dendritic web substrates of 140 µm thickness and 4 cm² area (114). Further cell features include an Al (1000 Å) back surface reflector, 100 Å passivating oxide thickness, TiPdAg contacts plated with 8 µm Ag, and a 430 Å ZnS/1000 Å MgF_2 DLAR coating. Measurements gave J_{SC} of 35.6 mA/cm², V_{OC} of 597 mV, FF of 0.781 and efficiency of 16.6%.

It can shortly be summarized, that high conversion efficiencies up to approximately 17% are in principle achievable with actual unconventionally crystallized silicon substrates. However, it is necessary to transfer such high efficiencies to the module level. Innovative concepts for high efficiency modules will be given in the next section.

5 Solar Generators and Application

There was significant progress in the field of photovoltaic concentrator systems. However, since the most exciting improvements have been made on the concentrator cells which were thoroughly discussed in a previous section, this application will be omitted here. An overview of recent photovoltaic concentrator research can be found in (151). The following statements will completely focus on flat-plate applications.

5.1 Solar Generators for Medium and High Power Application

A solar generator or photovoltaic array is consisting of one or more photovoltaic modules mounted on a support structure, and additional other components such as diodes or switches. To fabricate a flat-plate module, the silicon solar cells are interconnected to strings and then laminated between protective sheets. Finally, a frame is mounted and sealed. To achieve highly reliable modules with low cost to energy output ratio, the necessary module design and mass production processes had to be investigated.

The interconnection of the solar cells is routinely performed by thin tabs of copper or aluminium that are attached to the cells using soldering, ultrasonic welding or resistance welding.

The major functional components and the used approved materials of nowadays standard modules are for the laminate (starting from the top side): a cover of tempered glass sheet with high transmission, an encapsulant of ethylene vinyl acetate (EVA) foils, a scrim of fiberglass mat (optionally) and a back cover/substrate of polyester film, fluorocarbon film, adhesive laminated plastic film + metal foil combinations or another glass sheet (151 - 155). A stainless steel or Al frame is mounted and sealed with rubber. For example, the module production sequence of AEG (155) is given in Table 3..

There are many factors that cause the difference between cell, module and array efficiency, as pointed out in Fig. 27 for three older prototype arrays operated by the Florida Solar Energy Center (FSEC) (156). To achieve higher module efficiencies in present commercial modules, the packing factor was enhanced to a typical value of 90% by using rectangular shaped larger cells and an increased module area up to 2 sqm, typically 0.4 - 1 sqm. Thus, module efficiencies are distinctly over 10% for commercial modules measured under AM 1.5 (100 mW/cm^2, 25 °C) conditions (154).

Another crucial factor is the power loss due to increased temperature under operating conditions. Beside the use of BSR solar cells and transparent or reflecting back covers, the introduction of the bifacial grid contact cell (157, 158) obviously reduces the amount of absorbed IR radiation in connection with transparent back covers. In addition, if properly designed, this cell can use the scattered light which penetrates the module back side with a conversion efficiency almost as high as for the front side. The spectral response of such a bifacial cell is shown in Fig. 46. Depending on the surrounding, the application of this bifacial light sensitive cell can largely boost the module power putput under operating conditions.

An innovative approach to achieve high efficiency modules has recently been investigated by TELEFUNKEN electronic GmbH (149). This approach is concerned with a new interconnection scheme for the solar cells. As schematically shown in Fig. 47, many solar cells are two-dimensionally interconnected to a large-area solar cell array with a 'shingle-roof' pattern. Interconnection of cells in series is performed by marginally overlapping and

Table 3: Module manufacturing sequence of AEG (after Boller and Wandel /155/)

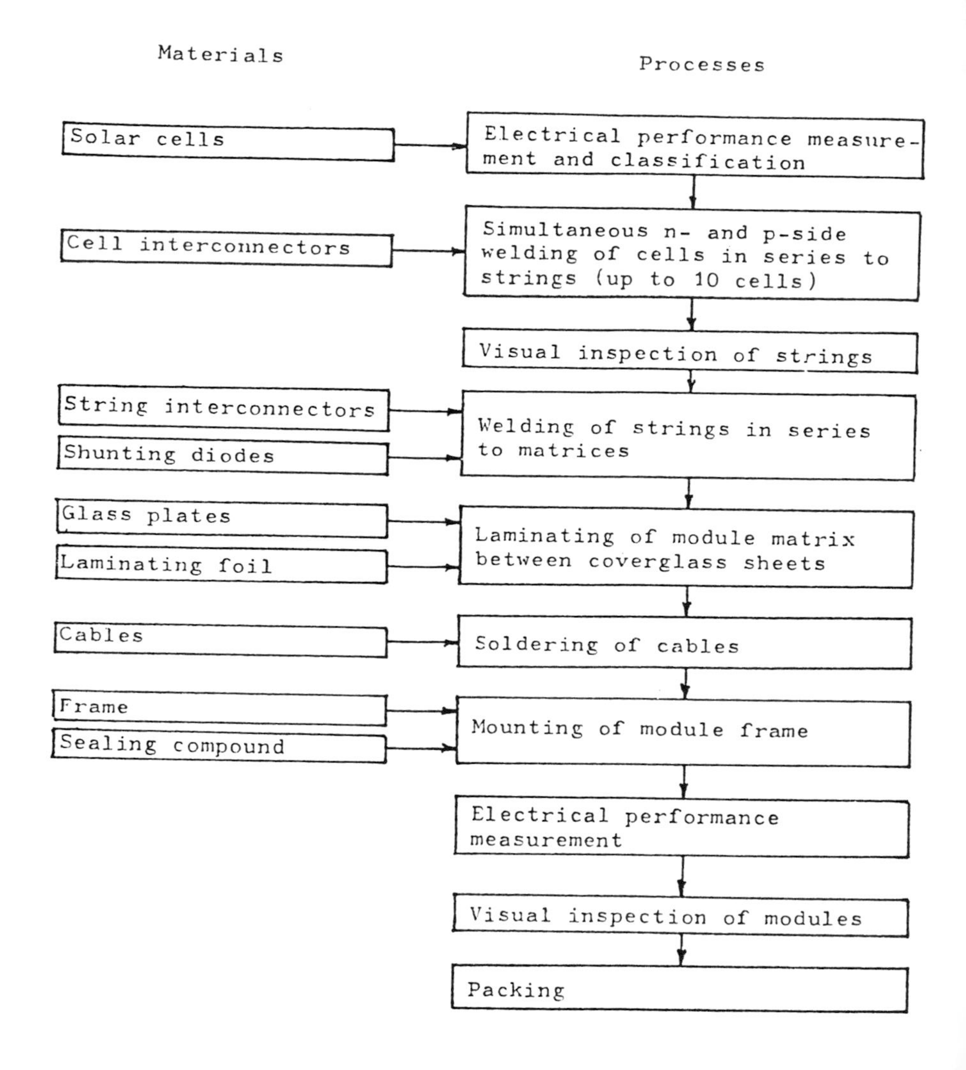

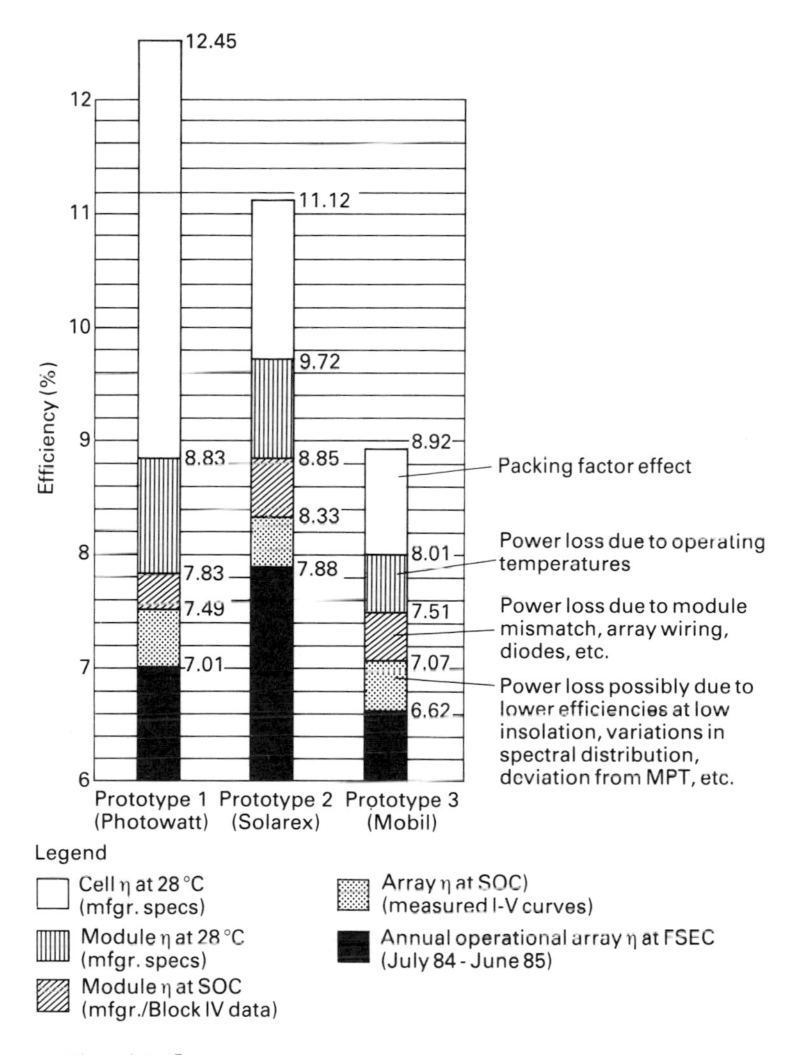

Fig. 13.45.

attaching the back contact of the upper cell to the front busbar of the lower cell, e.g. by soldering. Interconnection of cells in parallel is performed by dislocating the cells in the adjacent row. Large cell arrays of over 1000 cm² have been fabricated in such a way. They can be interconnected to strings by the usual methods as applied for single solar cells and eventually encapsulated in a module.

This new approach obviously enhances module efficiency because the distance between the cells in the array (conventionally some

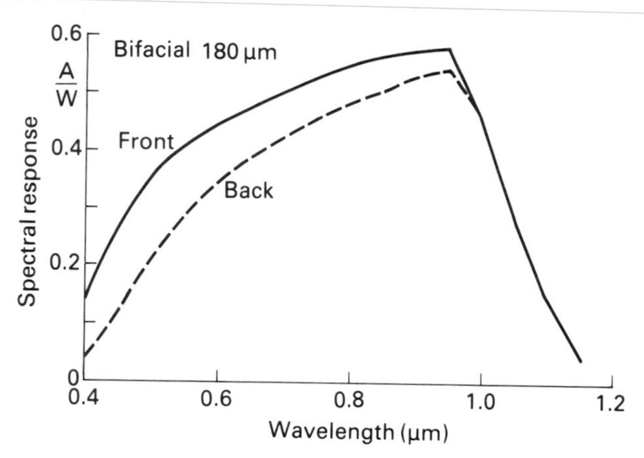

Fig. 13.46.

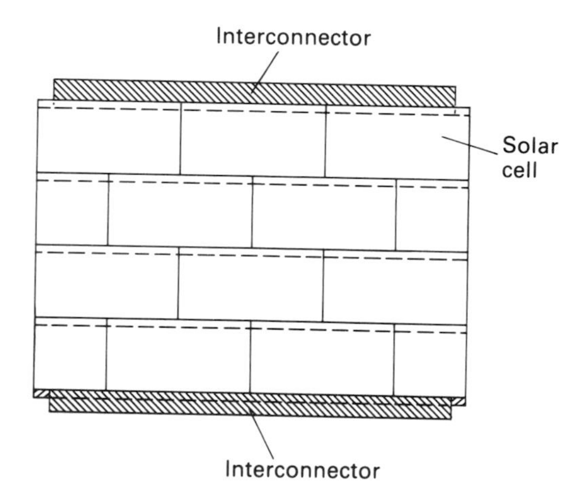

Fig. 13.47.

mm) can be reduced to negligible values. Furthermore the shading busbar of many cells is overlapped by active cell area. These two facts enhance the active module area and therefore the efficiency by 8 - 10 % relative to present commercial module designs. Under operating conditions, the hot spot problem is reduced since each cell is in good thermal connection to the other cells of the array.

Additionally, it should be pointed out that the interconnector stripe is saved within the array and that two cells can be interconnected in one step. This implies a further cost reduction, since cell array fabrication can be automized easily.

High efficiency test modules were fabricated at TELEFUNKEN electronic GmbH with usual sizes of about 0.4 m^2. Module efficiencies of about 17% (AM 1.5) with single crystal and 13% with polycrystalline silicon solar cells have been achieved (149), the highest reported up to now.

Presently, the commercially available photovoltaic generators can compete with other energy sources mainly in applications for remote locations. Some examples are

- Traffic systems: warning lights, street lighting, light houses, buoyes
- Telecommunications: transmitter stations and radio repeater stations
- Agriculture: water pumps, irrigation systems and cold-storage houses
- Public Health: refrigerators for vaccines, medical equipment
- Cathodic corrosion protection
- Power supply for residential houses, villages and desalination systems

For the power supply of houses, photovoltaic roof tiles have been developed. A useful new application is the development of medical containers. They can be brought in remote areas, serving as mobile hospitals with photovoltaic generators on the roof as power supplies.

An interesting application in future is the production of hydrogen by hydrolysis of water. This enables favourable storage and transport of solar energy. Demonstration projects are now under construction (159).

5.2 Solar Generators for Low Power (Consumer) Application

Representative applications for this type are radios and wall clocks. For those kinds of application, the photovoltaic generator in generally becoming an integral part of a product. Therefore new criterias for the design of the photovoltaic generators are relevant: optical appearance, available space, technique of integration and surrounding conditions. For limited space, the application of highly efficient expensive cells could be imperative. The shingle arrangement of cells as described in section 5.1 is also beneficial in this case. On the other hand, the generator can be fixed in a case behind a protective sheet or the product is normally not used under extreme weather conditions, allowing cheaper protective covering methods.

The indoor application as in a wall clock under low intensity insulation from different light sources has been investigated thoroughly (160, 161). Analysis of the diode characteristics revealed, that shunt resistance and space charge recombination current are dominant at 1 mW/sqcm and below. However, proper cell design allows the low-cost production of multicrystalline silicon solar cells useful down to illuminance values of 20 lux (fluorescent light) as shown in Fig. 48 and 49. The result is even much better for diffuse daylight or especially incandescent light where such cells are superior to commercially available amorphous silicon ones. In order to fabricate solar modules, the cells are generally mounted on a PC board, applying usual chip-on-board technology.

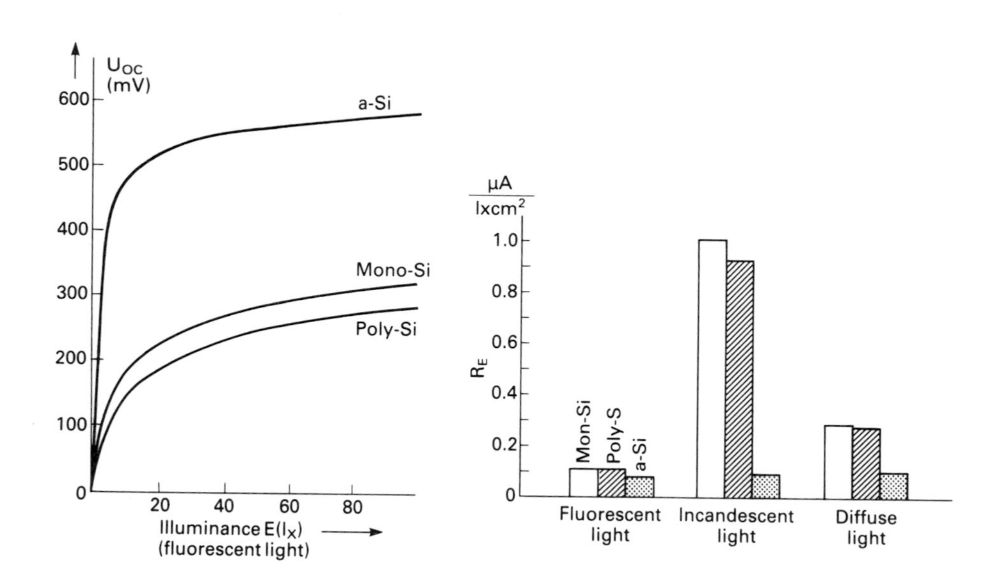

Figs. 13.48 and 13.49.

An interesting application at high intensity level is the solar roof for cars that powers a fan to cool the car at sunny days. Further examples of low power application include the wide range of toys, alarm systems, lamps, battery chargers and other small power devices, where a solar generator together with an accumulator can avoid the wasting of non-rechargeable batteries. In each case, the solar generator has to be well designed electrically and mechanically to fulfill the requirements at lowest fabrication costs.

6 Outlook

Becoming aware of the tremendous progress made in silicon solar cell efficiency during the last years, two questions immediately arise. First, what is the uppermost efficiency that can ultimately be achieved? And second, what efficiency values are expected to be attainable under commercial production conditions?

Upper Efficiency Limits

To answer the first question, a highly idealized silicon solar cell is considered, where only fundamental loss mechanisms as Auger recombination, radiative recombination and free carrier absorption are taken into account. Using the most recent Auger recombination coefficients (from (162)), then the limiting efficiencies have been calculated to be about 36% for concentrator cells (162, 163) and about 29% for one-sun cells (164, 165). More realistic performance values are provided, if technologically feasible values for recombination at the surfaces and in the adjacent doped regions are taken into account.

Experimental interface trap density and capture cross section values for the silicon/silicon oxide interface have been measured by Eades and Swanson (166) using deep level transient spectroscopy (DLTS). The results for oxides grown under three different conditions are shown in Fig. 50 and 51. It is noteworthy, that the capture cross section for electrons are about two orders of magnitude higher than those for holes. The data presented in Fig. 50 and 51 have been used to calculate the surface recombination velocity. The lowest values were obtained for the oxide grown in dry oxygen and pulled in argon. For this case, the results are shown in Fig. 52 as a function of minority carrier injection level and with substrate doping as parameter. For low injection levels, the different capture cross sections cause surface recombination velocity values that are about two orders of magnitude higher on p-type substrates than on n-type ones. The surface recombination velocity also increases with doping level. Under high injection conditions hovever, for all oxide growth conditions and on both substrate types, surface recombination velocities between 1 and 3 cm/s are predicted. These values were confirmed by measurements (166).

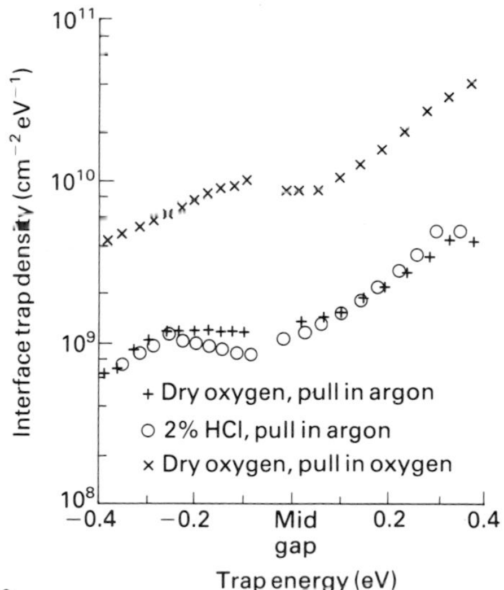

Fig. 13.50.

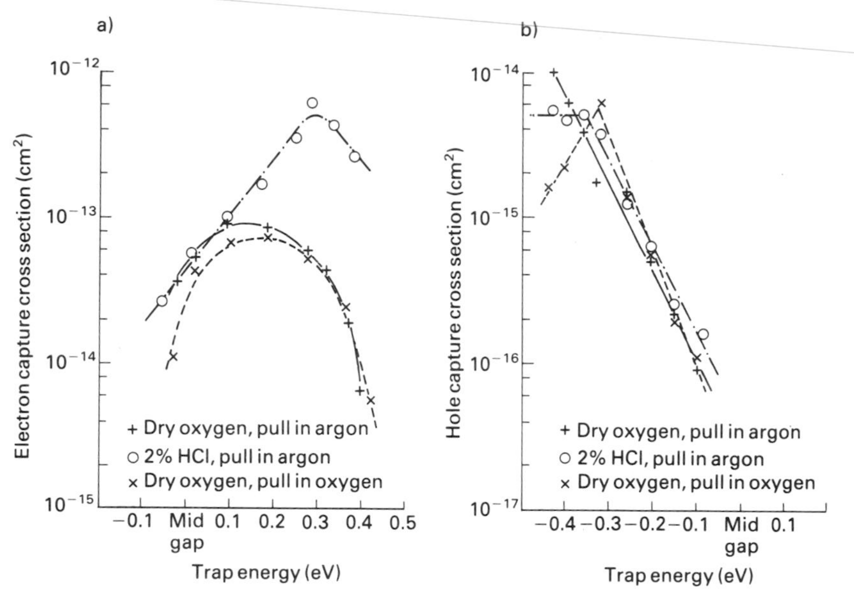

Fig. 13.51a and b.

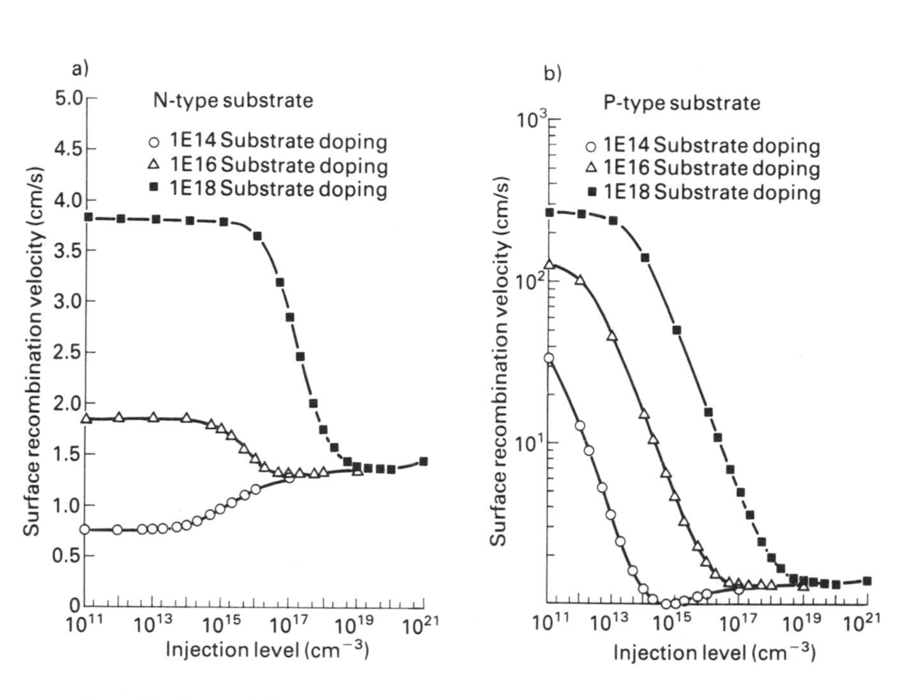

Fig. 13.52a and b.

Measurements of the surface recombination velocity on p-type silicon under low injection conditions were performed by King et al (167). They reported values that are about 70% lower as those presented in Fig. 52. The measured values range from 100 cm/s to 34 cm/s for boron concentrations from 1.3 x 10^{17} to 4 x 10^{15} atoms/cm^3.

Emitter saturation currents were measured by Kane and Swanson using photoconductivity decay methods (168). The results for a variety of phosphorus- and boron-doped emitters are presented in Table 4 and 5. These results indicate that boron-doped surface diffused regions exhibit lower saturation current density in contact regions, but larger ones in oxide passivated regions. The minimum contribution to saturation current density is 10^{-14} A/cm^2 and 5 x 10^{-14} A/cm^2 for phosphorus- and boron-doped samples respectively, with a surface doping concentration of 10^{18} cm^{-3} and oxide passivation.

The aforementioned values for surface and emitter recombination contributions have been applied to calculate the ultimately achievable efficiency. The results indicate that silicon solar cell efficiencies of about 31% under 360 suns concentration (58) and about 24% under AM 1.5 conditions (167) should be attainable on high-quality substrates.

Production Efficiency Goals

The same reported values for surface and emitter recombination together with other realistic parameter values have been used by Basore (169) to design and evaluate silicon solar cells that can be placed into commercial production within the next five years. Figure 53 shows the concentrator cell design intended for use at 100 suns, and Fig. 54 the one-sun cell design respectively. The concentrator cell is characterized by a front metal covering 50% of cell area to minimize series resistance, and to enhance light trapping in connection with a quarter-wavelength thick dielectric layer that separates most of the metal contact from the silicon. A grooved cover is applied to diffract the incident light away from the grid. The base is n-type to achieve low surface recombination in undoped regions. In contacted regions, the silicon is heavily doped. The one-sun cell design of Fig. 54 is similar to that of the laser grooved, buried contact cell approach shown in Fig. 8. The sheet resistance of the lightly doped layers is assumed to be 1000 Ω/square for the n-type and 500 Ω/square for the p-type layers. The front grid obscures about 3.5% of the surface area. The back contact design unites effective light trapping and low surface recombination features. For the calculations, the modelling program PC-1D (55) was used to predict the efficiency as a function of cell thickness for low-resistivity and high-resistivity substrates with lifetime as parameter. The lifetimes used are representative of FZ material (highest value), CZ material and semicrystalline silicon material (lowest value).

The calculated efficiencies of concentrator cells on high-resistivity 100 Ωcm and low resistivity 0.1 Ωcm substrates are shown in Fig. 55. The main result is that efficiency values of 30% at 100 suns can be expected for both substrate resistivities at a cell thickness of 50 - 100 μm. Even for CZ material,

Table 4: Saturation current density of phosphorus-doped surface diffused regions on high lifetime n-type (10^{13} cm^{-3}) <100> FZ silicon substrates (after Kane and Swanson /168/)

Sheet resistivity (Ω/square)	Junction depth (μm)	Surface doping concentration (cm^{-3})	Saturation current density (pA/cm^2)	
			Oxide passivated	Al-contacted
9.0	5.8	5.0×10^{19}	0.45	0.50
57.2	3.6	1.2×10^{19}	0.10	1.10
75.3	1.8	2.0×10^{19}	0.13	1.70
370	1.2	2.5×10^{18}	0.08	1.70
890	1.3	1.0×10^{18}	0.01	7.00

Table 5: Saturation current density of boron-doped surface diffused regions on high lifetime n-type (10^{13} cm^{-3}) <100> FZ silicon substrates (after Kane and Swanson /168/)

Sheet resistivity (Ω/square)	Junction depth (μm)	Surface doping concentration (cm^{-3})	Saturation current density (pA/cm^2)	
			Oxide passivated	Al-contacted
5.3	4.8	7.0×10^{19}	0.35	0.42
37	2.1	3.6×10^{19}	0.07	0.57
63	3.0	1.3×10^{19}	0.09	1.10
240	1.1	5.4×10^{18}	0.13	1.00
463	1.2	5.4×10^{18}	0.06	1.10

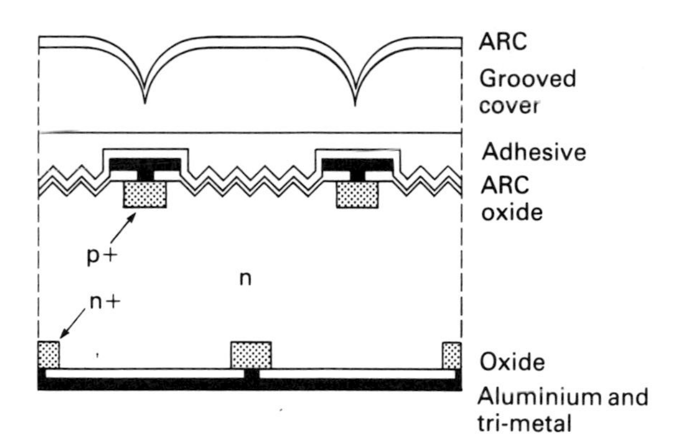

Fig. 13.53.

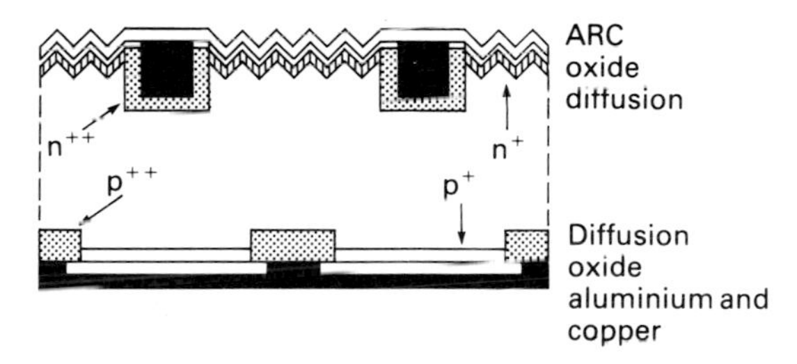

Fig. 13.54.

reasonably high efficiency values are predicted on such thin substrates.

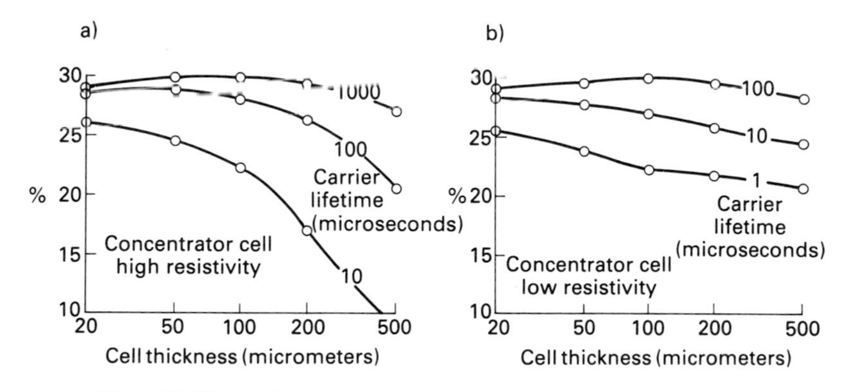

Fig. 13.55a and b.

The calculated efficiencies of non-concentrating large-area cells on high-resistivity 200 Ωcm and low-resistivity 0.2 Ωcm substrates are plotted in Fig. 56. The highest efficiencies, exceeding 21%, are found for cells less than 50 μm thick. And even with semi-crystalline materials, efficiencies exceeding 16% are expected on thin low-resistivity substrates. This later result is confirmed by another performance simulation investigation (170). In a first experimental approach, silicon solar cells less than 50 μm thick have been realized by the epitaxial growth of silicon layers on a low-cost conductive substrate (ceramic) (171), but their efficiencies are still modest.

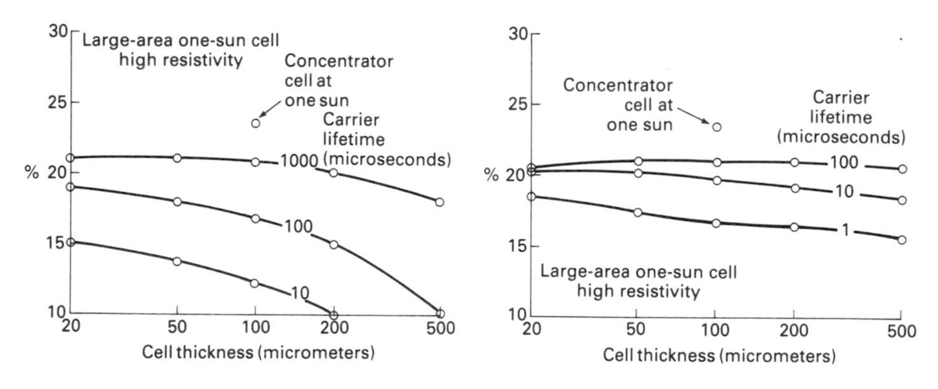

Fig. 13.56a and b.

Cost Goals

Based on the performance predictions for silicon solar cells as outlined before, it looks feasible to fabricate commercial concentrator modules with over 20% efficiency and flat-plate modules with over 15% efficiency. To be able to compete with conventional electricity generation, the price goal of about 0.12 US$/kWh has to be reached (172). For example, this corresponds to US$ 165/sqm of 15% efficient flat-plate modules and 195 (277) US$/sqm of 20 (25) % efficient concentrator modules (172). This price goal does not take into account the social costs of conventional electricity generation due to external effects such as environmental damages, depletion of resources, public subsidies and others. Conservative estimations of the additional net social costs for the conditions in the Federal Republic of Germany gave values of 0.07 to 0.17 DM/kWh (173). This means that solar electricity costs have to be reduced to about 0.3 - 0.4 DM/kWh (about 0.15 - 0.20 US$/kWh) in order to make photovoltaic systems competitive.

Even these price goals require the development of new low-cost and probably thin crystalline silicon substrate material. As soon as thin material is available, production technologies and equipments for solar cells and modules have to be adapted and optimized to fabricate high efficiency low-cost solar generators. The final breakthrough in market introduction of photovoltaic systems will surely require considerable sums of R & D money to be invested in large scale production plants. Only large scale production will allow to realize the full potential of cost reductions due to automization and fine optimization, taking benefit from the learning curve effect.

7. Summary

During the last years, many new technologies have been developed and improved to fabricate cost-effectively silicon solar cells and modules in industrial production lines. Thereby, production costs have come down to a level, where silicon photovoltaic generators became competitive with other energy sources in special application fields. However, the longterm goal is to fall short of the cost level of conventionally generated electrical energy from the grid. Therefore, experimental and theoretical work has been started to exploit the full potential of crystalline silicon solar cells. As a result, laboratory concentrator cell efficiency over 28% and one-sun cell efficiency over 21% have been achieved. Calculations based on confirmed experimental data, have indicated that these efficiencies should largely be transferable to commercially fabricated cells within the next years. To realize that, strong efforts are necessary to further develop the silicon base material, to find and improve alternative cell and module processing, evaluate the appropriate production equipment and perhaps find other alternative cell designs.

References

1. D. Pulfrey, "Photovoltaic Power Generation", Van Norstrand Co., 1978.
2. A.L. Fahrenbruch and R. H. Bube, "Fundamentals of Solar Cells" Academic Press, 1983.
3. M.A. Green, "Solar Cells, Operating Principles, Technology and System Applications", Prentice-Hall, New Jersey, 1982.
4. R.F. Van Overstraten, R.P. Mertens, "Physics, Technology and Use of Photovoltaics", Adam Hilger, Bristol, 1986.
5. Conference Record of the IEEE Photovoltaic Specialist Conference (PVSC) Series.
   Proceedings of the E.C. Photovoltaic Solar Energy Conference (PVSEC) Series.
   Technical Digest of the International Photovoltaic Science and Engineering Conference (PVSEC) Series.
6. K.-D. Rasch et al, Proc. 4th E.C. PVSEC, Stresa, 919 (1982).
7. S. Chitre, J. Donon, Proc. 3rd E.C. PVSEC, Cannes, 608 (1980).
8. Proc. of the Flat Plate Solar Array Project Research Forum on High Efficiency Crystalline Silicon Solar Cells, JPL Publication 85 - 38, 1985.
9. M.A. Green, "High Efficiency Silicon Solar Cells", Trans Tech Publications, Switzerland, 1987.
10. R.B. Godfrey et al, Conf. Rec. 16th IEEE PVSC, San Diego, 892 (1982).
11. H. Morita et al, Techn. Digest Int. PVSEC-1, Kobe, Japan, 31 (1984).
12. G. Cheek et al, Proc. 4th E.C. PVSEC, Stresa, 926 (1982).

13. Y. Tahara et al, Conf. Rec. 18th IEEE PVSC, Las Vagas, 792 (1985).
14. A. Usami et al, Conf. Rec. 18th IEEE PVSC, Las Vagas, 797 (1985).
15. W. Schmidt, K.-D. Rasch, Proc. 4th E.C. PVSEC, Stresa, 999 (1982).
16. A.R. Kirkpatrick, J.A. Minucci, A.C. Greenwald, Conf. Rec. 14th IEEE PVSC, San Diego, 820 (1980).
17. J.C. Muller et al, Proc. 6th E.C. PVSEC, London, 985 (1985).
18. A. Shibata et al, Proc. 2nd Int. PVSEC, Beijing, 265 (1986).
19. H.D. Hecht et al, Proc. 5th E.C. PVSEC, Athens, 1143 (1983).
20. H. Morita et al, Techn. Digest Int. PVSEC-1, Kobe, 817 (1984).
21. Ma Ho Ting, Proc. 2nd Int. PVSEC, Beijing, 325 (1986).
22. J.H. Wohlgemuth et al, Conf. Record 16th IEEE PVSC, San Diego, 809 (1982).
23. K. Kimura, Techn. Digest Int. PVSEC-1, Kobe, 37 (1984).
24. K. Roy, K.-D. Rasch, Conf. Rec. 18th IEEE PVSC, Las Vagas, 1133 (1985).
25. J.B. Milstein, Y.S. Tsuo, Conf. Rec. 17th IEEE PVSC, Orlando, 248 (1984).
26. A. Rohatgi, Conf. Rec. 18th IEEE PVSC, Las Vagas, 7 (1985).
27. M.A. Green, S.R. Wenham, A.W. Blakers, Conf. Rec. 19th IEEE PVSC, New Orleans, 6 (1987).
28. P. Campbell, M.A. Green, J. Appl. Physics 62 (1), 243 (1987).
29. K.-D. Rasch, K. Roy, Proc. Europ. Symp. PV Generator in Space, ESA SP-140, 15 (1978).
30. R. Hulstom, R. Bird, C. Riorda, Solar Cells, 15, 365 (1985).
31. A. Goetzberger, Conf. Rec. 15th IEEE PVSC, Orlando, 867 (1981).
32. P. Campbell, M.A. Green, Conf. Rec. 19th IEEE PVSC, New Orleans, 912 (1987).
33. E.M. Gaddy, Techn. Digest Int. PVSEC-3, Tokyo, 199 (1987).
34. H.J. Hovel, "Solar Cells, Semiconductors and Semimetals", Academic Press (1975).
35. J. Lindmayer, J.F. Allison, COMSAT Technical Review 3, 1 (1973).
36. M.A. Shibib, F.A. Lindholm, F. Cherez, IEEE Trans. El. Dev. ED-26, 959 (1959).
37. J.G. Fossum, E.L. Burgess, Appl. Phys. Lett. 33, 238 (1978).
38. F.A. Lindholm et al, Solid-State Electronics 23, 967 (1980).
39. M.A. Green et al, Conf. Rec. 18th IEEE PVSC, Las Vagas, 39 (1985).
40. M.A. Green et al, Conf. Rec. 15th IEEE PVSC, Orlando, 1405 (1981).
41. A.W. Blakers, M.A. Green, Appl. Phys. Lett. 47 (8), 818 (1985).
42. T. Uematsu, Y. Kida, T. Saitoh, Techn. Digest Int. PVSEC-3, Tokyo, 71 (1987).
43. M.A. Green et al, IEEE Trans. El. Dev. ED-31 (5), 679 (1984).
44. M.A. Green et al, Conf. Rec. 17th IEEE PVSC, Orlando, 386 (1984).
45. A.W. Blakers, M.A. Green, Appl. Phys. Lett. 48 (3), 215 (1986).
46. F.A. Lindholm et al, IEEE El. Dev. Lett. EDL-6, 363 (1985).
47. N.G. Tarr, IEEE El. Dev. Lett. EDL-6, 655 (1985).
48. E. Yablonovitch et al, Appl. Phys. Lett. 47 (11), 1211 (1985).

49. P. Verlinden et al, Proc. 7th E.C. PVSEC, Sevilla, 885 (1986).
50. R. M. Swanson, Conf. Rec. 18th IEEE PVSC, Las Vagas, 604 (1985).
51. R.A. Sinton, R.M. Swanson, IEEE Trans. El Dev. ED-34, 2116 (1987).
52. M.D. Lammert, R.J. Swartz, IEEE Trans. El. Dev. ED-24, 337 (1977).
53. M. Wolf, IEEE Trans. El. Dev. ED-28, 566 (1981).
54. J.L. Gray, R.J. Swartz, Conf. Rec. 17th IEEE PVSC, Orlando, 1297 (1984).
55. D.T. Rover, P.A. Basore, G.A. Thorson, Conf. Rec. 18th IEEE PVSC, Las Vagas, 70 (1985).
56. R. Ruckteschler, J. Knobloch, Proc. 7th E.C. PVSEC, Sevilla, 1094 (1986).
57. T. Saitoh et al, Techn. Digest Int. PVSEC-3, Tokyo, 83 (1987).
58. R.A. Sinton, R.M. Swanson, Conf. Rec. 19th IEEE PVSC, New Orleans, 1201 (1987).
59. P. Verlinden et al, Conf. Rec. 18th IEEE PVSC, Las Vagas, 55 (1985).
60. P. Verlinden et al, Conf. Rec. 19th IEEE PVSC, New Orleans, 405 (1987).
61. R.A. Sinton et al, Proc. 8th E.C. PVSEC, Florence (1988), to be published.
62. R.A. Sinton et al, IEEE El. Dev. Lett., EDL-7 (10), 567 (1986).
63. P. Verlinden et al, Proc. 8th E.C. PVSEC, Florence (1988), to be published.
64. M.A. Green et al, IEEE El. Dev. Lett. EDL-7 (10), 583 (1986).
65. M.A. Green, Proc. 7th E.C. PVSEC, Sevilla, 681 (1986).
66. M.A. Green et al, Conf. Rec. 19th IEEE PVSC, New Orleans, 49 (1987).
67. M.A. Green, Proc. 8th E.C. PVSEC, Florence (1988), to be published.
68. M.A. Green, Techn. Digest Int. PVSEC-3, Tokyo, 153 (1987).
69. T. Saitoh et al, Conf. Rec. 19th IEEE PVSC, New Orleans, 1518 (1987).
70. D.B. Bickler, W.T. Callaghan, Conf. Rec. 19th IEEE PVSC, New Orleans, 1424 (1987).
71. R.F. Wood, R.D. Westbrook, G.E. Jellison Jr., Conf. Rec. 19th IEEE PVSC, New Orleans, 519 (1987).
72. R.F. Wood, R.D. Westbrook, G.E. Jellison Jr., IEEE Electron Dev. Lett. EDL-8 (5), 249 (1987).
73. M.B. Spitzer et al, Conf. Rec. 17th IEEE PVSC, Orlando, 398 (1984).
74. M.B. Spitzer, C.J. Keavney, Conf. Rec. 18th IEEE PVSC, Las Vagas, 43 (1985).
75. M.B. Spitzer, C.J. Keavney, L.M. Geoffrey, Solar Cells 17, 135 (1986).
76. M.B. Spitzer et al, Solar Cells, 20, 333 (1987).
77. S.P. Tobin et al, Conf. Rec. 19th IEEE PVSC, New Orleans, 70 (1987).
78. A. Cuevas, M.A. Balbuena, R. Galloni, Conf. Rec. 19th IEEE PVSC, New Orleans, 918 (1987).
79. A. Cuevas, M.A. Balbuena, Proc. 8th E.C. PVSEC, Florence (1988), to be published.
80. B. Authier, Adv. in Solid State Physics XVIII, 1 (1978).

81. C. Gessert, D. Helmreich, M. Peterat, Proc. 6th E.C. PVSEC, London, 891 (1985).
82. J. Lindmayer, Conf. Rec. 12th IEEE PVSC, Baton Rouge, (1976).
83. J. Lindmayer, Z.C. Putney, Conf. Rec. 14th IEEE PVSC, San Diego, 208 (1980).
84. C.P. Khattak, F. Schmid, Am. Ceram. Soc. Bull., 57, 609 (1978).
85. J. Fally, C. Guignot, L. Goeffon, Proc. 6th E.C. PVSEC, London, 961 (1985).
86. K. Kaneko et al, Techn. Digest Int. PVSEC-3, Tokyo, 810 (1987).
87. T. Matsuyama et al, Conf. Rec. 19th IEEE PVSC, New Orleans, 738 (1987).
88. J. Geissler et al, Proc. 6th E.C. PVSEC, London, 976 (1985).
89. H. Fischer, W. Pschunder, IEEE Trans. El. Dev. ED-24 (4), 438 (1977).
90. K. Roy, K.-D. Rasch, H. Fischer, Conf. Rec. 14th IEEE PVSC, San Diego, 897 (1980).
91. W. Schmidt, G. Friedrich, K.-D. Rasch, Proc. 3rd E.C. PVSEC, Cannes, 664 (1980).
92. K.A. Dumas, C.P. Khattak, F. Schmid, Conf. Rec. 15th IEEE PVSC, Orlando, 954 (1981).
93. T. Daud, M. Koliwad, F.G. Allen, Appl. Phys. Lett. 33 (12), 1009 (1978).
94. A.K. Ghosh, C. Fishman, T. Feng, J. Appl. Phys. 51 (1), 446 (1980).
95. G.M. Storti et al, Conf. Rec. 14th IEEE PVSC, San Diego, 191 (1980).
96. S.M. Johnson, J. Culik, Conf. Rec. 16th IEEE PVSC, San Diego, 548 (1982).
97. S.M. Johnson, C. Winter, Conf. Rec. 17th IEEE PVSC, Orlando, 1121 (1984).
98. J. Culik, K. Grimes, Conf. Rec. 17th IEEE PVSC, Orlando, 1137 (1984).
99. D. Leung et al, Conf. Rec. 17th IEEE PVSC, Orlando, 264 (1984).
100. J. Fossum, F. Lindholm, IEEE Trans. El. Dev. ED-27, 692 (1980).
101. J. Fossum, R. Sundaresan, IEEE Trans. El. Dev. ED-29 (8), 1185 (1982).
102. A. Neugroschel, J. Mazer, IEEE Trans. El. Dev. ED-29 (2), 225 (1982).
103. D. Redfield, J. Appl. Phys. 40 (2), 163 (1982).
104. L.L. Kazmerski, Conf. Rec. 17th IEEE PVSC, Orlando, 379 (1984).
105. F.J. Stutzler, H.J. Queisser, J. Appl. Phys. 60 (11), 3910 (1986).
106. D. Helmreich, M. Peterat, Proc. 7th E.C. PVSEC, Sevilla, 703 (1986).
107. J. Fally, D. Guignot, L. Geoffon, Proc. 7th E.C. PVSEC, Sevilla, 754 (1986).
108. C.P. Khattak et al, Int. PVSEC-1, Kobe, Japan (1984).
109. J.H. Wohlgemuth et al, Conf. Rec. 19th IEEE PVSC, New Orleans, 1524 (1987).
110. T.F. Ciszek, Journal of Crystal Growth 66, 655 (1984).
111. F.V. Wald in "Crystals: Growth, Properties and Applications", Vol. 5, J. Grabmeier (editor), Springer, Berlin, 1981, pp 147 - 98.
112. F.V. Wald, Conf. Rec. 19th IEEE PVSC, New Orleans, 514 (1987).

113. C.S. Duncan et al, Conf. Rec. 14th IEEE PVSC, San Diego, 25 (1980).
114. D.L. Maier et al, Conf. Rec. 14th IEEE PVSC, New Orleans, 506 (1987).
115. I.A. Schwirtlich et al, Proc. 7th E.C. PVSEC, Sevilla, 736 (1986).
116. H.A. Aulich et al, Proc. 6th E.C. PVSEC, London, 951 (1985).
117. J. Dietl, C. Holm, Proc. 7th E.C. PVSEC, Sevilla, 726 (1986).
118. J. Dietl, Proc. 8th E.C. PVSEC, Florence (1988), to be published.
119. T. Saitoh et al, Conf. Rec. 14th IEEE PVSC, San Diego, 912 (1980).
120. K. Leo et al, Proc. 7th E.C. PVSEC, Sevilla, 1070 (1986).
121. A. Rohatgi et al, Conf. Rec. 14th IEEE PVSC, San Diego, 908 (1980).
122. G.J. Vendura Jr., T.M. Taverner, Conf. Rec. 17th IEEE PVSC, Orlando, 1368 (1984).
123. M. Poitevin et al, Proc. 7th E.C. PVSEC, Sevilla, 1127 (1986).
124. R. Sundaresan et al, J. Appl. Phys. 55, 1162 (1984).
125. S. Martinuzzi et al, Conf. Rec. 18th IEEE PVSC, Las Vagas, 1127 (1965).
126. S. Martinuzzi et al, Proc. 2nd Int. PVSEC, Beijing (China), 191 (1986).
127. H. Poitevin et al, Proc 8th E.C. PVSEC, Florence (1988), to be published.
128. C.H. Seager, D.S. Ginley, Appl. Phys. Lett. 34 (5), 337 (1979).
129. P.H. Robinson, R.V. D'Aiello, Appl. Phys. Lett. 39 (1), 63 (1981).
130. C.H. Seager et al, J. Vac. Sci. Technol. 20 (3), 430 (1982).
131. W. Schmidt, K.-D. Rasch, K. Roy, Conf. Rec. 16th IEEE PVSC, San Diego, 597 (1982).
132. C. Dubè, J.I. Hanoka, D.B. Sandstrom, Appl. Phys. Lett. 44 (4), 425 (1984).
133. M. Casini et al, Proc. 6th E.C. PVSEC, London, 1001 (1985).
134. J.C. Muller et al, Solar Cells 17, 201 (1986).
135. D.L. Meier et al, Conf. Rec. 17th IEEE PVSC, Orlando, 427 (1984).
136. N. Lewalski, R. Schindler, B. Voss, Conf. Rec. 19th IEEE PVSC, New Orleans, 1059 (1987).
137. S. Martinuzzi et al, Conf. Rec. 19th IEEE PVSC, New Orleans, 1069 (1987).
138. H.L. Grasser, K.A. Münzer, D.E. Claeys, Proc. 7th E.C. PVSEC, Sevilla, 855 (1986).
139. J.E. Johnson, J.I. Hanoka, J.A. Gregory, Conf. Rec. 18th IEEE PVSC, Las Vagas, 1112 (1985).
140. K.-D. Rasch, private communication.
141. J.M.E. Harper, J.J. Cuomo, H.R. Kaufman, Ann. Rev. Mat. Sci. 13, 413 (1983).
142. L.L. Kazmerski, Conf. Rec. 17th IEEE PVSC, Orlando, 379 (1984).
143. L.L. Kazmerski, Conf. Rec. 18th IEEE OVSC, Las Vagas, 993 (1985).
144. L.L. Kazmerski, Proc. 2nd Int. PVSEC, Beijing, China, 655 (1986).
145. L.L. Kazmerski et al, Conf. Rec. 19th IEEE PVSC, New Orleans 944 (1987).

146. K. Shirasawa et al, Techn. Digest Int. PVSEC-3, Tokyo, 97 (1987).
147. N. Takamori et al, Jap. J. Appl. Phys. 25 (12), L985 (1986).
148. Photovoltaic Insiders Report, Vol. VI (10), 6 (1987).
149. K.-D. Rasch et al, to be presented at the 20th IEEE PVSC, Las Vegas (1988).
150. S. Narayanan, S.R. Wenham, M.A. Green, Appl. Phys. Lett. 48 (13), 873 (1986).
151. D.E. Arvizu, Proc. 7th E.C. PVSEC, Sevilla, 771 (1986).
152. R.G. Little, M.J. Nowlan, Techn. Digest Int. PVSEC-1, Kobe, 651 (1984).
153. H. Nakano et al, Techn. Digest Int. PVSEC-1, Kobe, 657 (1984).
154. M.I. Smokler et al, Conf. Rec. 18th IEEE PVSC, Las Vegas, 1150 (1985).
155. H.W. Boller, G. Wandel, Proc. 6th E.C. PVSEC, London, 328 (1985).
156. G.H. Atmaram et al, Conf. Rec. 18th IEEE PVSC, Las Vegas, 1255 (1985).
157. M.J. Calleja et al, Proc. 5th E.C. PVSEC, Athens, 1138 (1983).
158. G. Strobl et al, Proc. 18th IEEE PVSC, Las Vegas, 454 (1985).
159. H. Steeb, A. Brinner, Proc. 8th E.C. PVSEC, Florence (1988), to be published.
160. W. Schmidt, G. Friedrich, K.-D. Rasch, unpublished results.
161. K. Heidler, S. Kunzelmann, Proc. 7th E.C. PVSEC, Sevilla, 201 (1986).
162. R.A. Sinton, S.M. Swanson, IEEE Trans. Elec. Dev. ED-34 (7) 1380 (1987).
163. P. Campbell, M.A. Green, IEEE Trans. Elec. Dev. ED-33, 234 (1986).
164. M.A. Green, IEEE Trans. Elec. Dev. ED-31 (5), 671 (1984).
165. T. Tiedje et al, IEEE Trans. Elec. Dev. ED-31 (5), 711 (1984).
166. W.D. Eades, R.M. Swanson, J. Appl. Phys. 58 (11), 4268 (1985).
167. R.R. King et al, Conf. Rec. 19th IEEE PVSC, New Orleans, 1168 (1987).
168. D.E. Kane, R.M. Swanson, Conf. Rec. 18th IEEE PVSC, Las Vegas, 578 (1985).
169. P.A. Basore, Conf. Rec. 19th IEEE PVSC, New Orleans, 905 (1987).
170. K. Matsukama et al, Techn. Digest Int. PVSEC-3, Tokyo, 223 (1987).
171. A.M. Barnett, Proc. 8th E.C. PVSEC, Florence (1988), to be published.
172. E.C. Boes, Proc. 8th E.C. PVSEC, Florence (1988), to be published.
173. O. Hohmeyer, 'Social Costs of Energy Consumption', Springer Verlag, Berlin-Heidelberg, 1988.

Chapter 14

PILOT AND DEMONSTRATION PROJECTS

F. C. TREBLE
Consulting Engineer, Farnborough, UK

14.1 Introduction

Pilot and demonstration projects are an important link in the process of bringing future applications of photovoltaics to the market place. Both have the primary aim of demonstrating the feasibility and usefulness of the technology to entrepreneurs, decision makers and the general public. But there is a distinction. Pilot projects are part of the process of developing efficient, reliable and durable systems and components. They enable different technological approaches to be compared and bring to light any problems in the installation, operation and maintenance of the plants. From this experience, lessons can be learned for the future guidance of system designers.

Demonstration projects, on the other hand, are the link between successful R&D and commercial exploitation. They are full-scale installations with the prospect of early economic viability. Ideally, they should be constructed from fully proven components to reduce the risk of failure to a minimum. Costs should be kept as low as possible. By demonstrating the applications of photovoltaics in this way and showing that, on a life-cycle cost basis, solar power can often be the cheapest and most reliable option, it is hoped to encourage others to construct similar plants. The likelihood of successful replication is an important criterion in the selection of demonstration projects.

Since the late 1970s, many pilot and demonstration plants have been constructed all over the world, with varying degrees of success. In this chapter, we shall look at the projects sponsored by the Commission of the European Communities (CEC).

14.2 CEC Pilot Projects

As part of their second 4-year solar energy R&D programme (1979-1983), the CEC's Directorate-General for Sciences, Research and Development (DG XII) sponsored sixteen photovoltaic pilot plants in the intermediate range of 30 to 300 kWp. The plants, some standalone and some grid-connected, serve a wide range of applications and are based on a number of different design approaches. All embody flat-plate arrays at a fixed inclination. They were constructed by consortia of private companies, electricity authorities, universities, government organisations and regional agencies from all over the Community. The total cost was about 22M European Units of Account (roughly $ 22M), of which about one third was borne by the Commission. The rest was contributed by industry, member Governments and other institutions.

The programme involved nine European manufacturers in the production of over 1 MWp of photovoltaic modules, in sizes ranging from 19 to 125 Wp. For design qualification, samples of each type were subjected at the CEC Joint Research Centre, Ispra, Italy to rigorous performance and environmental tests in accordance with CEC Specification No. 502. (See Chapter 4).

As soon as possible after the completion of each plant, official acceptance tests were carried out by a visiting team from JRC. The array and other components were visually inspected, functional tests were carried out on the whole installation and the rated power of the array was determined from on-site measurements. Array losses from module mismatch, cabling and diodes were estimated.

Every participant in the programme was required to record hourly values of global irradiation, the irradiation at the inclination of the array, wind speed and direction, ambient temperature, mean module temperature, array output energy and the energy fed to each load. Where applicable, they were also required to record the hourly input and output energies to and from the battery, the power conditioning units and the grid, together with any input from the back-up generator. The recordings were required to be made in a standard format on tape or disc and sent every month to JRC for processing and analysis. Finally, a log book was to be kept to record failures and accidents, remedial action, severe weather, maintenance operations, sensor calibrations, etc. Such analytical monitoring is an essential ingredient of any pilot project programme.

There follows a brief description of each of the sixteen plants.

Aghia Roumeli

A 50 kWp pv generator with 480 kWh battery supplies ac power for homes, shops, hotels and street lighting in a remote village of 105 inhabitants on the south coast of Crete. The village (Fig. 17.1) is accessible only from the sea or by a 5-hour walk through the Samaria Canyon, a well known beauty spot. The array, which is installed on the stony beach, consists of 720 France-Photon 72 Wp crystalline silicon modules. A 50 kVa diesel generator provides back-up but cannot be operated in parallel with the pv system.

Chevotogne

Fig. 14.1. 50 kWp array, Aghia Roumeli, Crete.

A 63 kWp array, set on a grassy hillside in a large camping and sports centre near Rochefort, Belgium, provides 40 kWp for the pumps of a solar-heated outdoor swimming pool and 23 kWp for lighting and auxiliary services. The array consists of 1984 IDE 33 Wp modules. The battery capacity is 275 kWh. Surplus power is fed to the grid, which also provides back-up.

Fota

A 50 kWp array, comprising 2775 AEG 19.2 Wp modules, together with a 160 kWh battery, supply ac power to a dairy farm on an island south of Cork (Fig. 14.2). The loads consist of milking machines, coolers and washing apparatus. The demand, peaking at 15 kW, is well matched to the solar input, as in Southern Ireland about 16 times more milk is produced in summer than in winter. Excess power is fed to the grid, which also provides back-up. The space below the single-plane array is divided into bays, two of which serve as the control and battery rooms and the others as cowsheds.

Giglio

Installed on a small island off the west coast of Italy, this plant comprises two systems - a 30 kWp array and 60 kWh battery for an agricultural cold store and a 15 kWp array with a 200 kWh battery for a water disinfection plant. The two arrays are constructed from 868 Pragma 54.5 Wp modules. A novel feature of the

Fig. 14.2. 50 kWp array, Fota Island, Cork, Ireland.

cold store system, which embodies a 5-cylinder compressor, is that the motor speed and the number of compressor cylinders in operation are automatically controlled to match the available pv power. Back-up is provided by the grid.

Hoboken

A 30 kWp array, consisting of 912 IDE 33 Wp modules, is installed on the roof of a warehouse in a metallurgical factory near Antwerp to provide dc power for the production of hydrogen by the electrolysis of water. There is no battery. The hydrogen is used for metallurgical processing. Excess power is used for water pumping. Grid back-up is provided through a 15 kW rectifier.

Kaw

A 35 kWp system with 480 kWh battery supplies ac power for a remote village of 70 inhabitants in French Guyana. The electricity is used for homes, a school, a church and street lighting The array consists of 492 France-Photon 72 Wp modules. A diesel generator provides back-up.

Kythnos

A 100 kWp array, comprising 800 Siemens 125 Wp modules, augments the power from an existing diesel power station and an experiment wind park on the Greek island of Kythnos. It saves fuel and enhances the reliability of the supply to the islanders. The battery capacity is 630 kWh.

Marchwood

A 30 kWp array (960 BP Solar 33 Wp modules), with an 88 kWh battery, installed at Marchwood power station, near Southampton, can be operated experimentally in standalone or grid-connected modes, with or without the battery. The array tilt is adjustable. When operated in standalone mode, the plant supplies selected individual loads.

Mont Bouquet

A 50 kWp array and 176 kWh battery, installed on the summit of Mont Bouquet, near Nimes in Southern France, partially powers three radio and three television transmitters with a total power consumption of between 35 and 40 kVA. The array is constructed from 710 Photowatt 72 Wp modules. The grid provides back-up.

Nice

A 50 kWp array, similar to that at Mont Bouquet, mounted on the roof of a building at Nice International Airport, supplies power for electronic equipment in the control tower and the public address system in the Main Hall. The sophisticated electronics in the control tower require a continuous 5 kVA supply, free from distortion and micro-cuts - a requirement which can be admirably met by photovoltaics. The grid provides back-up.

Pellworm

This 300 kWp system with a 2770 kWh battery, the largest and most northerly of the CEC pilot plants, is shown in Fig. 14.3. It supplies all the electrical requirements of an enclosed recreational centre on Pellworm, a German island in the North Sea. The array comprises 17568 AEG 19.2 Wp modules. The load consists of stoves and other items in the restaurant, heat pumps for the indoor swimming pool, lighting, TV, radio, etc. As the centre is used mostly in the summer, the load is well matched to the available solar radiation. Surplus electricity is fed to the local grid. The ground beneath the arrays is used for grazing sheep.

Rondulinu

A 44 kWp standalone pv plant supplies ac power for lighting, water pumping and refrigeration to a small hamlet in the mountains of Corsica. The array comprises 1220 France-Photon 36 Wp modules. The battery capacity is 430 kWh. A 25 kVA diesel generator provides back-up and can work in parallel with the pv system.

Fig. 14.3. 300 kWp array, Pellworm Island, Germany.

Terschelling

A pv/wind hybrid system, consisting of a 50 kWp array, 30 kW wind generator and a 180 kWh battery, provides about 95% of the electrical power needed by a marine training school on Terschelling, a large island off the Dutch coast. The array consists of 2610 AEG 19.2 Wp modules. Surplus power is fed to the grid.

Tremiti

A 65 kWp standalone system with a 500 kWh battery, installed on a cliff top on a small island in the Adriatic, is designed to desalinate seawater by the reverse osmosis process, producing about 4000 m^3 of potable water every year. The array consists of 189 Siemens 125 Wp modules and 1254 Ansaldo 33 Wp modules. The average energy requirement of the reverse osmosis plant is 13 kWh/m^3. The high pressure variable-flow pumps are driven by ac induction motors, which are supplied through separate variable-frequency inverters. Formerly, all drinking water for the island had to be imported by tanker from the mainland. There is no back-up generator. The load is reasonably well matched to the solar input, as summer visitors increase the demand for drinking water.

Vulcano

An 80 kWp system with a 550 kWh battery, installed on a volcanic island just north of Sicily, can feed power into the island's grid in parallel with the existing diesel station or, alternatively, supply a dedicated load of about 40 isolated houses. The array comprises 1210 Ansaldo 33 Wp modules and 740 Pragma 54.5 Wp modules.

Zambelli

A 70 kWp photovoltaic pumping station, powered by 1286 Pragma 54.5 Wp modules, raises drinking water up to a village reservoir in the mountains near Verona. Two commercial piston pumps, each capable of maintaining a flow of 30 m³/h, are driven by 35 kW ac motors supplied through a variable-frequency inverter.

14.3 CEC Demonstration Projects

The CEC's Directorate-General for Energy (DG XVII) launched its Solar Energy Demonstration Programme in 1978. In the early years, photovoltaics played only a minor role but, between 1983 and 1986, 48 pv projects were selected for implementation, representing a total installed capacity of 730 kWp. The Commission contributed about 40% of the total cost of 30M European Units of Account.

Table 14.1 lists the applications, locations and array powers of these plants. Most are standalone systems, providing power for telecommunications, lighthouses, water pumping, seawater desalination and the electrification of isolated homes, farmhouses and mountain refuges.

Table 14.1 CEC DEMONSTRATION PROJECTS

Application	Location	Power kWp
1983		
Fire prevention and high water alarm	L'Aquila & Pescara, Italy	0.4
Isolated farmhouse	Baden-Wurttemberg, Alsace	5
Water pumping for irrigation	Carpathos Is., Greece	10
Two remote villages	Gavdos Is., Greece	20
Groups of isolated houses	Antikythira Is., Greece	35
Remote village	Arki Is., Greece	25
40 isolated houses	Southern France	32
Ground water level measurement at water works	Hamburg, Germany	0.3
Isolated relay station	Antikythira Is., Greece	20
1984		
Three isolated houses	Baden-Wurttemberg, Alsace	2.6
Bird and weather station on remote island	Scharhoern, Germany	4.1
PV/wind hybrid system for dwelling houses	Milton Keynes, UK	4.2
Integration of pv generator in mains-connected house	Cork, Ireland	5

Table 14.1 CEC DEMONSTRATION PROJECTS (cont'd)

Application	Location	Power kWp
PV/wind cogeneration plant for remote village	Hasle, Bornholme, Denmark	36
PV/micro-hydro hybrid system for isolated farm	Campoligure, Italy	2.1
PV/wind hybrid system for unmanned lighthouse	Lithari, Skyros Is. Greece	2.6
PV/wind hybrid system for unmanned lighthouse	Methoni, Sapienza, Greece	2.6
Lighthouse	Palmaiola Is., Italy	3.9
Lighting of Etruscan archeological sites	Cetano & Sorano, Italy	26.1
Alpine lodgings and huts	French Alps	5.3
PV/wind supply for various purposes in recreational areas	Freiburg im Breisgau, Germany	2.1
1985		
Electrification of nature reserve	Camargue, France	12
PV/diesel supply for an isolated microwave relay	Toulon, France	1.9
TV and FM emitter	French Guyana	20
30 isolated houses	Eolian Islands, Italy	9
Desalination, refrigeration and lighting	Milos Is., Greece	31.9
Mountain dairy farm	Ambruzzo, Italy	6.3
Fog detection system for high risk highway	Verona, Italy	14.3
Housing complex	Poros Is., Greece	4.2
Grid-connected low power system	Saarbrucken, Germany	8
Mains-connected house	Berlin, Germany	5.1
29 mountain refuges	French Alps and Pyrenees	8.35
7 dairy farms in high mountains	Liguria, Italy	27.4
Lighthouse	Sicily	18.2
Water pumping in rural areas	Corsica	3.8
1986		
Houses, TV repeater, harbour beacon, sterilise r	Corsica	9.8
Three telephone exchanges	Guadaloupe	12.4
Isolated rural dwellings	Portugal	35
Seawater desalination (reverse osmosis)	Isla del Moro, Spain	23.1
Light buoys	Atlantic & Mediterranean	24
PV power plant	Tabarca Is., Spain	100
57 isolated houses	Sierra de Segura, Spain	28.2
PV power for cheese cooperative	Trebujena, Spain	19.7
30 isolated houses	Ivares, Oden Lleida, Spain	18.9
Seawater desalination (PV/wind)	Fuerteventura Is., Spain	20
Airport signalling lights	Lucca - Tassignano, Italy	4.8
Passive/thermo/PV hybrid system for climatic control and electrification of isolated buildings	Palombara Sabina, Italy	3
PV equipment programme of the International Activity Park	Valbonne, Sophia Antipolis France	8.3

All the plants are being monitored over a period of at least two years, in accordance with guidelines laid down by JRC, Ispra.

With the simpler plants, the monitoring is limited to recording hourly values of the global irradiation, array output current, the energy to all loads and, where applicable, the energy to and from the grid or from a back-up generator. A record is also kept of the hours during which the load is connected to the pv generator. The more complex systems are monitored in the same way as are pilot plants.

Towards the end of 1986, a data base, SESAME, was set up by the Commission to publicise and promote successful energy demonstrations. It contains technical, operational, administrative and financial information relating to each project and can be accessed via European data processing networks.

The current 4-year demonstration programme in the energy field started in 1986 and has a total budget of 360 MECU. Invitations for project proposals are published annually in the Official Journal of the European Communities.

14.4 Pilot project experience

The collection, processing and analysis of data from pilot and demonstration plants is a lengthy process and it will be some years before the lessons learned from the CEC programmes are fully collated and disseminated. However, experience from the pilot projects has already provided some important guidelines for the future.

The crystalline silicon modules have proved to be the most reliable part of the systems, despite the fact that those used for the pilot plants were "first generation" designs. Relatively few have had to be replaced. The most common fault is cracked glass windows, which does not affect the electrical performance. This damage has been attributed to a severe hailstorm (Giglio), vandalism (Terschelling) and design faults (Fota, Terschelling, Pellworm). The modules used in the last three plants are clamped to the structure by washered bolts, which tend to loosen after a time, causing slippage. At Fota, insufficient clearance between the glass backing of the modules and the galvanised supporting beam resulted in high stress points and eventual breakage in some cases. The thin pigtails used for interconnecting these modules are easily damaged and difficult to stow neatly. The manufacturer has corrected these faults in later designs.

To date, there has been no evidence to indicate that modules backed with Tedlar-coated foil are any more or less reliable than those with a glass backing. One manufacturer's glass/pvb/glass modules have shown signs of delamination but this is attributed to a faulty lamination process, which has now been corrected.

No trouble from hot-spot failures has been reported. This is probably due to the fact that the effect was well known before the European programmes started and so designers provided adequate protection in the form of by-pass diodes.

Module efficiencies in the pilot plants, based on the total module area, ranged from 6.5% to 8.5%. This is low by modern standards. No comparable data are yet available from the demonstration plants but it is likely that most recent modules are in the 10% to 11% bracket.

All the arrays in the pilot plants fit in well with their surroundings, thanks to the involvement of architects in the design process - a Commission requirement. In many cases, however, this has meant a considerable amount of ground clearance and site preparation. In one case (Vulcano), removal of the natural ground cover exposed a thick layer of fine volcanic ash, which had then to be covered with concrete to prevent array contamination. The projects with the lowest site preparation costs were those on existing structures, e.g. the roof at Hoboken and the concrete hard-standing at Marchwood, or where it was possible to leave the contours and ground cover undisturbed, as at Chevetogne and Pellworm. The cowsheds at Fota and the sheep grazing at Pellworm are examples of how a second use can be found for the space covered by an array field.

More designers opted for galvanised steel support structures but aluminium was used in two cases and, at Pellworm, half of the array field is supported on hardwood frames. Here, the module clamping bolts have had to be tightened once a year to counteract the shrinkage of the timber. Galvanised steel fastening screws proved to be a problem at Hoboken, where heavy corrosion led to an expensive renewal campaign.

At Fota, where the array is in a single plane, a "cherry-picker" platform is required whenever one wants to inspect or replace the higher modules. This could have been avoided by the provision of laddered gangways between sections of the array, as in the roof-mounted installation at Hoboken.

So far, the only reports of lighning damage have come from Kythnos, where the data monitoring system suffered, and Vulcano, where one module had to be replaced. At Kythnos, the lightning protection system consists of ten 10m steel masts, erected at intervals over the array field, while, at Giglio, three single-span earthed cables are suspended over the arrays from masts positioned outside the field. In all other plants, reliance has been placed on a good earthing system, with varistors to conduct lightning-induced current surges to ground. At Mont Bouquet and Nice, the selected varistors proved to be incapable of withstanding the maximum system voltage under normal operation. Excessive leakage currents caused progressive degradation and in July 1983 the varistors in both plants failed catastrophically and caused fires, which damaged the power conditioning equipment. After repair in 1986, Mont Bouquet had the misfortune to suffer another fire, this time from the surrounding woodland. Much of the array field was destroyed.

In the power conditioning subsystems, some designers opted for commercially-available inverters, with minor modifications to improve low-load efficiency, while others chose new designs, embodying high power transistors, in the quest for better performance. Again, some decided on single inverters, while others sought to reduce losses by using multiple units, switched in and out to suit the load and irradiance conditions. Instantaneous inverter efficiencies in the pilot plants ranged from 90% to 96% at full load, depending on the design and the acceptable degree of harmonic distortion in the output waveform. But, at 10% load, some efficiencies dropped below 80%. Since the inverters are working for most of the time under less than full load conditions, the energy loss over a period of time can be considerable. The energy efficiency can be reduced still further by no-load losses, unless the inverter is automatically switched off when not in use. The energy efficiencies of the pilot plant inverters have been found to range from 60% to 85%.

Inverter malfunction has been one of the principal causes of shutdown in the pilot plants. Electronic faults, synchronisation failures, high starting currents and poor frequency control (in variable frequency types) have been reported and some units have been affected by high temperatures, dust and humidity. These troubles have sometimes been compounded by shortages of spare parts and skilled maintenance personnel.

Only a minority of the pilot plant designers decided to use maximum power point trackers. Giglio and Marchwood have single units but Tremiti has three and Kythnos four. At Kythnos, the MPPTs are also used as part of the control system to reduce the pv energy flow, when necessary, without switching out sections of the array. At Terschelling, there are no less than 29 MPPTs - one for each of the subarrays. The arguments for this complexity were (a) that it enabled the requirements for subarray matching to be relaxed and (b) that it could accommodate the effects of partial shadowing from the nearby building and differing rates of degradation. It has not yet been possible to determine whether the MPPTs have increased the net energy output of any of the arrays and, if so, by how much. But experience has given some support to the view that, in plants of this size, any energy bonus is not worth the added expense and complication. At Terschelling, the MPPTs have proved unreliable, mainly due to transistor faults.

The most common battery problem has been excessive discharge, which probably caused 22 cells to fail at Pellworm, 12 at Aghia Roumeli and a further 8 at Kaw. This reflects the difficulties of charge control, which were discussed in Chapter 7. Battery energy efficiencies have proved to be somewhat lower than those predicted from manufacturer's data. At Fota, Giglio, Tremiti and Vulcano, monitoring has shown the battery to be undersized, thus aggravating the problem of overcharge protection.

At Fota, Pellworm and Terschelling, the battery bank is divided into two parts, each of which can be connected to either the main bus or an auxiliary bus. This arrangement gives great flexibility and enables one battery to be fully charged over a long period without discharge interruption, while the other supplies the load during the night and periods of low irradiance. It was expected to extend the life of the batteries - an assumpion that now appears to be in some doubt. Moreover, computer simulation studies at Fota have indicated that the contribution of the pv plant to the dairy farm load would be increased by over 60% if the two batteries were combined in the conventional arrangement. This charge would also simplify the control software and reduce the amount of switching, thus enhancing the reliability of the system.

All the pilot plants have some type of power management and control system. Aghia Roumeli, Kaw, Mont Bouquet, Nice, Rondulinu and Vulcano have simple systems, which have proved reliable but have the disadvantage of relatively poor battery management. Chevetogne, Fota and Pellworm have more sophisticated systems embodying microprocessors. These perform better and the software can be improved progressively in the light of experience. Significant improvements have, in fact, been made at Fota as a result of the continuous analysis of monitored data. At the other plants, the control system is of similar complexity but it is modular in design. This involves extra equipment but it has the advantage that the control algorithms for each module can be used in other systems employing the same unit.

At Giglio, Hoboken and Mont Bouquet, control is achieved by load matching. In other cases, when load requirements are being met and the battery is fully charged, the control system limits the power available to avoid overcharging the battery. This is done by feeding surplus power to the grid (Chevetogne, Fota, Marchwood, Pellworm and Terschelling), by disconnecting the entire array (Aghia Roumeli, Kaw and Rondulinu) or by disconnecting the requisite number of subarrays (Fota, Nice, Pellworm, Terschelling, Tremiti and Vulcano). At Rondulinu, battery overcharge protection is effected by switching in a resistor network to move the operating point in the array.

The matching of the pv system performance with the load has caused problems at some of the pilot plants. At Aghia Roumeli and Pellworm, there were difficulties due to the uncontrolled addition of loads after the plants were commissioned. At Terschelling, the load was underestimated, while at Chevetogne and Rondulinu the loads turned out to be less than expected. However, only minor problems have been encountered in cases where the load profile was well defined or could be controlled by the plant operator.

In general, the pilot plants have required very little maintenance. At none of the sites has manual cleaning of the modules proved to be necessary, as it has been found that deposits of dust and dirt are cleared away periodically by rain. Losses from window contamination are estimated to be generally less than 5%, although at Hoboken, a very dirty site, they have amounted to 10%. The textured glass surface of some modules tends to retain the dirt and is a feature to be avoided in future. Batteries have not needed to be topped up very frequently. The maximum frequency is twice a year (Terschelling) and the minimum once every two years (Pellworm).

Fota, Marchwood and Pellworm are fully automatic plants, needing no operators. At the other end of the scale, the desalination plant at Tremiti requires one full-time operator and three extra staff in summer.

An important photovolatic performance parameter is the mean

monthly efficiency of the array, η_a. This is defined as:-

$$\eta_a = \frac{\text{mean monthly array output energy (kWh)}}{\left[\begin{array}{l}\text{mean monthly irradiation in the}\\ \text{plane of the array (kWh.m}^{-2}\text{)}\end{array}\right] \times \left[\text{array area (m}^2\text{)}\right]} \times 100\%$$

Data analysis has shown that, in the pilot plants, η_a varies from 0.95% to 7.39%. The poor performance at the lower extreme can be attributed, inter alia, to poor load matching, inadequate battery capacity, plant malfunction and non-optimised control. Of course, the performance would be improved all round by more efficient modules of the best type now available (13%).

The capacity factor, CF, defined as:-

$$CF = \frac{\text{array output energy over a specified period (kWh)}}{[\text{nominal rated power (kWp)}]\times[\text{no. of hours in the period}]} \times 100\%$$

ranges from just over 5% for Marchwood and Fota (taken over a full year) to 10% at Vulcano (taken over the period July - November 1985). In their best months, Vulcano achieved a CF of 20% and Kythnos 17%.

In 1987, the European Commission (DG XII) invited all pilot plant contractors to submit proposals for cost-sharing contracts to improve their plants in the light of experience. The response was encouraging and a programme of modifications designed to improve performance and reliability is now under way.

14.5 Lessons learned

The lessons learned so far from the pilot plants may be summarised as follows:-

1. If full value is to be obtained from the investment in a pilot plant, detailed arrangements should be made at the outset for the ownership, operation, maintenance, repair and monitoring of the plant over a specified number of years. These arrangements should be spelt out clearly in the contract.

2. A manual should be prepared for each plant, giving instructions for the operation, maintenance and repair of the plant and the data monitoring system.

3. System reliability is all-important, particularly in remote areas. The design should therefore be kept as simple as possible, even if this involves some loss of efficiency.

4. A single battery and bus system is preferable to a dual system.

5. In designing the power management and control system, computer simulation studies, based on the best available data, should be carried out to determine the arrangement most

likely to produce maximum system efficiency. Later, the model should be up-dated as more realistic data become available.

6. The existing load and likely future changes should be assessed as accurately as possible, with a view to optimum matching.

7. Adequate shielding should be installed to prevent interference with control and data acquisition systems.

8. In choosing the site and designing the array, special attention should be paid to the need to minimise the costs of transportation, ground clearance, site preparation and installation.

9. As a general rule, only components which have been design qualified in laboratory type tests should be used in a pilot plant. This applies also to load appliances.

10. High efficiency modules, now available on the market, should be used in future projects, in the interests of lower balance-of-system costs.

11. Modules should be mounted through fixing holes to give positive location, rather than by clamping. Interconnecting cables should be properly supported and protected against damage from animals and birds. In large, single-plane arrays, provision should be made to facilitate the cleaning, inspection and replacement of modules.

12. More effort is needed to improve the low-load efficiency and reliability of inverters.

13. Maximum power point trackers should be used only when careful computer simulation studies have indicated that they will produce a worthwhile gain. These devices need to be made more reliable.

14. Further investigation is necessary to determine the best way to protect pv systems from lightning damage.

15. Steps should be taken to ensure that varistors used to conduct lightning-induced currents to ground are capable of withstanding, with a reasonable safety factor, the highest voltages they are likely to be subjected to in normal operation.

These lessons are already being applied by the designers of CEC photovoltaic demonstration plants. The Commission has also organised expert groups to study common problems, such as battery charge control, system modelling, inverter efficiency and lightning protection, and seek solutions.

Although this chapter has rightly laid stress on problems and difficulties, it should be said in closing that the pilot and demonstration projects have made and are making an invaluable contribution to the progress of photovoltaic system technology.

They gave not only proved the technical feasibility of generating electricity by the direct conversion of solar energy but have also shown the social benefits and social acceptability of photovoltaics.